Advances in shape memory polymers

The Textile Institute and Woodhead Publishing

The Textile Institute is a unique organization in textiles, clothing and footwear. Incorporated in England by a Royal Charter granted in 1925, the Institute has individual and corporate members in over 90 countries. The aim of the Institute is to facilitate learning, recognize achievement, reward excellence and disseminate information within the global textiles, clothing and footwear industries.

Historically, The Textile Institute has published books of interest to its members and the textile industry. To maintain this policy, the Institute has entered into partnership with Woodhead Publishing Limited to ensure that Institute members and the textile industry continue to have access to high calibre titles on textile science and technology.

Most Woodhead titles on textiles are now published in collaboration with The Textile Institute. Through this arrangement, the Institute provides an Editorial Board which advises Woodhead on appropriate titles for future publication and suggests possible editors and authors for these books. Each book published under this arrangement carries the Institute's logo.

Woodhead books published in collaboration with The Textile Institute are offered to Textile Institute members at a substantial discount. These books, together with those published by The Textile Institute that are still in print, are offered on the Woodhead web site at: www.woodheadpublishing.com. Textile Institute books still in print are also available directly from the Institute's web site at: www.textileinstitutebooks.com.

A list of Woodhead books on textiles science and technology, most of which have been published in collaboration with the Textile Institute, can be found towards the end of the contents pages.

We are always happy to receive suggestions for new books from potential editors. To enquire about contributing to our Textiles series, please send your name, contact address and details of the topic/s you are interested in to sarah.lynch@woodheadpublishing.com. We look forward to hearing from you.

The Woodhead team responsible for publishing this book
Commissioning Editor: Kathryn Picking
Publications Coordinator: Lynsey Gathercole
Project Editor: Cathryn Freear
Editorial and Production Manager: Mary Campbell
Production Editor: Richard Fairclough
Freelance Project Manager: Annette Wiseman
Copyeditor: Jo Egré
Proofreader: Don Cleeter
Cover Designer: Terry Callanan

Woodhead Publishing Series in Textiles: Number 146

Advances in shape memory polymers

Jinlian Hu

The Textile Institute

WOODHEAD PUBLISHING

Oxford Cambridge Philadelphia New Delhi

Published by Woodhead Publishing Limited in association with The Textile Institute
Woodhead Publishing Limited, 80 High Street, Sawston,
Cambridge CB22 3HJ, UK
www.woodheadpublishing.com
www.woodheadpublishingonline.com

Woodhead Publishing, 1518 Walnut Street, Suite 1100, Philadelphia,
PA 19102-3406, USA

Woodhead Publishing India Private Limited, G-2, Vardaan House, 7/28 Ansari Road,
Daryaganj, New Delhi – 110002, India
www.woodheadpublishingindia.com

First published 2013, Woodhead Publishing Limited

British Library Cataloguing in Publication Data
A catalogue record for this book is available from the British Library.

Library of Congress Control Number: 2013931492

ISBN 978-0-85709-852-8 (print)
ISBN 978-0-85709-854-2 (online)
ISSN 2042-0803 Woodhead Publishing Series in Textiles (print)
ISSN 2042-0811 Woodhead Publishing Series in Textiles (online)

The publishers' policy is to use permanent paper from mills that operate a sustainable forestry policy, and which has been manufactured from pulp which is processed using acid-free and elemental chlorine-free practices. Furthermore, the publishers ensure that the text paper and cover board used have met acceptable environmental accreditation standards.

Typeset by RefineCatch Limited, Bungay, Suffolk

Contents

Woodhead Publishing Series in Textiles

1 **Watson's textile design and colour Seventh edition**
Edited by Z. Grosicki
2 **Watson's advanced textile design**
Edited by Z. Grosicki
3 **Weaving Second edition**
P. R. Lord and M. H. Mohamed
4 **Handbook of textile fibres Volume 1: Natural fibres**
J. Gordon Cook
5 **Handbook of textile fibres Volume 2: Man-made fibres**
J. Gordon Cook
6 **Recycling textile and plastic waste**
Edited by A. R. Horrocks
7 **New fibers Second edition**
T. Hongu and G. O. Phillips
8 **Atlas of fibre fracture and damage to textiles Second edition**
J. W. S. Hearle, B. Lomas and W. D. Cooke
9 **Ecotextile '98**
Edited by A. R. Horrocks
10 **Physical testing of textiles**
B. P. Saville
11 **Geometric symmetry in patterns and tilings**
C. E. Horne
12 **Handbook of technical textiles**
Edited by A. R. Horrocks and S. C. Anand
13 **Textiles in automotive engineering**
W. Fung and J. M. Hardcastle
14 **Handbook of textile design**
J. Wilson
15 **High-performance fibres**
Edited by J. W. S. Hearle
16 **Knitting technology Third edition**
D. J. Spencer
17 **Medical textiles**
Edited by S. C. Anand
18 **Regenerated cellulose fibres**
Edited by C. Woodings
19 **Silk, mohair, cashmere and other luxury fibres**
Edited by R. R. Franck

20 **Smart fibres, fabrics and clothing**
Edited by X. M. Tao
21 **Yarn texturing technology**
J. W. S. Hearle, L. Hollick and D. K. Wilson
22 **Encyclopedia of textile finishing**
H-K. Rouette
23 **Coated and laminated textiles**
W. Fung
24 **Fancy yarns**
R. H. Gong and R. M. Wright
25 **Wool: Science and technology**
Edited by W. S. Simpson and G. Crawshaw
26 **Dictionary of textile finishing**
H-K. Rouette
27 **Environmental impact of textiles**
K. Slater
28 **Handbook of yarn production**
P. R. Lord
29 **Textile processing with enzymes**
Edited by A. Cavaco-Paulo and G. Gübitz
30 **The China and Hong Kong denim industry**
Y. Li, L. Yao and K. W. Yeung
31 **The World Trade Organization and international denim trading**
Y. Li, Y. Shen, L. Yao and E. Newton
32 **Chemical finishing of textiles**
W. D. Schindler and P. J. Hauser
33 **Clothing appearance and fit**
J. Fan, W. Yu and L. Hunter
34 **Handbook of fibre rope technology**
H. A. McKenna, J. W. S. Hearle and N. O'Hear
35 **Structure and mechanics of woven fabrics**
J. Hu
36 **Synthetic fibres: Nylon, polyester, acrylic, polyolefin**
Edited by J. E. McIntyre
37 **Woollen and worsted woven fabric design**
E. G. Gilligan
38 **Analytical electrochemistry in textiles**
P. Westbroek, G. Priniotakis and P. Kiekens
39 **Bast and other plant fibres**
R. R. Franck
40 **Chemical testing of textiles**
Edited by Q. Fan
41 **Design and manufacture of textile composites**
Edited by A. C. Long
42 **Effect of mechanical and physical properties on fabric hand**
Edited by H. M. Behery
43 **New millennium fibers**
T. Hongu, M. Takigami and G. O. Phillips

44 **Textiles for protection**
Edited by R. A. Scott
45 **Textiles in sport**
Edited by R. Shishoo
46 **Wearable electronics and photonics**
Edited by X. M. Tao
47 **Biodegradable and sustainable fibres**
Edited by R. S. Blackburn
48 **Medical textiles and biomaterials for healthcare**
Edited by S. C. Anand, M. Miraftab, S. Rajendran and J. F. Kennedy
49 **Total colour management in textiles**
Edited by J. Xin
50 **Recycling in textiles**
Edited by Y. Wang
51 **Clothing biosensory engineering**
Y. Li and A. S. W. Wong
52 **Biomechanical engineering of textiles and clothing**
Edited by Y. Li and D. X-Q. Dai
53 **Digital printing of textiles**
Edited by H. Ujiie
54 **Intelligent textiles and clothing**
Edited by H. R. Mattila
55 **Innovation and technology of women's intimate apparel**
W. Yu, J. Fan, S. C. Harlock and S. P. Ng
56 **Thermal and moisture transport in fibrous materials**
Edited by N. Pan and P. Gibson
57 **Geosynthetics in civil engineering**
Edited by R. W. Sarsby
58 **Handbook of nonwovens**
Edited by S. Russell
59 **Cotton: Science and technology**
Edited by S. Gordon and Y-L. Hsieh
60 **Ecotextiles**
Edited by M. Miraftab and A. R. Horrocks
61 **Composite forming technologies**
Edited by A. C. Long
62 **Plasma technology for textiles**
Edited by R. Shishoo
63 **Smart textiles for medicine and healthcare**
Edited by L. Van Langenhove
64 **Sizing in clothing**
Edited by S. Ashdown
65 **Shape memory polymers and textiles**
J. Hu
66 **Environmental aspects of textile dyeing**
Edited by R. Christie
67 **Nanofibers and nanotechnology in textiles**
Edited by P. Brown and K. Stevens

68 **Physical properties of textile fibres Fourth edition**
W. E. Morton and J. W. S. Hearle
69 **Advances in apparel production**
Edited by C. Fairhurst
70 **Advances in fire retardant materials**
Edited by A. R. Horrocks and D. Price
71 **Polyesters and polyamides**
Edited by B. L. Deopura, R. Alagirusamy, M. Joshi and B. S. Gupta
72 **Advances in wool technology**
Edited by N. A. G. Johnson and I. Russell
73 **Military textiles**
Edited by E. Wilusz
74 **3D fibrous assemblies: Properties, applications and modelling of three-dimensional textile structures**
J. Hu
75 **Medical and healthcare textiles**
Edited by S. C. Anand, J. F. Kennedy, M. Miraftab and S. Rajendran
76 **Fabric testing**
Edited by J. Hu
77 **Biologically inspired textiles**
Edited by A. Abbott and M. Ellison
78 **Friction in textile materials**
Edited by B. S. Gupta
79 **Textile advances in the automotive industry**
Edited by R. Shishoo
80 **Structure and mechanics of textile fibre assemblies**
Edited by P. Schwartz
81 **Engineering textiles: Integrating the design and manufacture of textile products**
Edited by Y. E. El-Mogahzy
82 **Polyolefin fibres: Industrial and medical applications**
Edited by S. C. O. Ugbolue
83 **Smart clothes and wearable technology**
Edited by J. McCann and D. Bryson
84 **Identification of textile fibres**
Edited by M. Houck
85 **Advanced textiles for wound care**
Edited by S. Rajendran
86 **Fatigue failure of textile fibres**
Edited by M. Miraftab
87 **Advances in carpet technology**
Edited by K. Goswami
88 **Handbook of textile fibre structure Volume 1 and Volume 2**
Edited by S. J. Eichhorn, J. W. S. Hearle, M. Jaffe and T. Kikutani
89 **Advances in knitting technology**
Edited by K-F. Au
90 **Smart textile coatings and laminates**
Edited by W. C. Smith
91 **Handbook of tensile properties of textile and technical fibres**
Edited by A. R. Bunsell

92 **Interior textiles: Design and developments**
Edited by T. Rowe
93 **Textiles for cold weather apparel**
Edited by J. T. Williams
94 **Modelling and predicting textile behaviour**
Edited by X. Chen
95 **Textiles, polymers and composites for buildings**
Edited by G. Pohl
96 **Engineering apparel fabrics and garments**
J. Fan and L. Hunter
97 **Surface modification of textiles**
Edited by Q. Wei
98 **Sustainable textiles**
Edited by R. S. Blackburn
99 **Advances in yarn spinning technology**
Edited by C. A. Lawrence
100 **Handbook of medical textiles**
Edited by V. T. Bartels
101 **Technical textile yarns**
Edited by R. Alagirusamy and A. Das
102 **Applications of nonwovens in technical textiles**
Edited by R. A. Chapman
103 **Colour measurement: Principles, advances and industrial applications**
Edited by M. L. Gulrajani
104 **Fibrous and composite materials for civil engineering applications**
Edited by R. Fangueiro
105 **New product development in textiles: Innovation and production**
Edited by L. Horne
106 **Improving comfort in clothing**
Edited by G. Song
107 **Advances in textile biotechnology**
Edited by V. A. Nierstrasz and A. Cavaco-Paulo
108 **Textiles for hygiene and infection control**
Edited by B. McCarthy
109 **Nanofunctional textiles**
Edited by Y. Li
110 **Joining textiles: Principles and applications**
Edited by I. Jones and G. Stylios
111 **Soft computing in textile engineering**
Edited by A. Majumdar
112 **Textile design**
Edited by A. Briggs-Goode and K. Townsend
113 **Biotextiles as medical implants**
Edited by M. King and B. Gupta
114 **Textile thermal bioengineering**
Edited by Y. Li
115 **Woven textile structure**
B. K. Behera and P. K. Hari

116 **Handbook of textile and industrial dyeing. Volume 1: Principles, processes and types of dyes**
Edited by M. Clark
117 **Handbook of textile and industrial dyeing. Volume 2: Applications of dyes**
Edited by M. Clark
118 **Handbook of natural fibres. Volume 1: Types, properties and factors affecting breeding and cultivation**
Edited by R. Kozłowski
119 **Handbook of natural fibres. Volume 2: Processing and applications**
Edited by R. Kozłowski
120 **Functional textiles for improved performance, protection and health**
Edited by N. Pan and G. Sun
121 **Computer technology for textiles and apparel**
Edited by J. Hu
122 **Advances in military textiles and personal equipment**
Edited by E. Sparks
123 **Specialist yarn and fabric structures**
Edited by R. H. Gong
124 **Handbook of sustainable textile production**
M. I. Tobler-Rohr
125 **Woven textiles: Principles, developments and applications**
Edited by K. Gandhi
126 **Textiles and fashion: Materials design and technology**
Edited by R. Sinclair
127 **Industrial cutting of textile materials**
I. Vilumsone-Nemes
128 **Colour design: Theories and applications**
Edited by J. Best
129 **False twist textured yarns**
C. Atkinson
130 **Modelling, simulation and control of the dyeing process**
R. Shamey and X. Zhao
131 **Process control in textile manufacturing**
Edited by A. Majumdar, A. Das, R. Alagirusamy and V. K. Kothari
132 **Understanding and improving the durability of textiles**
Edited by P. A. Annis
133 **Smart textiles for protection**
Edited by R. A. Chapman
134 **Functional nanofibers and applications**
Edited by Q. Wei
135 **The global textile and clothing industry: Technological advances and future challenges**
Edited by R. Shishoo
136 **Simulation in textile technology: Theory and applications**
Edited by D. Veit
137 **Pattern cutting for clothing using CAD: How to use Lectra Modaris pattern cutting software**
M. Stott

138 **Advances in the dyeing and finishing of technical textiles**
M. L. Gulrajani
139 **Multidisciplinary know-how for smart textiles developers**
Edited by T. Kirstein
140 **Handbook of fire resistant textiles**
F-S. Kilinc-Balci
141 **Handbook of footwear design and manufacture**
A. Luximon
142 **Textile-led design for the active ageing population**
J. McCann and D. Bryson
143 **Optimizing decision making in clothing management using artificial intelligence (AI): From production to fashion retail**
W. K. Wong, Z. X. Guo and S. Y. S. Leung
144 **Mechanisms of flat weaving technology**
V. Choogin, P. Bandara and E. Chepelyuk
145 **Innovative jacquard textile design using digital technologies**
F. Ng and J. Zhou
146 **Advances in shape memory polymers**
J. Hu
147 **Clothing manufacture management: A systematic approach to planning, scheduling and control**
J. Gersak
148 **Anthropometry, apparel sizing and design**
D. Gupta and N. Zakaria

Acknowledgements

The content of this book is based mainly on the original work produced by my research group (Shape Memory Textile Center, Hong Kong Polytechnic University). A number of people have contributed to and assisted with the preparation of the manuscript. I am especially grateful to Qinghao Meng who helped with editing the manuscript in the first draft and You Wu who helped with editing the later versions. I am deeply indebted to my research students, who have contributed the major part of the book. Acknowledgements for assistance and support for specific chapters are as follows:

- Chapters 1, 9, 11 and 12: Qinghao Meng;
- Chapters 2 to 4: Fenglong Ji;
- Chapters 5 to 8: Shaojun Chen;
- Chapters 10 and 11: Lap Yan Yeung.

The original data and part of the content in Chapters 5, 6, 7 and 8 are provided from the PhD thesis of Shaojun Chen. Much of the data reported in Chapters 2, 3 and 4 are from Fenglong Ji's PhD thesis. You Wu provided additional material to Chapter 5. Lap Yan Yeung contributed the work related to textile applications of shape memory polymers in Chapters 10 and 11. In addition to his help with the manuscript as a whole, Qinghao Meng made a particular contribution to Chapters 1, 9, 11 and 12.

I also wish to express my appreciation to many of my students who contributed to the research underpinning this book through their excellent work on shape memory fiber preparation and testing. I have really enjoyed working with these outstanding students whose work has made this book possible. In addition, I would like to acknowledge the generous support of the Hong Kong textile and apparel industry as well as the Hong Kong Government for funding support from the Innovation and Technology Commission over the past four years. From these industry-guided projects on the development of shape memory garments/fabrics and fibers, we can appreciate the significance of shape memory polymers to the textile and apparel industry as well as to the academic world.

Preface

Since the first reported shape memory effect of T_g-type shape memory polymers three decades ago, many shape memory polymers based on different switching mechanisms have been developed. The research in shape memory polymers has been extensive and intensive, particularly in the past five years. This book does not intend to cover all aspects of shape memory polymer research. It aims to give a brief introduction to some of the important advances. The book is mainly based on recent research of the Shape Memory Textile Center at the Hong Kong Polytechnic University, whilst referring to the research results obtained in other distinguished research groups around the world.

The research highlights covered in this book include:

1. the relationship between morphological structures and shape memory properties;
2. high performance T_g-type and T_m-type shape memory polymers;
3. shape memory polymers with supramolecular switches;
4. the thermally-active shape memory effect of supermolecular shape memory polymers;
5. the moisture-active shape memory effect of supermolecular shape memory polymers;
6. shape memory polymers based on phase change raw materials;
7. the shape memory effect achieved by indirect heating;
8. shape memory polymer fibers.

Chapter 1 briefly introduces the mechanisms of the shape memory effect and shape memory polymers, summarizing key research in the subject. It also introduces one of the most important shape memory polymers, shape memory polyurethane, and its unique properties. Since a melting transition temperature (T_m) or a glass transition temperature (T_g) can act as the switch transition for shape memory polyurethane, it can be called either a T_m-type or a T_g-type shape memory polyurethane. Chapters 2 and 3 discuss T_m-type and T_g-type shape memory polymers. The effects of deformation temperature, deformation amplitude, shape fixing temperature and pre-deformation are discussed in terms of their influences on shape memory properties.

Chapter 4 discusses a series of shape memory polymer polyurethane-ureas chain extended with aromatic diamines. The structures and properties of the aromatic and non-aromatic shape memory polyurethane-(ureas) are compared.

The results indicate that it is possible to enhance the shape recovery and recovery stress without sacrificing much shape fixity, by using rigid hard segments such as ureas, provided that the hard segment content is properly controlled. Chapter 4 also introduces the development of T_g-type shape memory polyurethane using a modified two-step copolymerization method. In the polyurethane prepared by the two-step method, the urethane chains can be regarded as soft segments. With urethane chains as soft segments, the degree of phase separation can be greatly enhanced. The shape memory polyurethane has a narrow shape recovery region and high recovery stress. The switching temperature and shape memory properties of the shape memory polyurethane can be adjusted using the urethane segment as the soft segment.

Shape memory functionality and supramolecular chemistry have a significant relationship in terms of shape memory effect and molecular interaction. Supramolecular structures can act as promising shape switches in shape memory polymers. Chapter 5 reviews the relationship between the shape memory polymers and supramolecules and the potential of shape memory polymers employing supramolecular switches.

Supramolecular structures of hydrogen bonding have been widely used to fabricate functional polymer. Chapter 6 reviews the synthesis of supramolecular shape memory polyurethane containing pyridine moieties. Chapter 7 discusses the thermally-active shape memory effect of supermolecular shape memory polymers. The pyridine ring superamolecular structure is the key element in determining shape recovery. Chapter 8 introduces the moisture-sensitive shape memory effect of supermolecular shape memory polymers. The hydrogen bonding present in the pyridine ring serves as moisture-sensitive 'switch', whilst the hard phase formed via hydrogen bonding present in the urethane groups acts as physical net points for the moisture-sensitive shape memory effect. Finally, potential applications of supermolecular shape memory polymers in smart windows and reshape applications are reviewed.

The shape memory effect in thermally-active shape memory polymers is usually induced by directly heating the shape memory polymer to a temperature above the switching temperature. Chapter 9 reviews the shape memory effect achieved by the indirect heating, for example by using electric fields. When electrically-conductive materials are incorporated into thermally-active shape memory polymers, the current passing through the SMPs can induce Joule heating and trigger shape recovery. When ferromagnetic fillers are employed in shape memory polymers, a magnetically-active shape memory effect can also be achieved by induced heating as a result of the magnetic field. Light absorbed by a deformed shape memory polymer can also increase the polymer's temperature to the switch transition temperature to trigger shape recovery.

Shape memory polymer solutions or emulsions can be used as easy-care finishing agents on fabrics. The shape memory polymer layer on fabric can provide a recovery force during heating, which will result in a more flat appearance and crease retention upon thermal stimulation. Chapter 10 reviews studies of shape memory polyurethane treated cotton fabrics. The chapter also discusses combining shape memory treatments with existing durable-finishing agents and a newly developed pre-treatment method using liquid ammonia.

The penultimate chapter in the book reviews methods for manufacturing T_g and T_m shape memory polyurethane polymer fibers using melt and wet spinning techniques. It reviews synthesis strategies, analysis of molecular structure, mechanical and thermal properties. Finally, though there have been many achievements in the development of shape memory polymers, there are still many challenging problems left to overcome. Chapter 12 discusses future trends and challenges, together with several promising applications of smart shape memory polymers. Shape memory polymers can store or release heat energy over a narrow temperature range. They are therefore able to help the human body maintain its natural temperature across hot and cold environments. Amongst a range of applications, Chapter 12 discusses the use of shape memory polymers with temperature-regulating effects.

Jinlian Hu

1
Introduction to shape memory polymers

DOI: 10.1533/9780857098542.1

Abstract: This chapter provides a general introduction to stimuli-responsive shape memory polymers (SMPs). Many SMPs based on different switching mechanisms have been developed and advances in research into SMPs are highlighted. At the end of the chapter, a typical SMP, shape memory polyurethane (SMPU), is introduced because of its tremendous advantages compared with other SMPs in terms of applications.

Key words: shape memory polymer, shape memory polyurethane, shape memory effect, high performance shape memory polymer.

1.1 Introduction

Shape memory materials (SMMs) are those materials that have the capability of recovering their original shapes upon exposure to an external stimuli such as heat (Hu *et al.*, 2002), electricity (Asaka and Oguro, 2000), light (Jiang *et al.*, 2006), a magnetic field (Makhosaxana *et al.*, 2000) or moisture (Yang *et al.*, 2006). Examples of SMMs include shape memory alloys (SMAs), shape memory ceramics and shape memory polymers (SMPs). The most well-known SMMs currently are SMAs, which have outstanding properties such as high strength with wide technical applications. SMAs, such as gold-cadmium, nickel-titanium or copper-zinc-aluminum, were developed in the 1980s, while SMPs and ceramics were developed in the 1990s.

SMPs have several advantages over SMAs and shape memory ceramics. These include light weight, low cost, good processability, high deformability, high shape recoverability, soft 'handle' (soft feel when handled) and tailorable switching temperature (Hyashi, 1993; Kim *et al.*, 1996; Liang and Rogers, 1997; Lin and Chen, 1998a,b; Tobushi *et al.*, 1998; Wei *et al.*, 1998; Hu *et al.*, 2002, 2005a,b; Lendlein and Kelch, 2002; Yang *et al.*, 2003; Hayashi *et al.*, 2004; Hu and Mondal, 2005; Zhu *et al.*, 2006b, 2007b, 2008b, 2009a,b; Liu *et al.*, 2007a, 2008; Chen *et al.*, 2008; Gunes and Jana, 2008; Ratna and Karger-Kocsis, 2008; Rousseau, 2008; Meng *et al.*, 2009; Xie and Rousseau, 2009).

1.2 Defining shape memory polymers

Figure 1.1 (Hu and Chen, 2010) shows a flower shape made using a SMP. The original shape is an open flower. By increasing the temperature to above its switching temperature, for example 80°C, the petals of the flower are coiled by hand to make a closed flower, as shown in Fig. 1.1(a). After the flower is cooled

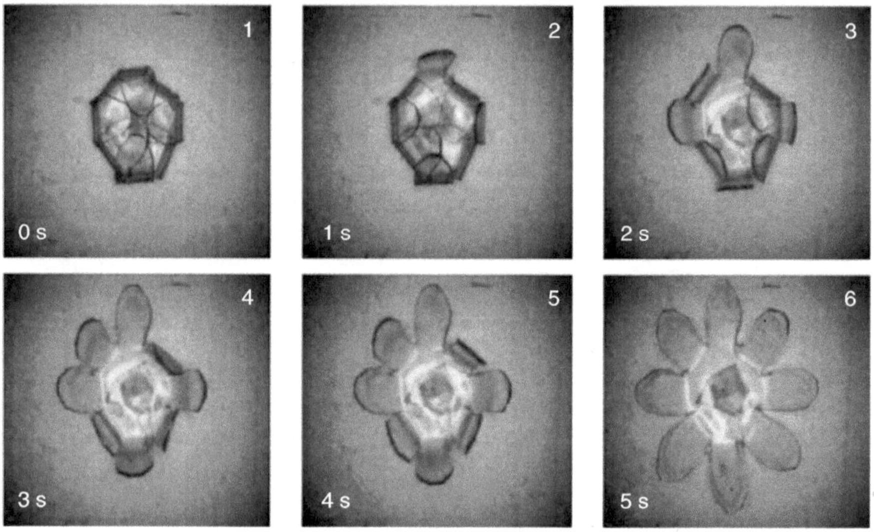

1.1 Opening process of shape memory flower at 80°C (Hu and Chen, 2010).

down to room temperature (T_{room}), the closed flower is fixed. Then if the closed flower is put into an oven at 80°C, the flower opens its petals within five seconds. Figure 1.2 shows the molecular mechanism of SMPs (Hu and Chen, 2010). SMPs consist of:

- cross-links or net points which determine the SMPU permanent shape (a SMP is required to maintain a stable network structure so as to recover its original shape);
- switchable segments which are used to create and maintain the temporary shape.

A SMP is deformed at a temperature above a switching transition temperature (T_{trans}) and fixed into a temporary shape by being cooled down to a temperature below the T_{trans} (Kim *et al.*, 1996; Takahashi *et al.*, 1996; Lin and Chen, 1998a; Lendlein and Kelch, 2002). Heating the SMP above its T_{trans} results in it recovering its permanent shape. Because of their unique shape memory effect (SME), high elongation at break (usually >100%), low cost and light weight, SMPs have attracted increasing attention from industry (Lendlein and Kelch, 2002). The superior properties of SMPs can be applied to the textile industry and a variety of shape memory textiles have been developed.

In polymer networks with a SME, the chemical or physical cross-links play the role of network conjunctions that stabilize the network during a series of thermo-mechanical processes (Reyntjens *et al.*, 1999; Lendlein *et al.*, 2001, 2005; Liu

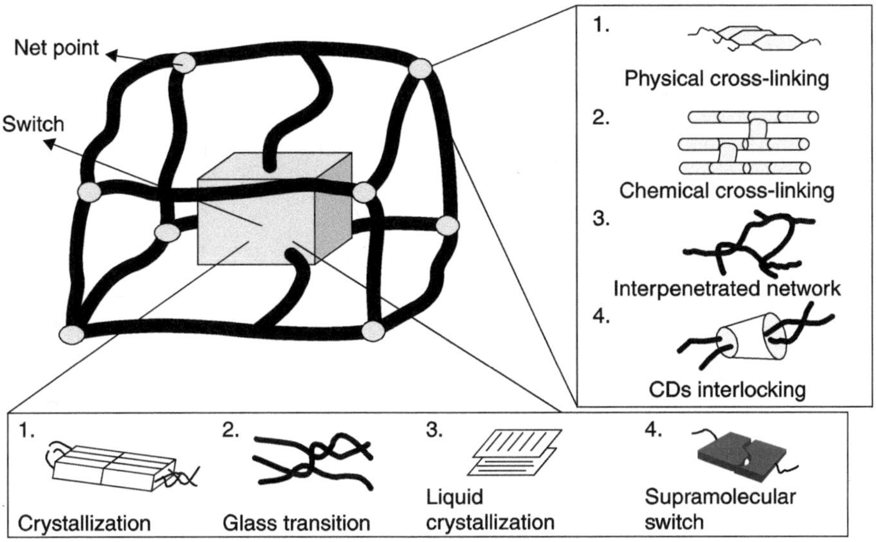

1.2 Molecular mechanism of shape memory polymers (Hu and Chen, 2010).

et al., 2002a, 2004; Zhu *et al.*, 2003, 2005; 2007a,b; Ji *et al.*, 2007; Chen *et al.*, 2007b; Meng *et al.*, 2007a,b; Mondal and Hu, 2007a). The polymeric network chains play the role of 'switching segments', whose thermal T_{trans} essentially serves as the T_{trans} in triggering the SME.

The molecular mobility of these switching segments changes significantly above and below T_{trans} and the materials' modulus can thus change by at least one or two orders of magnitude in a narrow temperature range around T_{trans}. The network chain segments are flexible at a temperature above T_{trans} but rigid at a temperature below T_{trans} where the mobility of the chains is frozen or at least limited. The polymer materials can therefore develop large deformation at a temperature above T_{trans} and can then be fixed into a temporary shape at a temperature below T_{trans}.

It is worthwhile clarifying the SME of polymers compared to SMPs. Liu *et al.* (2002a) have pointed out that all polymers, even those with no apparent network structure, intrinsically show SMEs on the basis of rubber elasticity, but with varied shape memory performance. If a polymer is deformed and cooled down to a frozen state quickly, so that the stress relaxation is effectively avoided, the elastic stress generated in the deformation can mostly be preserved. When the material is reheated to a high temperature, the stored stress is released, resulting in some recovery of its original shape.

The major difference between traditional polymeric elastomers and SMPs lies in the different thermal T_{trans} of their polymeric network chains. A traditional polymeric elastomer can be fixed in a temporary shape if it is cooled down to a sufficiently low temperature. In other words, traditional polymeric elastomers can show a SME under 'particular conditions'. However, the SME under these 'particular conditions' is not useful from the point of view of practical applications. Hayashi *et al.* (1995) developed a series of polyurethane SMPs with a T_{trans} located in the range from –30 to 65°C (Liang *et al.*, 1997). However, it is the SMPs which can hold a temporary shape around T_{room} that will be more useful for practical applications. Therefore the T_{trans} of SMPs is mostly above T_{room}.

1.3 Types of shape memory polymers

1.3.1 Morphological structures and shape memory properties

The network chains of the SMPs can be either crystalline or amorphous and therefore the thermal transition for triggering the SME can be either a melting transition or a glass transition (Kim *et al.*, 1996, 2000; Jeong *et al.*, 2000c; Lendlein and Kelch, 2002; Liu *et al.*, 2002a; Alteheld *et al.*, 2005). Correspondingly, SMPs can have either a melting temperature (T_m) or a glass transition temperature (T_g) as T_{trans}. Where $T_{trans} = T_m$, strain-induced crystallization of the switching segments can be obtained by cooling the deformed materials down to a temperature below T_m. The crystallites prevent shape recovery until the material is reheated above T_m (Kim *et al.*, 1996; Li *et al.*, 1996b). Where $T_{trans} = T_g$, the micro-Brown motions of the polymer are frozen and the switching segments are set into the glassy state when the material is cooled down to below T_g (Takahashi *et al.*, 1996; Tobushi *et al.*, 1996). The material therefore cannot recover its original shape and remains in the non-equilibrium state until reheated to above T_{trans} where the micro-Brown motions are activated.

Research in SMPs based on conventional glass or melting transition has been conducted intensively (Irie, 1998; Mother *et al.*, 2000; Tupper *et al.*, 2001; Lake and Beavers, 2002; Abrahamson *et al.*, 2003; Khonakdar *et al.*, 2007; Mondal and Hu, 2007a; Rezanejad and Kokabi, 2007; Zhu *et al.*, 2008a). With glass or melting transitions as the switch, many polymer systems have been reported to possess SMEs. According to Liu *et al.* (2007a), SMPs based on conventional glass or melting transitions fall into four classes:

- class I: covalently cross-linked glassy thermoset networks (glass transition as a switch);
- class II: covalently cross-linked semi-crystalline networks (melting transition as a switch);
- class III: physically cross-linked glassy copolymers (glass transition or melting transition as a switch); and
- class IV: physically cross-linked semi-crystalline block copolymers (glass transition or melting transition as a switch).

The shape memory properties of SMPs are closely related to the morphological structures, it is necessary to study in depth this relationship. For example, the

SME of shape memory polyurethanes (SMPUs), the shape recoverability and shape fixing ability, are dependent on the polyurethane network structure and phase transition of the SMPU, respectively. The roles played by the reversible phase and fixed phase are complex. Recovery stress is essential for most of the applications for SMPs since in practical applications the shape recovery can be impeded by external stimuli.

It has been widely accepted that the programming technology of SMPs has a significant influence on the shape memory properties. The programming process, w, is also known as the thermo-mechanical cycle. Thermo-mechanical cycles are generally composed of three steps, i.e. deformation, shape fixing and shape recovery. In the first step, the SMP is deformed at a temperature which may be above or below its switching temperature. In the second step, the SMP is set into a temporary shape by cooling the polymer to a temperature below its switching temperature. In the third step, the SMP recovers its original shape. In this thermo-mechanical shape memory cycle, many thermo-mechanical parameters can affect the performance of SMPs. The typical parameters are deformation temperature, deformation time, temperature decreasing speed, temperature increasing speed and deformation amplitude, etc.

1.3.2 High performance T_m-type shape memory polymers

Research has shown that physical cross-linking can play a significant role in the properties of SMPs. The shape fixity of T_m-type SMPs is a result of the crystallization of the soft phase. Recovery stress results from the internal stress stored during the deformation process. It is also affected by the shape fixity. High hard segment content (HSC) and physical cross-linking can improve the recovery stress, but can decreases shape fixity (Bogart *et al.*, 1983; Kim *et al.*, 1996). Raising physical cross-linking while employing low HSCs may be effective in improving the shape memory properties of SMPUs.

SMPU-ureas have been demonstrated to have well separated phase structures and shape recovery stress because of their strong hydrogen bonding in the hard segment phase (Garrett *et al.*, 2000, 2002, 2003; Luo *et al.*, 1996, 1997).

1.3.3 High performance T_g-type shape memory polymers

T_g-SMPUs are usually synthesized using polyols with an average molecular weight in the range of 300 to 1000 as the soft segment by a two-step polymerization technique. First, polyols are capped with isocyanates at both ends. Then the prepolymers are chain extended with small-sized diols or diamines. Compared with T_m-type SMPs, T_g-type SMPs have tailorable switching temperatures by varying soft segment contents or molecular weight. Decreasing soft segment contents can raise switching temperature, but can also cause significant phase mixing. Employing low molecular weight polyols can improve the switching temperature, which also decreases phase separation.

This book introduces a modified two-step polymerization technique. Partial urethane chains are employed as the soft segment phase. Such modified two-step polymerization can improve the molecule flexibility of the SMPU. After the modified two-step polymerization, short polyols and rigid 4,4′-diphenylmethane diisocyanate (MDI) form the soft segment phase. To further improve the phase separation, rigid chain extenders can be used.

1.3.4 Shape memory polymers with supramolecular switches

Non-covalent bonding is the dominant type of inter-molecular force in supramolecular chemistry. These non-covalent interactions include ionic bond, hydrophobic interactions, hydrogen bonds, Van der Waals forces and dipole–dipole bonds.

At the molecular level, the SME of SMPs is a result of rapid elastic modulus change due to molecule mobility. The dynamic supermolecular structure can lead to the mechanical property changes of polymers; as a result, dynamic supermolecular structures can be employed as the switches of SMPs. For example, in some circumstances, inter-molecular forces can lead to the phase separation of polymers. It has been found that the hydrogen bonding in SMPs, especially in the hard segment phase, can promote the formation of that phase. Hydrogen bonding is highly sensitive to temperature, i.e. at high temperature, hydrogen bonding decreases, whilst at low temperature, hydrogen bonding increases.

Researchers have successively prepared SMPs by taking advantage of the thermal reversibility of hydrogen bonding. By using hydrogen bonding as thermal reversible switches, many SMPs have been fabricated. Li *et al.* (2007) employed a thermal reversible switch associating a quadruple H-bonding structure as the switch for SMPs. The SMP consists of a lightly cross-linked network which is bonded to an associating ureidopyrimidinone (UPy) moiety. The UPy moiety forms a quadruple hydrogen bonding interaction. The molecular structure of the SMP network is presented in Fig. 1.3(a). The shape memory mechanism with thermal reversible hydrogen bonding as the switch is shown in Fig. 1.3(b). As can be observed from Fig. 1.3(b), during the fixity process, the thermal reversible hydrogen bonding forms new net points, which ‘pin’ the elastomer elastic recovery. As a result, the deformed shape is fixed. Upon heating to a high temperature, the net points formed by hydrogen bonding break. The material recovers its original shape as a result of entropy elasticity.

Chen *et al.* (2009a) and Zhu *et al.* (2009a) also incorporated self-complementary quadruple hydrogen bonding units into SMPUs and studied the SME. Other thermal reversible supermolecular interactions, such as ion–ion and ion–dipole, may also be used as the switch for SMPs. By employing the supermolecular inclusion between α-CD or γ-CD and polycaprotolactone and polyethylene glycol, Zhang *et al.* (2008) fabricated SMPs with the α-CD or γ-CD inclusion

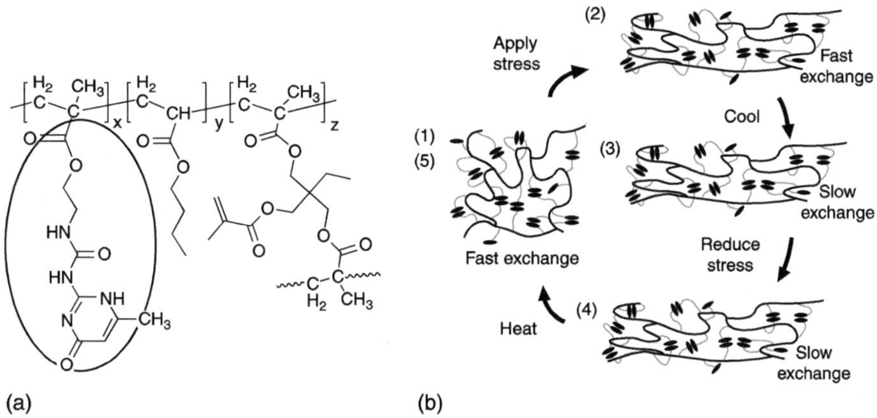

1.3 (a) Lightly cross-linked shape memory polymers containing pendent ureidopyrimidinone side-groups; (b) shape memory mechanism of the shape memory polymers with thermo-reversible H-bonding. The ≡ represent H-bonding groups in the hot and cold states, and the darker lines represent the lightly cross-linked covalent network (Li *et al.*, 2007).

crystallites, with PEG or PCL as the fixing phase and the polycaprotolactone and polyethylene glycol as the reversible phase.

Previous research on SMPs mainly focused on T_g- or T_m-type SMPs, which used crystallization and vitrification processes to fix a temporary shape. They recover their original shapes if they are heated to a temperature above their melting or T_g. With glass or melting transition as the switch, many polymer systems have been reported to possess SMEs.

Though recent research has demonstrated the feasibility of fabricating supramolecular SMPs, the studies on the properties of supramolecular SMPs are not systemic. The supramolecular structures significantly affect the mechanical and thermo-mechanical properties of the SMPs. In this book, the influence of hydrogen bonding as the shape memory switch on thermally-active SMPs is introduced. A type of supramolecular SMP, SMPU containing pyridine moieties, is presented. The relation between the thermally-active SME of the SMP and the non-covalent bonding is discussed.

The most studied SMPs are thermally activated. This means that the shape recovery is triggered by heating the polymer to a temperature above its switching temperature. Many researchers have achieved the SME on conventionally thermally-active SMPs. Employing hydrophilic or water soluble ingredients in SMPs can accelerate the moisture/water-active shape recovery process. Chen *et al.* (2007a) developed a water-active shape memory biodegradable polymer from chitosan cross-linked with epoxy. The chitosan is relatively hydrophilic.

The supramolecular moieties in supramolecular SMPs can significantly affect the hydrophilicity of the polymers. It has been found that the hydrogen bonding in

BIN-SMPU can significantly affect the mobility of the SMPU molecules (Chen *et al.*, 2009b). As has been suggested by Jaczewska *et al.* (2007), the water absorption of poly(vinyl pyridine) is several times higher than that of polythiophene. The pyridine ring is also sensitive to moisture. Therefore, the BIN-SMPUs with pyridine may respond to moisture.

1.3.5 Phase change materials for SMPUs

When the temperature of the skin of the human body differs by more than 3.0°C from its ideal temperature, the person feels uncomfortable (Tao, 2001; Mattila, 2006). It would be desirable, if at high temperatures the excess heat produced could be stored in the clothing system and then released when the person starts to get cold. Phase change materials (PCMs) that have been used to regulate temperature fluctuations have this function (Han and Hubbell, 1997; Langer, 1990; Son *et al.*, 1991; Veronese *et al.*, 1996).

PEG is a solid–liquid phase change polymeric material. It has a T_m from around 3.2 to 68.7°C and a very high phase change enthalpy depending on its molecular weight (Son *et al.*, 1991; Bryant, 1999; Mattila, 2006). Several research groups (Liang and Guo, 1995; Ye and Ge, 2000; He and Zhang, 2001) have prepared solid–solid PCMs, by employing PEG as the phase change ingredient and another skeleton-forming ingredient to keep the material in a solid state after the melting of PEG.

Zhang *et al.* (1999, 2004) prepared fibers of PEG/polypropylene, poly(ethylene terephthalate) and ethylene-vinyl acetate by controlling suitable spinning parameters and component contents. The PCMs prepared via physical blending have a tendency to lose their phase change characteristics after several heating–cooling cycles due to the loss of PEG. Jiang *et al.* (2002) developed a network solid–solid PCM with rigid polymer cellulose diacetate (CDA) serving as a skeleton, and the PEG as a branch chain. However, because of the covalent network structure of the PEG-grafted CDA, the material is not suitable for fiber preparation. More recently, Li and Ding (2007) prepared a novel PEG/MDI (MDI)/pentaerythrito cross-linking copolymer via the condensation reaction of PEG with tetrafunctional pentaerythritol isocyanate. The phase transition enthalpy was more than 100 J/g, with a transition point at 58.68°C.

Since the SME of SMPs is the result of the so-called micro-phase, the separated heterogeneous structure is composed of the hard phase and the soft phase. The solid–solid PCMs also show obvious SMEs.

1.3.6 Shape memory effect by indirect heating

Conventionally, the SME of SMPs is induced by directly heating the polymer to a temperature above the switching temperature. In many practical applications, electrical power is more convenient to use to trigger the shape recovery process than by external heating. It has been demonstrated that the shape recovery of SMPs can be achieved by Joule heating after the SMPs are filled with conductive fillers such as carbon black and carbon nanotubes (Hilmarkoerner, 2004; Cho *et al.*, 2005).

A certain level of electrical conductivity of SMPs can be reached by incorporating electrical conductive ingredients. When a current passes through the conductive ingredient network within SMPs, the induced Joule heating may raise the internal temperature to above the switching T_{trans} of the polymer to trigger shape recovery (Koerner *et al.*, 2004; Cho *et al.*, 2005; Paik *et al.*, 2005, 2006; Goo *et al.*, 2007). The conductive ingredients which have been used as conductive fillers include carbon nanotubes, polypyrrole (PPy) (Sahoo *et al.*, 2005, 2007a,b), carbon black, and short carbon fiber (Leng *et al.*, 2007, 2008a–d; Lan *et al.*, 2008; Lv *et al.*, 2008). Light absorbed by a deformed SMP can also increase SMP temperature to trigger shape recovery (Small *et al.*, 2005a,b, 2007). To improve light absorbing efficiency, dyes such as indocyanine green and Epolight 4121 (Small *et al.*, 2005a,b), carbon blacks and carbon nanotubes may also be used (Laroche *et al.*, 2002; Koerner *et al.*, 2004; Langer and Tirrell, 2004).

By employing ferromagnetic fillers in SMPs, magnetic-responsive SMPs were also prepared. SMPs recovered their original shape as a result of electromagnetic field-induced heating (Buckley *et al.*, 2006; Mohr *et al.*, 2006; Varga *et al.*, 2006; Razzaq *et al.*, 2007a,b; Behl, 2008; Cuevas *et al.*, 2009; Weigel *et al.*, 2009; Yakacki *et al.*, 2009). Figure 1.4 shows the shape recovery effect of the electro-active, light-active and magnetic field-active SMPs.

1.3.7 Shape memory polymer fibers

Though there has been much research on SMPUs in the last two decades (Hyashi, 1993; Kim *et al.*, 1996; Kim and Lee, 1998; Lin and Chen, 1998a,b; Tobushi *et al.*, 1998; Gall *et al.*, 2002; Lendlein and Kelch, 2002; Tang and Stylios, 2006; Gunes *et al.*, 2008; Liu *et al.*, 2002b, 2007b; Xie and Rousseau, 2009; Zhang *et al.*, 2008, 2009), the study on shape memory fibers (SMFs) is still at the initial stage. The limited reports on SMFs may in part be due to the difficulty in making qualified SMPs for fiber spinning. Kaursoin and Agrawal (2007) used an MM-4510 polyester SMPU procured from DiAPLEX, to prepare SMFs by melt-spinning. The soft segment T_g at 40.88°C acted as the switch temperature. However, the detailed specification and synthesis technology of the SMPU are not given.

Zhu *et al.* (2006c) reported a SMPU fiber prepared by using a wet-spinning process. The SMF showed complete shape recovery during the thermo-mechanical cyclic tensile testing process, because the disorientation and thermal shrinkage of the SMF partially contributed to the shape recovery. Figure 1.5 shows the elastic modulus between SMFs and other synthetic fibers. The main difference between SMFs and conventional synthetic fiber is the variation of the elastic modulus at the temperature normally used. The elastic modulus decreases significantly in this temperature range. For other synthetic fibers, such as Lycra and Polyester, the elastic modulus is constant with small changes in temperature.

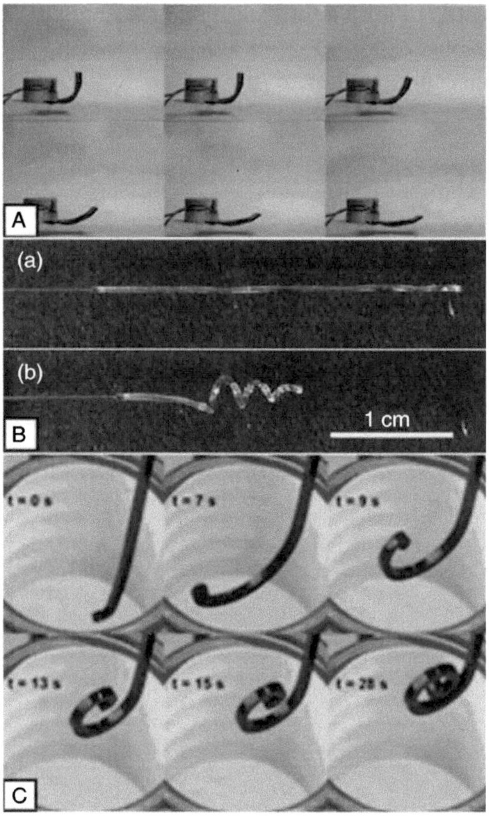

1.4 Shape memory effect induced by in-direction heating: (a) the electro-active shape recovery behavior of CNT/SMPU composites; (b) light-active shape recovery of an SMPU micro-actuator coupled to an optical fiber (A) temporary straight rod, (B) permanent corkscrew form and (C) magnetically-induced SME of an SMPU filled with magnetic nanoparticles inside a magnetic field of an inductor ring. Figure 1.4(A) reproduced with kind permission from Elsevier, Ltd., http://dx.doi.org/10.1016/j.compscitech.2008.08.016. Figure 1.4(B) reproduced with kind permission from Optical Society of America, http://dx.doi.org/10.1364/opex.13.008204. Figure 1.4(C) copyright (2006) National Academy of Sciences, USA, http://dx.doi.org/10.1073/pnas.0600079103.

At the switching temperature, the elastic modulus of the SMF decreases significantly due to the glass transition. The T_g could be in the range of 0 to 100°C. Other commonly used clothing synthetic fibers do not have the thermal transition in this range. The comparison studies between SMPU films and fibers show that SMFs more readily form aggregated hard segment domains.

SMFs have promising applications in the textile and clothing industry as they provide inspiration to create intelligent textiles with a self-regulating structure

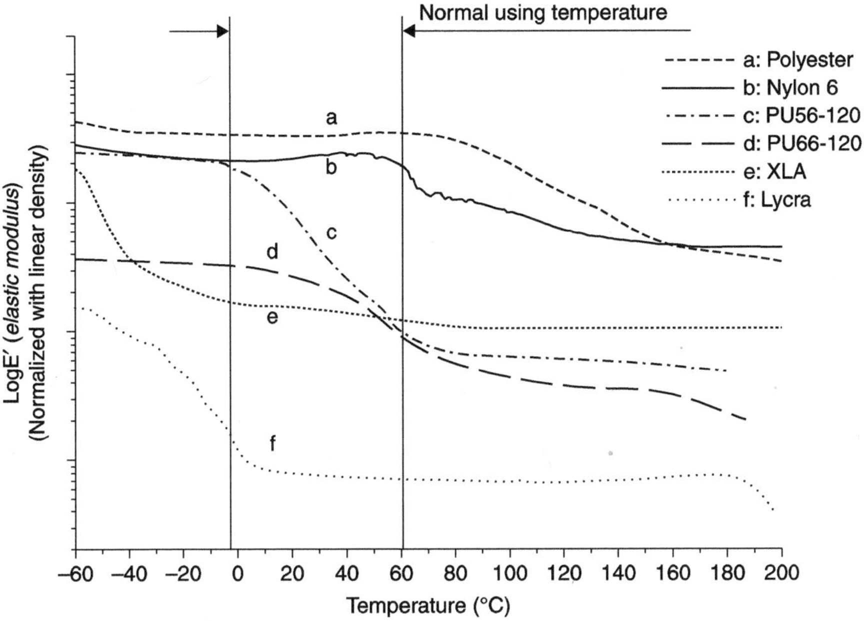

1.5 Comparison of elastic modulus between SMPU fiber and various man-made fibers (adapted from Zhu *et al.*, 2006c; Hu and Chen, 2010; Li *et al.*, 2007).

and performance in response to external stimulus. Compared with their bulk counterparts, SMFs have high mechanical properties and shape recovery force because of molecular orientation. Comparison with Spandex and Polyester fiber, shape memory fabrics have better capabilities for 3-D textiles due to their good shape fixity. Furthermore, the fabrics made of SMFs may fit wearers well, as a result of good deformability, retention capacity and partial elasticity of SMFs. Because of the stimulus sensitive effect, the SMFs in this special fiber format may also find applications in biomedical materials, high performance sensors, actuators and microgrippers.

Because of their molecule orientation, SMFs have outstanding mechanical properties. More importantly, compared with shape memory films without molecule orientation, the fiber shape recovery stress may be much higher.

1.4 A typical shape memory polymer: shape memory polyurethanes (SMPUs)

1.4.1 Chemistry of shape memory polyurethanes

Generally, SMPU composed of a soft segment and a hard segment is a stimuli-sensitive block copolymer, having the ability to change its shape at a temperature above its T_{tans} (Choi *et al.*, 2006). In the T_m-type SMPU, a high crystallinity (X_c) of the soft phase at T_{room} and the formation of stable hard segment domains acting as physical net points in the temperature above the T_m are the two conditions for the segmented PU with SME. Accordingly, the soft segment with good crystallizability can be the reversible phase of SMPU.

So far, PCL (Hu *et al.*, 2005b; Zhu *et al.*, 2003, 2007b), PBA (Ding *et al.*, 2006), PEA (Ding *et al.*, 2006), PTMG (Lee *et al.*, 2001) and PHA (Ding *et al.*, 2006) were all reported to synthesize SMPU with diisocyanate such as MDI and chain extenders such as 1,4-butandiol (BDO). In particular, the PCL-based SMPU and the PTMG-based SMPU were extensively researched. For example, Li *et al.* (1996a,b) investigated PCL-based SMPU by varying the soft segment length (SSL) and HSC, and proposed the above critical conditions for the segmented copolymer with SMEs. They also studied the dependency of SSL and HSC on its response temperature, final shape recovery and speed of that recovery.

Furthermore, Lee *et al.* (2001) investigated the structure and thermo-mechanical properties of PTMG-based SMPU with various HSCs. They also studied the influence of SSL and HSC on the shape memory behavior of PTMG-based SMPU (Lin and Chen, 1998a,b). Because PHA had a higher crystallizability than that of PEA and PBA, PHA-based SMPU was found to exhibit better shape memory behavior when compared with the PCL-based-, PEA-based- and PBA-based SMPUs (Ding *et al.*, 2006). Furthermore, the cross-linked SMPU (Hu *et al.*, 2005b), SMPU ionomer (Zhu *et al.*, 2006a,b) and SMPU composite blended with either resin (Jeong *et al.*, 2001), or functional inorganic particles (Li *et al.*, 2000) were recently developed from PCL or PTMG. In addition, a great deal of research conducted on the water vapor permeability and various applications of SMPU was carried out in many fields (Jeong *et al.*, 2000a; Mondal and Hu, 2006, 2007b).

Since their T_g can be improved to above the ambient temperature, many polyols with low Mn are also used for the T_g-type SMPU, such as PBAG with Mn of 600, and PTMG with Mn of 250 or 650. In these systems, higher HSC is usually required to obtain a T_{trans} above T_{room}; the HSC is usually above 60 wt.% in the PBA600-based SMPU. HSC should be above 55 wt.% in the PPG400-based SMPU, and the HSC is usually above 65 wt.% in the PTMG650-based SMPU. It seems that the polyols with higher Mn need higher HSC. Hence, an SMPU of desired T_{trans} can be synthesized by controlling the SSL and HSC.

Shape memory properties of SMPU are adjustable by controlling the SSLs and HSCs. In the T_m-type SMPU, as the SSL increases, the X_c of the soft segment will increase. Consequently, the T_{trans} increases and the shape fixity will be improved. However, as the HSC increases, the X_c will decrease. The T_{trans} and shape fixity will also decrease. The shape recovery is usually higher in the SMPU with higher HSCs. In the T_g-type SMPU, the T_g will move to a lower temperature range as the SSL increases or the HSC decreases. Hence, the T_{trans} decreases with the increase of SSL and the decrease of HSC.

1.4.2 Phase separation of shape memory polyurethanes

Phase separation is one of the most important criteria for classifying whether a polyurethane will have SMEs. Extensive research studies on the parameters that affect phase separation have been conducted (Chu *et al.*, 1992; Tao *et al.*, 1994; Mclean and Sauer, 1997; Velankar and Cooper, 1998, 2000a,b; Yontz, 1999; Jeong *et al.*, 2000b; Garrett *et al.*, 2000; Janik *et al.*, 2003). The soft-segment content has been proven as one of the major parameters which affect the phase separation process. When the soft-segment content is too low, the soft segment will be embedded into the hard segment domain and vice-versa. Typically, 30 to 70% of the soft segment content will be taken at phase separation.

Li and Cooper (1990) examined PTMO-based PUs through the use of a high-voltage electron microscopy (HVEM) to reduce electron damage. Their experiments showed that no significant phase separation exists when the HSC is 26 wt.%. Some short cylinders are found when HSC increases to 33 wt.%. Furthermore, lamellar morphology exists when HSC increases to 50 wt.%, which indicates a complete process of phase separation. The PU sample with 50 wt.% HSC was then annealed at 125°C for 24 hours. The HVEM micrograph showed longer lamellae with uncharged inter-lamellae spaces, which indicates a more complete phase separation process due to the rearrangement of segregation below the hard-segment T_m.

Valankar and Cooper (1998, 2000a,b) studied the phase separation and rheological properties of PU melts in terms of block length and incompatibility. A series of PUs were synthesized by soft segment with different molecular weights (830, 1250, 2000 and 3000). The results identified that phase separation does not occur in molecular weights of the 830 and 1250 PU series. Furthermore, the molecular weights of the 2000 and 3000 PU series show an increasing phase mixing structure with increasing temperature. This result confirms that copolymer melts have tenacity of phase mixing at high temperatures, unlike the polymer films investigated by Li and Cooper (1990) with HVEM. By controlling the ionic group in the hard segment of the PU (quaternization), PU ionomers can be synthesized. Such PU ionomers are similar to traditional PUs, but are more effective in controlling the phase separation, as the inter-urethane hydrogen bonding between carbonyl and N-H groups is one of the major driving forces for microphase separation.

1.5 Conclusions

SMPs are a class of stimuli-responsive materials with the capability of changing their shape upon exposure to external stimuli (Tobushi *et al.*, 1996; Wei *et al.*, 1998; Poilane *et al.*, 2000; Lendlein and Kelch, 2002). Because of their unique SME, the SMPs have drawn increasing attention in the technical community. In this chapter, comprehensive literature pertaining to SMPs is reviewed and the research highlights in the area are introduced:

1. morphological structures and shape memory properties;
2. high performance T_m-type SMPs;
3. high performance T_g-type SMPs;
4. SMPs with supramolecular switches;
5. phase change materials based on SMPUs;
6. moisture-active SME of polymers with supermolecular structures;
7. SME by indirect heating;
8. SMFs.

One of the typical SMPs, i.e. SMPU, including chemistry and phase of SMPUs, is introduced since most research in this book employs SMPUs to study the relationship between the functions and structures of SMPs.

1.6 References

Abrahamson, E. R., Lake, M. S., Gall, K. (2003), Shape memory mechanics of an elastic memory composite resin, *Journal of Intelligent Material System and Structures*, **14**, 623–32.

Alteheld, A., Feng, Y., Kelch, S., Lendlein, A. (2005), Biodegradable, amorphous copolyester-urethane networks having shape-memory properties, *Angew. Chem. Int. Ed.*, **44**, 1188.

Asaka, K., Oguro, K. (2000), Bending of polyelectrolyte membrane platinum composites by electric stimuli, *Journal of Electro-analytical Chemistry*, **480**, 186–98.

Behl, M. (2008), Shape-memory polymers, *Third International Conference on Nanotechnology and Smart Textiles for Industry, Healthcare and Fashion*, 19 March 2008, The Royal Society, London.

Bogart, V., John, W. C., Gibson, P. E., Cooper, S. L. (1983), Structure–property relationships in polycaprolactone-polyurethanes, *J. Polym. Sci., Polym. Phys. Ed.*, **21**, 65–95.

Bryant, Y. G. (1999), Melt spun fibers containing microencapsulated phase change material, *Advances in Heat and Mass Transfer in Biotechnology – The ASME International Mechanical Engineering Congress and Exposition*, Nashville, TN.

Buckley, P. R., Mckinley, G. H., Wilson, T. S., Iv, W. S., Benett, W. J., *et al.* (2006), Inductively heated shape memory polymer for the magnetic actuation of medical devices, *IEEE Trans. Biomed. Eng.*, **53**, 2075–2083.

Chen, M. C., Tsai, H. W., Chang, Y., Lai, W. Y., Mi, F. L., *et al.* (2007a), Rapidly self-expandable polymeric stents with a shape-memory property, *Biomacromolecules*, **8**, 2774–80.

Chen, S. J., Hu, J. L., Liu, Y. Q., Liem, H. M., Zhu, Y., Meng, Q. H. (2007b), Effect of molecular weight on shape memory behavior in polyurethane films, *Polym. Int.*, **56**, 1128–34.

Chen, S., Hu, J., Zhuo, H., Zhu, Y. (2008), Two-way shape memory effect in polymer laminates, *Mater. Lett.*, **62**, 4088–90.

Chen, S., Hu, J., Yuen, C. W., Chan, L. (2009a), Novel moisture-sensitive shape memory polyurethanes containing pyridine moieties, *Polymer*, doi: 10.1016/j.polymer.2009.07.031.

Chen, S. J., Hu, J. L., Yuen, C. W., Chan, L. K. (2009b), Supramolecular polyurethane networks containing pyridine moieties for shape memory materials, *Mater. Lett.*, **63**, 1462–4.

Cho, J. W., Kim, J. W., Jung, Y. C., Goo, N. S. (2005), Electroactive shape-memory polyurethane composites incorporating carbon nanotube, *Macromol. Rapid Commun.*, **26**, 412–16.

Choi, N. Y., Kelch, S., Lendlein, A. (2006), Synthesis, shape-memory functionality and hydrolytical degradation studies on polymer networks from poly(rac-lactide) b-poly(propylene oxide)-b-poly(rac-lactide) dimethacrylates, *Adv. Eng. Mater.*, **85**, 439–45.

Chu, B., Gao, T., Li, Y., Wang, J., Desper, C. R., Byrne, C. A. (1992), Microphase separation kinetics in segmented polyurethanes: Effects of soft segment length and structure, *Macromolecules*, **25**, 5724–9.

Cuevas, J. M., Alonso, J., German, L., Iturrondobeitia, M., Laza, J. M., *et al.* (2009), Magneto-active shape memory composites by incorporating ferromagnetic microparticles in a thermo-responsive polyalkenamer, *Smart Materials and Structures*, **18**, 075003.

Ding, X. M., Hu, J. L., Tao, X. M., Hu, C. R. (2006), Preparation of temperature-sensitive polyurethanes for smart textiles, *Textile Research Journal*, **76**, 406–13.

Gall, K., Dunn, M. L., Liu, Y., Finch, D., Lake, M., Munshi, N. A. (2002), Shape memory polymer nanocomposites, *Acta Mater.*, **50**, 5115–26.

Garrett, J. T., Runt, J., Lin, J. S. (2000), Microphase separation of segmented poly(urethane urea) block copolymers, *Macromolecules*, **33**, 6353–9.

Garrett, J. T., Lin, J. S., Runt, J. (2002), Influence of preparation conditions on microdomain formation in poly(urethane urea) block copolymers, *Macromolecules*, **35**, 161–8.

Garrett, J. T., Xu, R., Cho, J., Runt, J. (2003), Phase separation of diamine chain-extended poly(urethane) copolymers: FT-IR spectroscopy and phase transitions, *Polymer*, **44**, 2711–19.

Goo, N. S., Paik, I. H., Yoon, K. J. (2007), The durability of a conducting shape memory polyurethane actuator, *Smart Materials and Structures*, **16**, N23–N26.

Gunes, I. S., Jana, S. C. (2008), Shape memory polymers and their nanocomposites: A review of science and technology of new multifunctional materials, *J. Nanosci. Nanotechnol.*, **8**, 1616–37.

Gunes, I. S., Cao, F., Jana, S. C. (2008), Evaluation of nanoparticulate fillers for development of shape memory polyurethane nanocomposites, *Polymer*, **49**, 2223–34.

Han, D., Hubbell, J. (1997), Synthesis of polymer network scaffolds from lactide and poly(ethylene glycol) and their interaction with cells, *Macromolecules*, **30**, 6077–83.

Hayashi, S., Kondo, S., Kapadia, P., Ushioda, E. (1995), Room-temperature-functional shape-memory polymers, *Plast. Eng.*, **51**, 29.

Hayashi, S., Tasaka, Y., Hayashi, N., Akita, Y. (2004), Development of smart polymer materials and its various applications, *Mitsubishi Heavy Industries, Ltd., Technical Review*, **41**, 1–3.

He, Q., Zhang, W. (2001), A study on latent heat storage exchangers with the high-temperature phase-change material, *International Journal of Energy Research*, **25**, 331–41.

Hilmarkoerner, G. P., Nathan, A., Pearce, M. A., Vaia, R. A. (2004), Remotely actuated polymer nanocomposites – Stress-recovery of carbon-nanotube-filled thermalplastic elastomers, *Natural Material*, **3**, 115–20.

Hu, J. L., Chen, S. (2010), A review of actively moving polymers in textile applications, *J. Mater. Chem.*, **20**, 3346–55.

Hu, J. L., Ding, X. M., Tao, X. M. (2002), Shape memory polymers and their applications to smart textile products, *Journal of China Textile University*, **19**, 89–93.

Hu, J. L., Mondal, S. (2005), Structural characterization and mass transfer properties of segmented polyurethane: Influence of block length of hydrophilic segments, *Polym. Int.*, **54**, 764–71.

Hu, J. L., Ji, F. L., Wong, Y. W. (2005a), Dependency of the shape memory properties of a polyurethane upon thermomechanical cyclic conditions, *Polym. Int.*, **54**, 600–5.

Hu, J. L., Yang, Z. H., Ji, F. L., Liu, Y. Q. (2005b), Cross-linked polyurethanes with shape memory properties, *Polym. Int.*, **54**, 854–9.

Hyashi, S. (1993), Properties and applications of polyurethane, *International Progress in Urethanes*, **6**, 90–115.

Irie, M. (1998), *Shape Memory Polymers, In Shape Memory Materials*, Cambridge: Cambridge University Press.

Jaczewska, J., Budkowski, A., Bernasik, A., Raptis, I., Raczkowska, J., *et al.* (2007), Humidity and solvent effects in spin-coated polythiophene-polystyrene blends, *J. Appl. Polym. Sci.*, **105**, 67–79.

Janik, H., Pałys, B., Petrovic, Z. S. (2003), Multiphase-separated polyurethanes studied by micro-Raman spectroscopy, *Macromol. Rapid. Commun.*, **24**, 265–8.

Jeong, H. M., Ahn, B. K., Cho, S. M., Kim, B. K. (2000a), Water vapor permeability of shape memory polyurethane with amorphous reversible phase, *Journal of Polymer Science: Part B: Polymer Physics*, **38**, 3009–17.

Jeong, H. M., Ahn, B. K., Kim, B. K. (2000b), Temperature sensitive water vapour permeability and shape memory effect of polyurethane with crystalline reversible phase and hydrophilic segments, *Polym. Int.*, **49**, 1714–21.

Jeong, H. M., Lee, S. Y., Kim, B. K. (2000c). Shape memory polyurethane containing amorphous reversible phase, *J. Mater. Sci.*, **35**, 1579.

Jeong, H. M., Ahn, B. K., Kim, B. K. (2001), Miscibility and shape memory effect of thermoplastic polyurethane blends with phenoxy resin, *Eur. Polym. J.*, **37**, 2245–52.

Ji, F. L., Hu, J. L., Li, T. C., Wong, Y. W. (2007), Morphology and shape memory effect of segmented polyurethanes. Part I: With crystalline reversible phase, *Polymer*, **48**, 5133–45.

Jiang, Y., Ding, E., Li, G. (2002), Study on transition characteristics of PEG/CDA solid–solid phase change materials, *Polymer*, **43**, 117–22.

Jiang, H. Y., Kelch, S., Lendlein, A. (2006), Polymers move in response to light, *Adv. Mater.*, **18**, 1471–5.

Kaursoin, J., Agrawal, A. K. (2007), Melt spun thermoresponsive shape memory fibers based on polyurethanes: Effect of drawing and heat-setting on fiber morphology and properties, *J. Appl. Polym. Sci.*, **103**, 2172–82.

Khonakdar, H. A., Jafari, S. H., Rasouli, S., Morshedian, J., Abedini, H. (2007), Investigation and modeling of temperature dependence recovery behavior of shape-memory cross-linked polyethylene, *Macromol. Theory Simul.*, **16**, 43–52.

Kim, B. K., Lee, S. Y. (1998), Polyurethane ionomers having shape memory effects, *Polymer*, **39**, 2803–8.

Kim, B. K., Lee, S. Y., Xu, M. (1996), Polyurethane having shape memory effect, *Polymer*, **37**, 5781–93.

Kim, B. K., Shin, Y. J., Cho, S. M., Jeong, H. M. (2000), Shape-memory behavior of segmented polyurethanes with an amorphous reversible phase: The effect of block length and content, *J. Polym. Sci., Polym. Phys. Ed.*, **38**, 2652.

Koerner, H., Price, G., Pearce, N. A., Alexander, M., Vaia, R. A. (2004), Remotely actuated polymer nanocomposites – stress-recovery of carbon-nanotube-filled thermalplastic elastomers, *Natural Material*, **3**, 115–20.

Lake, M. S., Beavers, F. L. (2002), The fundamentals of designing deployable structures with elastic memory composites, Collection of Technical Papers – *AIAA/ASME/ASCE/AHS/ASC Structures, Structural Dynamics and Materials Conference*, vol., 3, 2050–63, Denver, CO.

Lan, X., Huang, W. M., Liu, N., Phee, S. Y., Leng, J. S., Du, S. Y. (2008), Improving the electrical conductivity by forming NI powder chains in a shape-memory polymer filled with carbon black, *Proceedings of SPIE – The International Society for Optical Engineering*, vol. 6927, *Electroactive Polymer Actuators and Devices (EAPAD) 2008*, 692–717, San Diego, CA.

Langer, R. (1990), New methods of drug delivery, *Science*, **249**, 1527–33.

Langer, R., Tirrell, D. A. (2004), Designing materials for biology and medicine, *Nature*, **428**, 487–92.

Laroche, F. E. F. G., Fiset, M., Mantovani, D. (2002), Shape memory materials for biomedical applications, *Advanced Engineering Materials*, **4**, 91–104.

Lee, B. S., Chun, B. C., Chung, Y.-C., Sul, K. I., Cho, J. W. (2001), Structure and thermomechanical properties of polyurethane block copolymers with shape memory effect, *Macromolecules*, **34**, 6431–7.

Lendlein, A., Kelch, S. (2002), Shape-memory polymers, *Angew. Chem. Int. Ed.*, **41**, 2034–57.

Lendlein, A., Schmidt, A. M., Langer, R. (2001), Ab-polymer networks based on oligo(ε-caprolactone) segments showing shape-memory properties, *PNAS*, **98**, 842.

Lendlein, A., Schmidt, A. M., Schroeter, M., Langer, R. (2005), Shape-memory polymer networks from oligo(ε-caprolactone) dimethacrylates, *J. Polym. Sci., Polym. Chem. Ed.*, **43**, 1369.

Leng, J., Lv, H., Liu, Y., Du, S. (2007), Electroactivate shape-memory polymer filled with nanocarbon particles and short carbon fibers, *Appl. Phys. Lett.*, **91**, 144105.

Leng, J., Lu, H., Liu, Y., Du, S. (2008a), Conductive nanoparticles in electro activated shape memory polymer sensor and actuator, *Proceedings of SPIE – The International Society for Optical Engineering*, vol. 6931, *Nanosensors and Microsensors for Bio-Systems 2008*, 693109, San Diego, CA.

Leng, J., Lv, H., Liu, Y., Du, S. (2008b), Synergic effect of carbon black and short carbon fiber on shape memory polymer actuation by electricity, *J. Appl. Phys.*, **104**, 104917.

Leng, J. S., Huang, W. M., Lan, X., Liu, Y. J., Du, S. Y. (2008c), Significantly reducing electrical resistivity by forming conductive NI chains in a polyurethane shape-memory polymer/carbon-black composite, *Appl. Phys. Lett.*, **92**, 204101–3.

Leng, J. S., Lan, X., Liu, Y. J., Du, S. Y., Huang, W. M., *et al.* (2008d), Electrical conductivity of thermoresponsive shape-memory polymer with embedded micron sized NI powder chains, *Appl. Phys. Lett.*, **92**, 014101–3.

Li, C., Cooper, S. L. (1990), Direct observation of the micromorphology of polyether polyurethanes using high-voltage electron microscopy, *Polymer*, **31**, 3–7.

Li, F. K., Hou, J. N., Zhu, W., Zhang, X., Xu, M., *et al.* (1996a), Crystallinity and morphology of segmented polyurethanes with different soft-segment length, *J. Appl. Polym. Sci.*, **62**, 631–8.

Li, F. K., Zhang, X., Hou, J. A., Xu, M., Luo, X. L., *et al.* (1996b), Studies on thermally stimulated shape memory effect of segmented polyurethanes, *J. Appl. Polym. Sci.*, **64**, 1511–16.

Li, F. K., Qi, L., Yang, J., Xu, M., Luo, X., Ma, D. (2000), Polyurethane/conducting carbon black composites: Structure, electric conductivity, strain recovery behavior, and their relationships, *J. Appl. Polym. Sci.*, **75**, 68–77.

Li, J., Viveros, J. A., Wrue, M. H., Anthamatten, M. (2007), Shape-memory effects in polymer networks containing reversibly associating side-groups, *Adv. Mater.*, **19**, 2851–5.

Li, W. D., Ding, E. Y. (2007), Preparation and characterization of cross-linking PEG/MDI/PE copolymer as solid–solid phase change heat storage material, *Solar Energy Materials & Solar Cells*, **91**, 764–8.

Liang, C., Rogers, C. A. (1997), One-dimensional thermomechanical constitutive relations for shape memory materials, *J. Intel. Mat. Syst. Str.*, **8**, 285–302.

Liang, C., Rogers, C. A., Malafeew, E. (1997), Investigation of shape memory polymers and their hybrid composites, *J. Intel. Mat. Syst. Str.*, **8**, 380–6.

Liang, X., Guo, Y. Q. (1995), Crystalline-amorphous phase transition of a poly(ethylene glycol)/cellulose blend, *Macromolecules*, **28**, 6551–5.

Lin, J. R., Chen, L. W. (1998a), Study on shape-memory behavior of polyether-based polyurethanes, Part I: Influence of the hard-segment content, *J. Appl. Polym. Sci.*, **69**, 1563–74.

Lin, J. R., Chen, L. W. (1998b), Study on shape-memory behavior of polyether-based polyurethanes, Part II: Influence of soft-segment molecular weight, *J. Appl. Polym. Sci.*, **69**, 1575–86.

Liu, C., Chun, S. B., Mather, P. T., Zheng, L., Haley, E. H., Coughlin, E. B. (2002a), Chemically cross-linked polycyclootene: Synthesis characterization and shape memory behavior, *Macromolecules*, **35**, 9868.

Liu, J., Ma, D. Z., Li, Z. (2002b), FT-IR studies on the compatibility of hard-soft segments for polyurethane-imide copolymers with different soft segments, *Eur. Polym. J.*, **38**, 661–5.

Liu, C., Qin, H., Mather, P. T. (2007a), Review of progress in shape-memory polymers, *J. Mater. Chem.*, **17**, 1543–58.

Liu, G., Ding, X., Cao, Y., Zheng, Z., Peng, Y. (2004), Shape memory of hydrogen-bonded polymer network/poly(ethyleneglycol) complexes, *Macromolecules*, **37**, 2228.

Liu, Y., Chung, A., Hu, J. L., Lu, J. (2007b), Shape memory behavior of SMPU knitted fabric, *Journal of Zhejiang University SCIENCE A*, **8**, 830–4.

Liu, Y., Lv, H., Lan, X., Leng, J., Du, S. (2008), Review of electro-active shape-memory polymer composite, Compscitech.2008.08.016. *Compos Sci Technol.*, **69**, 2064–8.

Luo, N., Wang, D. N., Ying, S. K. (1996), Crystallinity and hydrogen bonding of hard segments in segmented poly(urethane urea) copolymers, *Polymer*, **37**, 3577–83.

Luo, N., Wang, D. N., Ying, S. K. (1997), Hydrogen-bonding properties of segmented polyether poly(urethane urea) copolymer, *Macromolecules*, **30**, 4405–9.

Lv, H., Leng, J., Du, S. (2008), Electro-induced shape-memory polymer nanocomposite containing conductive particles and short fibers, *Proceedings of SPIE – The International Society for Optical Engineering*, vol. 6929, *Behavior and Mechanics of Multifunctional and Composite Materials 2008*, 69291L, 10–13 March 2008, San Diego, CA.

Makhosaxana, X. P., Filipcsei, G., Zrinyi, M. (2000), Preparation and responsive properties of magnetically soft poly(n-isopropylacrylamide) gels, *Macromolecules*, **33**, 1716–19.

Mattila, H. R. (2006), *Intelligent Textiles and Clothing*, vol. 3: Woodhead Publishing Limited.

Mclean, R. S., Sauer, B. B. (1997), Tapping-mode AFM studies using phase detection for resolution of nanophases in segmented polyurethanes and other block copolymers, *Macromolecules*, **30**, 8314–17.

Meng, Q. H., Hu, J. L., Yeung, L. Y. (2007a), An electro-active shape memory fibre by incorporating multi-walled carbon nanotubes, *Smart Materials and Structures*, **16**, 830–6.

Meng, Q. H., Hu, J. L., Zhu, Y., Lu, J., Liu, Y. (2007b), Polycaprolactone-based shape memory segmented polyurethane fiber, *J. Appl. Polym. Sci.*, **106**, 2515–23.

Meng, Q. H., Hu, J. L., Zhu, Y., Lu, J., Liu, B. H. (2009), Biological evaluations of a smart shape memory fabric, *Textile Research Journal*, **79**, 1522–33.

Mohr, R., Kratz, K., Weigel, T., Lucka-Gabor, M., Moneke, M., Lendlein, A. (2006), Initiation of shape-memory effect by inductive heating of magnetic nanoparticles in thermoplastic polymers, *Proceedings of the National Academy of Sciences*, **103**, 3540–5.

Mondal, S., Hu, J. L. (2006), Segmented shape memory polyurethane and its water vapor transport properties, *Des Monomers Polym.*, **9**, 527–50.

Mondal, S., Hu, J. L. (2007a), Studies of shape memory property on thermoplastic segmented polyurethanes: Influence of PEG 3400, *Journal of Elastomers and Plastics*, **39**, 81–91.

Mondal, S., Hu, J. L. (2007b), Water vapor permeability of cotton fabrics coated with shape memory polyurethane, *Carbohyd. Polym.*, **67**, 282–7.

Mother, P. T., Jeon, H. G., Haddad, T. S. (2000), Strain recovery in POSS hybrid thermoplastics, *Polymer Preprints*, **41**, 528–9.

Paik, I. H., Goo, N. S., Yoon, K. J., Jung, Y. C., Cho, J. W. (2005), Electric resistance property of a conducting shape memory polyurethane actuator, *Key Engineering Materials*, **297–300**, 1539–44.

Paik, I. H., Goo, N. S., Jung, Y. C., Cho, J. W. (2006), Development and application of conducting shape memory polyurethane actuators, *Smart Materials and Structures*, **15**, 147682.

Poilane, C., Delobelle, P., Lexcellent, C., Hayashi, S., Tobushi, H. (2000), Analysis of the mechanical behavior of shape memory polymer membranes by nanoindentation, bulging and point membrane deflection tests, *Thin Solid films*, **379**, 156.

Ratna, D., Karger-Kocsis, J. (2008), Recent advances in shape memory polymers and composites: A review, *J. Mater. Sci.*, **43**, 254–69.

Razzaq, M. Y., Anhalt, M., Frormann, L., Weidenfeller, B. (2007a), Mechanical spectroscopy of magnetite filled polyurethane shape memory polymers, *Materials Science and Engineering: A*, **471**, 57–62.

Razzaq, M. Y., Anhalt, M., Frormann, L., Weidenfeller, B. (2007b), Thermal, electrical and magnetic studies of magnetite filled polyurethane shape memory polymers, *Materials Science and Engineering: A*, **444**, 227–35.

Reyntjens, W. G., Prez, F. E. D., Goethals, E. J. (1999), Polymer networks containing crystallizable poly(octadecyl vinylether) segments for shape-memory materials, *Macromol. Rapid Commun.*, **20**, 251.

Rezanejad, S., Kokabi, M. (2007), Shape memory and mechanical properties of cross-linked polyethylene/clay nanocomposites, *Eur. Polym. J.*, **43**, 2856–65.

Rousseau, I. A. (2008), Challenges of shape memory polymers: A review of the progress toward overcoming SMP's limitations, *Polymer Engineering & Science*, **48**, 2075–89.

Sahoo, N. G., Jung, Y. C., Goo, N. S., Cho, J. W. (2005), Conducting shape memory polyurethane-polypyrrole composites for an electroactive actuator, *Macromol. Mater. Eng.*, **290**, 1049–55.

Sahoo, N. G., Jung, Y. C., Cho, J. W. (2007a), Electroactive shape memory effect of polyurethane composites filled with carbon nanotubes and conducting polymer, *Mater. Manuf. Process.*, **22**, 419–23.

Sahoo, N. G., Jung, Y. C., Yoo, H. J., Cho, J. W. (2007b), Influence of carbon nanotubes and polypyrrole on the thermal, mechanical and electroactive shape-memory properties of polyurethane nanocomposites, *Compos. Sci. Technol.*, **67**, 1920–9.

Small, W., Metzger, M. F., Wilson, T. S., Maitland, D. J. (2005a), Laser-activated shape memory polymer microactuator for thrombus removal following ischemic stroke: Preliminary *in vitro* analysis, *IEEE Journal of Selected Topics in Quantum Electronics*, **11**, 892–901.

Small, W., Wilson, T. S., Benett, W. J., Loge, J. M., Maitland, D. J. (2005b) Laser-activated shape memory polymer intravascular thrombectomy device, *Opt. Express*, **13**, 8204–13.

Small, W., Buckley, P., Wilson, T., Benett, W., Hartman, J. (2007), Shape memory polymer stent with expandable foam: A new concept for endovascular embolization of fusiform aneurysms, *IEEE Trans. Biomed. Eng.*, **54**, 1157–60.

Son, C., Moreshouse, H. L., Jeffrey, H. (1991). An experimental investigation of solid-state phase-change materials for solar thermal storage, *Journal of Solar Energy Engineering*, **113**, 244–9.

Takahashi, T., Hayashi, N., Hayashi, S. (1996), Structures and properties of shape-memory polyurethane block copolymers, *J. Appl. Polym. Sci.*, **60**, 1061.

Tang, S. L. P., Stylios, G. K. (2006), An overview of smart technologies for clothing design and engineering, *International Journal of Clothing Science and Technology*, **18**, 108–28.

Tao, H.-J., Meuse, C. W., Yang, X., Macknight, W. J., Hsu, S. L. (1994), A spectroscopic analysis of phase separation behavior of polyurethane in restricted geometry: Chain rigidity effects, *Macromolecules*, **27**, 7146–51.

Tao, X. M. (2001), *Smart Fibres, Fabrics and Clothing*, vol. 3: Woodhead Publishing Limited.

Tobushi, H., Hara, H., Yamada, E., Hayashi, S. (1996), Thermomechanical properties in a thin film of shape memory polymer of polyurethane series, *Smart. Mater. Struct.*, **5**, 483.

Tobushi, H., Hashimoto, T., Ito, N., Hayashi, S., Yamada, E. (1998), Shape fixity and shape recovery in a film of shape memory polymer of polyurethane series, *J. Intell. Mater. Syst. Struct.*, **9**, 127–36.

Tupper, M., Munshi, N., Beavers, F., Meink, T., Gall, K., Mikulas, M. J. (2001), Developments in elastic memory composite materials for spacecraft deployable structures, *IEEE Aerospace Conference Proceedings*, vol. 5, 52541–8, Big Sky, MT.

Varga, Z., Filipcsei, G., Zrínyi, M. (2006), Magnetic field sensitive functional elastomers with tuneable elastic modulus, *Polymer*, **47**, 227–33.

Velankar, S., Cooper, S. L. (1998), Microphase separation and rheological properties of polyurethane melts, Part I: Effect of block length, *Macromolecules*, **31**, 9181–92.

Velankar, S., Cooper, S. L. (2000a), Microphase separation and rheological properties of polyurethane melts, Part II: Effect of block incompatibility on the microstructure, *Macromolecules*, **33**, 382–94.

Velankar, S., Cooper, S. L. (2000b), Microphase separation and rheological properties of polyurethane melts, Part III: Effect of block incompatibility on the viscoelastic properties, *Macromolecules*, **33**, 395–403.

Veronese, M., Monfardini, C., Caliceti, P., Schiavon, O., Scrawen, M. D., Beer, D. (1996), Improvement of pharmacokinetic, immunological and stability properties of asparaginase by conjugation to linear and branched monomethoxy poly(ethylene glycol), *J. Control Release*, **40**, 199–209.

Wei, Z. G., Sandstrom, R., Miyazaki, S. (1998), Shape-memory materials and hybrid composites for smart systems, Part I: Shape-memory materials, *J. Mater. Sci.*, **33**, 3743–62.

Weigel, T., Mohr, R., Lendlein, A. (2009), Investigation of parameters to achieve temperatures required to initiate the shape-memory effect of magnetic nanocomposites by inductive heating, *Smart Materials and Structures*, **18**, 025011.

Xie, T., Rousseau, I. A. (2009), Facile tailoring of thermal transition temperatures of epoxy shape memory polymers, *Polymer*, **50**, 1852–6.

Yakacki, C. M., Satarkar, N. S., Gall, K., Likos, R., Hilt, J. Z. (2009), Shape-memory polymer networks with Fe_3O_4 nanoparticles for remote activation, *J. Appl. Polym. Sci.*, **112**, 3166–76.

Yang, B., Huang, W. M., Li, C., Li, L. (2006), Effects of moisture on the thermomechanical properties of a polyurethane shape memory polymer, *Polymer*, **47**, 1348–56.

Yang, J. H., Chun, B. C., Chung, Y.-C., Cho, J. H. (2003), Comparison of thermal/mechanical properties and shape memory effect of polyurethane block-copolymers with planar or bent shape of hard segment, *Polymer*, **44**, 3251–8.

Ye, H., Ge, X. (2000), Preparation of polyethylene-paraffin compound as a form-stable solid–liquid phase change material, *Solar Energy Materials & Solar Cells*, **64**, 37–44.

Yontz, D. J. (1999), *An analysis of molecular parameters governing phase separation in a reacting polyurethane system.* Thesis. University of Massachusetts, Amherst.

Zhang, H., Wang, H., Zhong, W., Du, Q. (2009), A novel type of shape memory polymer blend and the shape memory mechanism, *Polymer*, **50**, 1596–601.

Zhang, S., Yu, Z., Govender, T., Luo, H., Li, B. (2008), A novel supramolecular shape memory material based on partial A-CD-PEG inclusion complex, *Polymer*, **49**, 3205–10.

Zhang, X. X., Wang, X. C., Hu, L. (1999), Spinning and properties of PP/PEG composite fibers for heat storage and thermo-regulation, *Journal of Tianjin Institute Textiles Science and Technology*, **18**, 1–3.

Zhang, X. X., Zhang, H., Niu, J. J. (2004), *Thermoregulated fiber and its products*, China Patent ZL 00105837.1.

Zhu, G. M., Liang, G., Xu, Q., Yu, Q. (2003), Shape-memory effects of radiation cross-linked poly(epsilon-caprolactone), *J. Appl. Polym. Sci.*, **90**, 1589–95.

Zhu, G. M., Xu, Q. Y., Liang, G. Z., Zhou, H. F. (2005), Shape-memory behaviors of sensitizing radiation-cross-linked polycaprolactone with polyfunctional poly(ester acrylate), *J. Appl. Polym. Sci.*, **95**, 634.

Zhu, Y., Hu, J. L., Yeung, K. W., Fan, H. J., Liu, Y. Q. (2006a) Shape memory effect of PU ionomers with ionic groups on hard segments, *Chin. J. Polym. Sci.*, **24**, 173–86.

Zhu, Y., Hu, J. L., Yeung, K. W., Liu, Y. Q., Liem, H. M. (2006b), Influence of ionic groups on the crystallization and melting behavior of segmented polyurethane ionomers, *J. Appl. Polym. Sci.*, **100**, 4603–13.

Zhu, Y., Hu, J. L., Yeung, L. Y., Liu, Y., Ji, F. L., Yeung, K. W. (2006c), Development of shape memory polyurethane fiber with complete shape recoverability, *Smart Materials and Structures*, **15**, 1385–94.

Zhu, Y., Hu, J., Yeung, L.-Y., Lu, J., Meng, Q., *et al.* (2007a), Effect of steaming on shape memory polyurethane fibers with various hard segment contents, *Smart Materials and Structures*, **16**, 969–81.

Zhu, Y., Hu, J. L., Yeung, K. W., Choi, K. F., Liu, Y. Q., *et al.* (2007b), Effect of cationic group content on shape memory effect in segmented polyurethane cationomer, *J. Appl. Polym. Sci.*, **103**, 545–56.

Zhu, Y., Hu, J., Choi, K. F., Yeung, K. W., Meng, Q., Chen, S. (2008a) Crystallization and melting behavior of the crystalline soft segment in a shape-memory polyurethane ionomer, *J. Appl. Polym. Sci.*, **107**, 599–609.

Zhu, Y., Hu, J., Lu, J., Yeung, L. Y., Yeung, K.-W. (2008b), Shape memory fiber spun with segmented polyurethane ionomer, *Polym. Adv. Technol.*, **19**, 1745–53.

Zhu, Y., Hu, J., Liu, Y. (2009a), Shape memory effect of thermoplastic segmented polyurethanes with self-complementary quadruple hydrogen bonding in soft segments, *The European Physical Journal E*, **28**, 3–10.

Zhu, Y., Hu, J. L., Yeung, Y. (2009b), Effect of soft-segment crystallization, hard segment physical cross-link on shape memory function in antibacterial segmented PU ionomers, *Acta Biomaterials*, **5**, 3346–57.

2

T_m-type shape memory polymers

DOI: 10.1533/9780857098542.23

Abstract: With a melting transition as the switch, shape memory polymers can exhibit the shape memory effect (SME). This chapter demonstrates the experimental findings in the study of a series of T_m-SMPUs. The morphological variation with chemical composition and the influences of morphology and thermo-mechanical conditions including deformation strain, shape fixing speed and pre-deformation on the SME are discussed.

Key words: T_m-type shape memory polymers, phase separation, phase morphology, shape memory properties.

2.1 Introduction

The correlations between morphological structure, thermo-mechanical condition and shape memory properties are the key issues for the development of high-performance SMPUs. In previous reports, Kim *et al.* (1996) and Li *et al.* (1997)) investigated the thermal properties and structure of the T_m-SMPUs with DSC, DMA and WAXD. However, they could not reveal the effects of structure evolution with the change of chemical composition on shape memory properties. In addition, the morphology of segmented polyurethanes made from low-molecular-weight or semi-crystalline polyols has not been well investigated. As a novel physical property, the shape memory properties appear different from the conventional mechanical properties and are strongly related to thermo-mechanical conditions, which have not been studied in detail.

A set of T_m-SMPUs was prepared covering a wide composition range in which a series of morphological changes with the variation of chemical composition would occur. The formulations of the T_m-SMPUs are detailed in Table 2.1. The series of segmented polyurethanes are designated as PCL4000-B-xx where 'B' refers to the chain extender BD; 'xx' denotes the wt.% hard segment content (HSC). For example, PCL4000-B-30 means that the HSC of the segmented polyurethane is 30 wt.%.

The morphological structure was experimentally observed with multiple characterization techniques including DSC, TMA, tensile hysteresis and SAXS. The relationship between the morphological structure and shape memory properties of the T_m-SMPUs was explored. In addition, the influences of thermo-mechanical conditions such as deformation amplitude, stress relaxation and pre-deformation on the shape memory behaviors are discussed.

Table 2.1 Feed ratios and structural characteristics of T_m-SMPUs

Samples	HSC (MDI + BDO) (wt.%)	Density at 65°C (g/cm^3)	Theoretical volume fraction of hard segment	Average number of MDI per hard segment	Average number of BD per hard segment
PCL4000-B-10	10	1.090	0.076	1.41	0.41
PCL4000-B-15	15	1.106	0.124	2.22	1.22
PCL4000-B-20	20	1.120	0.167	2.97	1.97
PCL4000-B-25	25	1.133	0.211	3.91	2.91
PCL4000-B-30	30	1.143	0.256	4.98	3.98
PCL4000-B-35	35	1.155	0.302	6.32	5.32
PCL4000-B-40	40	1.170	0.349	7.66	6.66
PCL4000-B-45	45	1.182	0.397	9.36	3.36
PCL4000-B-50	50	1.199	0.446	11.4	10.4

2.2 Structure and properties of T_m-type shape memory polymers

2.2.1 Thermal properties of T_m-type shape memory polymers

The thermal properties of all the T_m-SMPUs were investigated through DSC tests at a scanning rate of 10°C/min. As shown in Fig. 2.1, all the segmented polyurethanes show an endotherm in the temperature range 40 to 50°C, and this is ascribed to the melting of the crystalline soft phase. The segmented polyurethanes with HSC $\geq$ 30% show an endothermic peak in the vicinity of 200°C, which is ascribed to the melting of the hard phase. This suggests that when HSC $\geq$ 30%, the hard phase exists in the segmented polyurethanes. With the increase of HSC, the intensity of the endothermic peak pertaining to the hard phase is increased while that corresponding to the soft phase is reduced, suggesting an increasing fraction of hard phase and a decreasing crystallinity (X_c) of the soft phase.

Table 2.2 presents the melting temperature T_m, heat of fusion (ΔH_m) and X_c of the two phases. The X_c of the PCL soft segments of the T_m-SMPUs was calculated according to the ΔH_m data obtained from the DSC tests, assuming the perfect PCL crystal has a melting enthalpy of 0.14 kJ/g (Ping *et al.*, 2005). It was found that T_m, ΔH_m and X_c of the soft phase decrease with the increase of HSC and drop abruptly when HSC $\leq$ 40%. This is ascribed to the increasing limitation imposed on soft segments. The T_m and ΔH_m of the hard phase increase with the increase of HSC, indicating an increase of fraction of the hard phase.

In the foregoing DSC tests, the segmented polyurethanes with HSC $\leq$ 20% clearly show no endotherm of hard phase. But the segmented polyurethane PCL4000-B-25 shows an indistinct endothermic behavior. To verify the change, a group of DSC tests at a higher scanning rate of 50°C/min were conducted

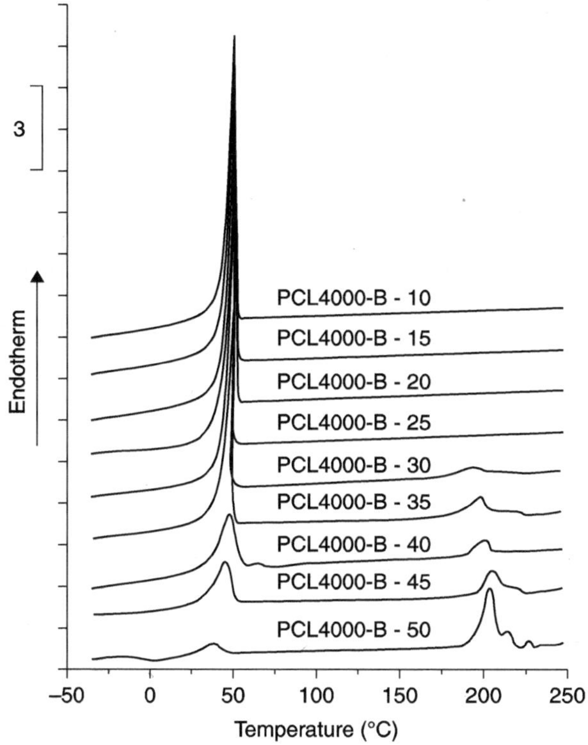

2.1 DSC traces of the T_m-SMPUs at scanning rate of 10°C/min.

Table 2.2 DSC experimental results of DSC with scanning rate of 10°C/min

Sample	T_m (°C)		Heat of melting	ΔH (J/g)	
	Soft segments	Hard segments	Soft segments	Hard segments	X_c of soft phase (%)
PCL4000-B-10	49.5	–	47.1	–	33.7
PCL4000-B-15	48.7	–	45.6	–	32.5
PCL4000-B-20	48.0	–	44.7	–	32.0
PCL4000-B-25	46.9	195.0	35.5	1.2	25.4
PCL4000-B-30	46.5	196.7	32.8	1.7	23.4
PCL4000-B-35	44.7	199.4	31.9	2.2	22.3
PCL4000-B-40	42.5	199.9	18.7	4.5	13.4
PCL4000-B-45	42.2	200.4	15.9	4.7	10.8
PCL4000-B-50	37.4	203.9/215.3/ 226.9	3.8	8.0/0.9/0.5	2.7

Source: Ping *et al.* (2005)

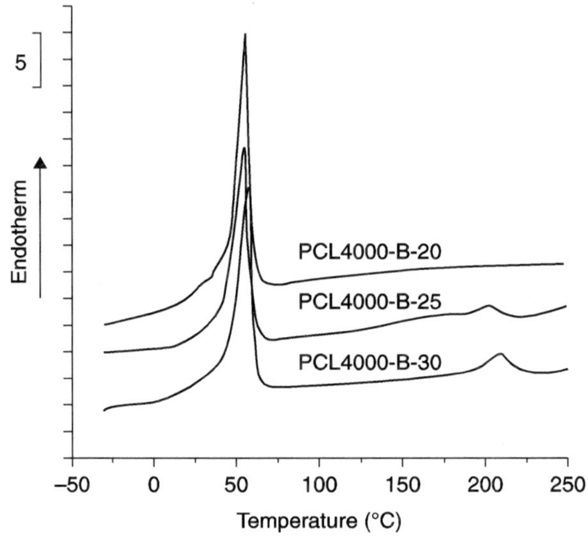

2.2 DSC traces of three T_m-SMPUs at scanning rate of 50°C/min.

on PCL4000-B-20, PCL4000-B-25 and PCL4000-B-30. As illustrated in Fig. 2.2, PCL4000-B-25 and PCL4000-B-30 show the endotherm of the hard phase while PCL4000-B-20 does not. This implies that no hard domains exist when HSC is less than 25%. However, TMA tests and SAXS tests reveal that there are still hard domains in PCL4000-B-15 and PCL4000-B-20. This may result from the limited sensitivity of the DSC technique in detecting the presence of the hard phase.

2.2.2 Thermo-mechanical properties of T_m-type shape memory polymers

The TMA tests of the segmented polyurethanes are shown in Fig. 2.3. The segmented polyurethanes exhibit two softening transitions. The transition emerging in the temperature range 40 to 50°C is ascribed to the softening of the soft phase, which is consistent with the DSC results. The transition arising in the higher temperature range 90 to 200°C is attributed to the softening of the hard phase. All the segmented polyurethanes except PCL4000-B-10 show a platform before the softening of the hard phase. PCL4000-B-10 does not exhibit a platform and the softening transition of the hard phase. This indicates that PCL4000-B-10 forms no hard domains, whereas the other polyurethanes do possess a hard phase. It is also found that the softening transition of the hard phase is shifted to higher temperatures with an increase of HSC. This can be ascribed to the increasing order and purity of the hard phase. PCL4000-B-45 and PCL4000-B-50 show no softening transition of the soft phase, which suggests that the hard phase changes from an isolated state to an interconnected state; or from a discontinuous phase to

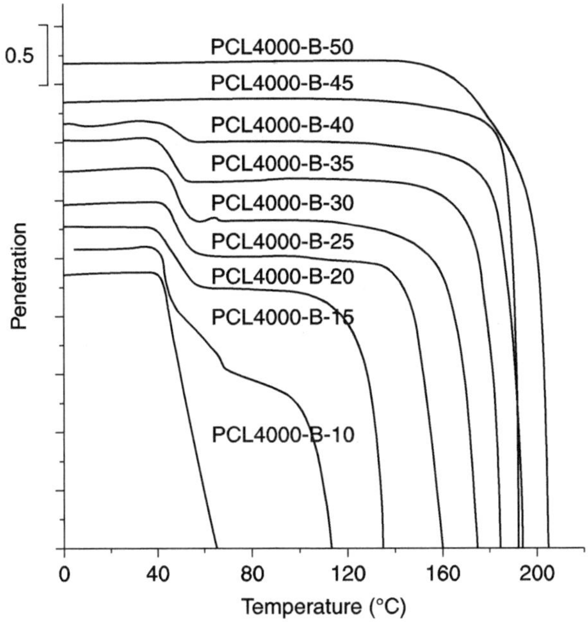

2.3 TMA traces of the T_m-SMPUs at the scanning rate of 10°C/min.

a continuous phase in this HSC range. The interconnected hard phase forms a rigid framework which prevents the materials from softening when the soft phase melts.

2.2.3 Tensile hysteresis of T_m-type shape memory polymers

To verify the change in morphology, tensile hysteresis tests were conducted on the segmented polyurethanes at the temperature $T_m + 20°C$ and the results are shown in Fig. 2.4. It can be seen that the percentage hysteresis for PCL4000-B-30, PCL4000-B-35 and PCL4000-B-40 is comparatively low, while that for PCL4000-B-45 is substantially raised. This is due to the hard phase changing from a discontinuous state to a continuous state. The interconnected state leads to more deformation occurring in the hard phase and consequently gives rise to a higher percentage hysteresis in the cases of PCL4000-B-45 and PCL4000-B-50.

2.2.4 Phase separation of T_m-type shape memory polymers

A great number of researchers have employed SAXS to investigate the morphology of segmented polyurethanes with an amorphous soft phase (Blundell *et al.*, 2002, 2004; Born *et al.*, 1982; Bras *et al.*, 1995; Chang *et al.*, 1998; Chu *et al.*, 1992; Elwell *et al.*, 1994; Garrett *et al.*, 2000, 2002; Krakovsky *et al.*, 1997; Li *et al.*,

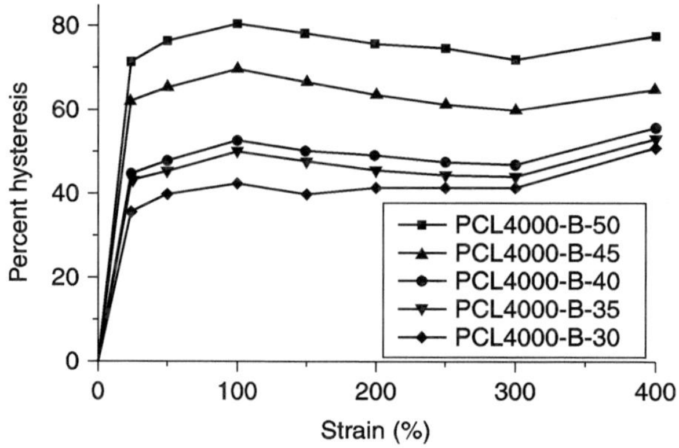

2.4 Tensile hysteresis tests of the T_m-SMPUs at 65°C.

1992a,b, 2002; Martin *et al.*, 2000; O'Sickey *et al.*, 2003; Petrovic *et al.*, 2005; Ryan *et al.*, 1991, 1992, 1993; Valenkar and Cooper, 1998, 2000). Few investigations have been made into the morphology of polyurethanes with a crystalline soft phase. Bogart *et al.* (1983) studied the morphology of a series of segmented polyurethanes with crystalline PCL diols as soft segments with SAXS at room temperature. The authors considered that the segmented polyurethanes could be treated as a 'pseudo two-phase system' due to the apparent difference between the soft and hard segments. The SAXS experimental results of the segmented polyurethanes were therefore mathematically analyzed based on two-phase models.

However, for the segmented polyurethanes with a crystalline soft phase, two-phase models are not appropriate because at room temperature the segmented polyurethanes contain a great deal of crystallites of soft segments which contribute to X-ray scattering. Hence the morphology of a hard phase cannot be properly derived. In this study, the SAXS tests of the segmented polyurethanes were performed at T_m + 20°C, in order to ensure the complete melt of the crystalline soft phase. The segmented polyurethanes mainly possess a hard phase and amorphous soft phase at T_m + 20°C. Thereby the segmented polyurethanes can be approximately regarded as a 'two-phase system'.

The SAXS profiles of the segmented polyurethanes are presented in Fig. 2.5. The segmented polyurethanes show scattering maxima or shoulders, except for PCL4000-B-10. This implies that PCL4000-B-10 is basically homogeneous while the other polyurethanes exhibit more or less phase separation. It can be concluded that all the segmented polyurethanes possess hard domains, except PCL4000-B-10. This is in agreement with the conclusions obtained from the TMA tests. The three-dimensional (3-D) interdomain spaces correspond to the maxima of the curves of I(q) vs q. With Lorentz correction, the one-dimensional (1-D) interdomain spaces are extracted from the maxima of the curves of $I(q)q^2$ vs q. Applying the model

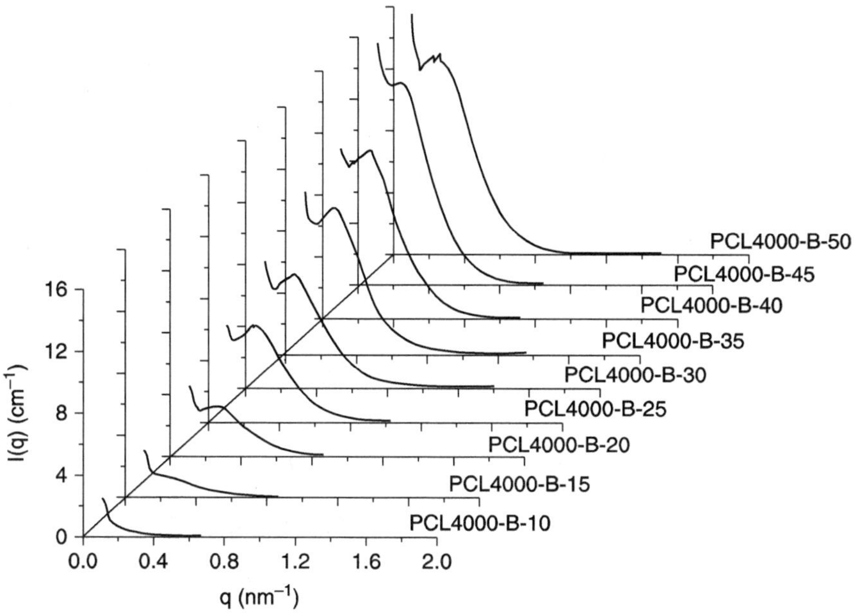

2.5 SAXS profiles of the T_m-SMPUs.

developed by Strobl *et al.* (1980), the 1-D interdomain spaces can also be obtained from the 1-D correlation function $\gamma_1(r)$, which is given by Eq. [2.1]:

$$\gamma_1(r) = \int_0^{\infty} I(q)q^2 \cos(qr)dq \qquad [2.1]$$

As indicated in Fig. 2.6, the 1-D interdomain spaces correspond to the primary maxima on the curves of the correlation function $\gamma_1(r)$. In addition, the domain sizes of the hard and soft phases can be extracted from the curves of the correlation function $\gamma_1(r)$. The 3- and 1-D inderdomain spaces of the segmented polyurethanes are presented in Table 2.3.

In general, the interdomain space of the segmented polyurethanes decreases with an increase of HSC. But at HSC = 45%, the interdomain space increases slightly. This may be attributed to the discontinuity–continuity transition of the hard phase. The decrease of the interdomain space with the increase of HSC suggests the increasing concentration of the hard domains, i.e., the physical cross-linking density of the polyurethane cross-linked networks.

Through the analysis based on Porod's law, the domain sizes of segmented polyurethanes can be estimated. Porod's law is given by Eq. [2.2] (Bogart *et al.*, 1983; Huang *et al.*, 2006; Koberstein *et al.*, 1992; Koberstein and Leung, 1992; Leung and Koberstein, 1985):

$$\lim_{q \to \infty} I(q) = \frac{K}{q^4} \qquad [2.2]$$

Table 2.3 Effect of HSC on interdomain spacings of the T_m-SMPUs

	3-dimensional (nm) Bragg's law	1-dimensional (nm)	
		Bragg's law	$\gamma_1(r)$
PCL4000-B-10	–	–	–
PCL4000-B-15	25.1	16.3	18.0
PCL4000-B-15	24.1	15.4	16.1
PCL4000-B-15	23.4	15.4	15.6
PCL4000-B-15	23.5	15.1	15.5
PCL4000-B-15	23.0	15.4	15.5
PCL4000-B-15	22.8	15.4	15.5
PCL4000-B-15	23.8	16.4	16.3
PCL4000-B-15	22.8	14.8	13.9
PCL4000-B-10	–	–	–
PCL4000-B-15	25.1	16.3	18.0
PCL4000-B-15	24.1	15.4	16.1
PCL4000-B-15	23.4	15.4	15.6

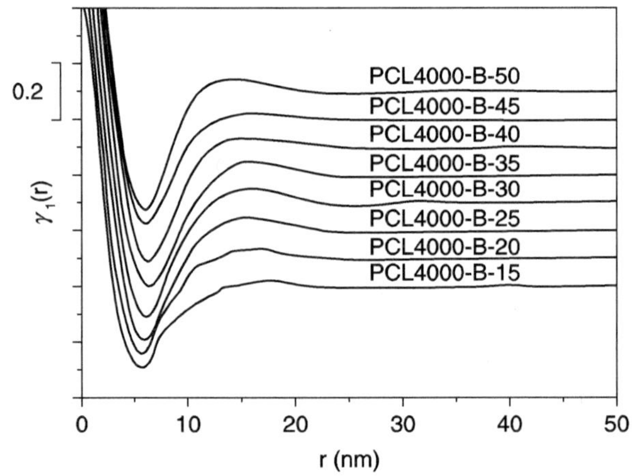

2.6 One-dimensional correlation function $\gamma_1(r)$ vs r.

where K is the Porod's law constant. Taking into account the influence of the background IB caused by phase mixing, the observed scattering intensity Iobs(q) is revised into (Chang *et al.*, 1998; Garrett *et al.*, 2000, 2002):

$$\lim_{q\to\infty} I_{obs}(q) = \lim_{q\to\infty} \frac{K}{q^4} + I_B \qquad [2.3]$$

The background corrected intensity is therefore equal to Iobs(q)-IB. By considering the effect of the diffuse boundary between the hard and soft domains, the background corrected intensity is written as (Chang *et al.*, 1998; Garrett *et al.*, 2000, 2002):

$$\lim_{q \to \infty} I(q) = \lim_{q \to \infty} \frac{K}{q^4} \exp(-\sigma^2 q^2) \qquad [2.4]$$

where σ is a measure of the interfacial boundary thickness. Porod's law constant K is related to the interfacial surface-to-volume ratio S/V:

$$\frac{S}{V} = \pi \cdot \phi_1 \phi_2 \frac{K}{\int_0^\infty I(q) q^2 dq} \qquad [2.5]$$

The average lengths of soft domains $\bar{l}_1$ and hard domains $\bar{l}_2$ can be calculated from the following equations:

$$\bar{l}_1 = 4\phi_1 / \left(\frac{S}{V}\right) \qquad [2.6]$$

$$\bar{l}_2 = 4\phi_2 / \left(\frac{S}{V}\right) \qquad [2.7]$$

where ϕ_1 and ϕ_2 refer to the theoretical volume fractions of the soft and hard phase, respectively. The domain sizes obtained from the analysis based on Porod's law and from $\gamma_1(r)$ curves are presented in Table 2.4. It is found that hard domain size generally increases with the increase of HSC. But at the point HSC = 45%, it slightly decreases. This is also due to the discontinuity–continuity transition of hard domains. In addition, the soft domain size decreases with the increase of HSC. This tendency is particularly noticeable in the low HSC region. In addition, the larger soft domain sizes indicate lower physical cross-linking density.

Table 2.4 Domain sizes of the T_m-SMPUs

	From Porod's law		From γ_1(r) lamellar model	
	$\bar{l}_2$ (nm)	$\bar{l}_1$ (nm)	d_2 (nm)	d_1 (nm)
PCL4000-B-15	2.6	18.9	3.1	13.3
PCL4000-B-20	2.6	13.4	3.3	11.0
PCL4000-B-25	2.8	10.8	3.7	9.8
PCL4000-B-30	2.9	8.7	4.1	9.5
PCL4000-B-35	3.4	8.1	4.3	8.8
PCL4000-B-40	3.6	6.9	4.4	9.1
PCL4000-B-45	3.5	5.5	4.0	9.2
PCL4000-B-50	3.6	4.9	3.9	8.0

2.3 Shape memory properties of T_m-type shape memory polymers

First, the shape memory effect (SME) was quantified via the set of thermo-mechanical cyclic tensile tests mentioned above. Figure 2.7 shows the SME of the segmented polyurethanes with varying HSC. It can be seen that the shape fixity decreases with an increase of HSC. When HSC $\leq$ 35%, the shape fixity of the segment is over 94%, while when HSC $\geq$ 40%, the shape fixity decreases abruptly.

The crystallization of the soft segments is considered to determine the shape fixing of the T_m-SMPU. In order to understand the effect of the crystallization on shape fixity, three sets of WAXD tests were performed on the segmented polyurethanes before extension, after shape fixing and after shape recovery to trace crystallization of the soft phase in shape memorization. Figure 2.8 illustrates the WAXD profiles of the segmented polyurethanes. Before extension, all the segmented polyurethanes with HSC $\leq$ 40% show distinct diffraction peaks, which are ascribed to the crystallization of the PCL soft phase. The intensity of the diffraction peaks decreases with an increase of HSC. This suggests that the X_c of the soft phase decreased gradually with an increase of HSC. When HSC reaches 45 and 50%, the diffraction peaks disappear, indicating that the soft segments are amorphous in PCL4000-B-45 and PCL4000-B-50.

As shown in Fig. 2.8(b), the intensity of the diffraction peaks of the segmented polyurethanes increase substantially after shape fixing. Even the polyurethanes PCL4000-B-45 and PCL4000-B-50 show diffraction peaks. Because the positions of the diffraction peaks are identical to those shown in the WAXD tests before extension, the peaks are still ascribed to the crystallization of the soft phase. The increase of the X_c is attributed to the strain-induced crystallization of the soft segments after deformation and shape fixing. The X_c of PCL4000-B-40,

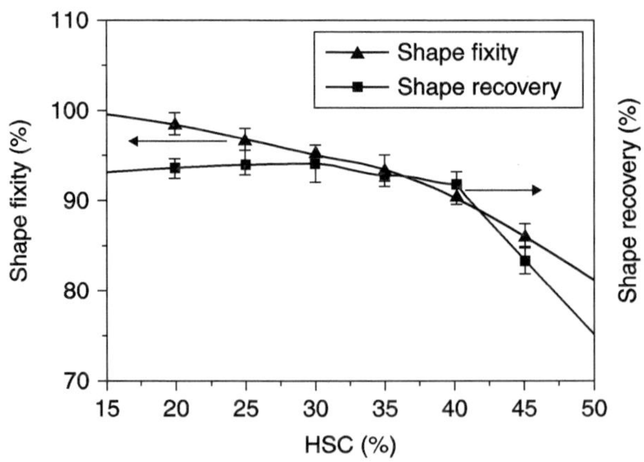

2.7 Effect of hard segment content on shape memory effect of the T_m-SMPUs.

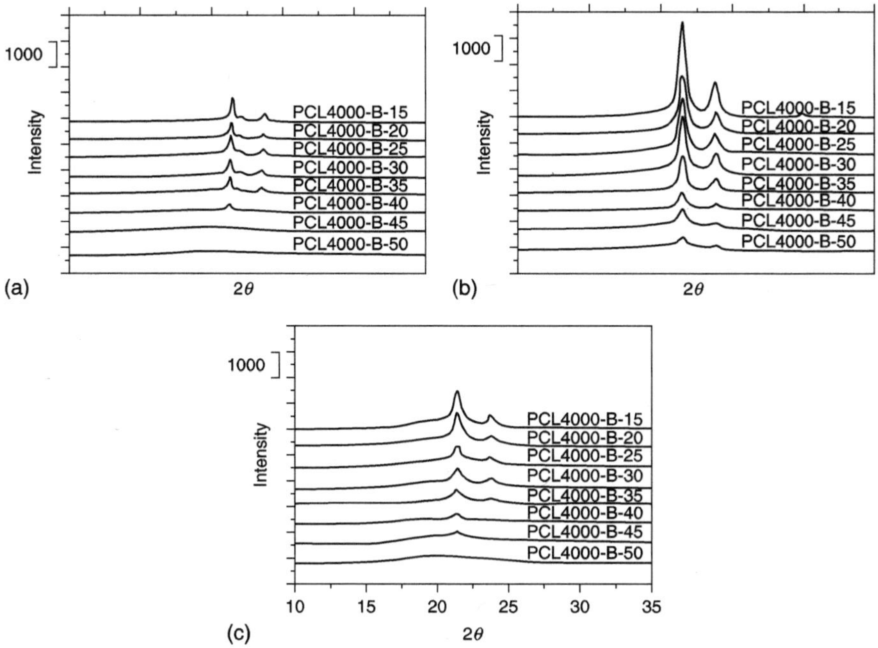

2.8 Tracing the structural changes in shape memorization by the WAXD tests: (a) before extension; (b) after shape fixing (being extended 100% and fixed in a temporary shape); (c) after shape recovery.

PCL4000-B-45 and PCL4000-B-50 is much lower than that of the other segmented polyurethanes. This is why the shape fixity descends abruptly when HSC $\geq$ 40%. In contrast, the high shape fixity of the segmented polyurethanes with HSC $\leq$ 35% arises from their high crystallizability. As shown in Fig. 2.8(c), the segmented polyurethanes after shape recovery show a little higher diffraction intensity as compared with the polyurethanes before deformation. This suggests that after a cycle of shape memorization the soft segments orientate partially, resulting in a slight increase of X_c, even though the polyurethanes almost recover to their original shape.

As mentioned above, the shape recovery of the segmented polyurethanes is dependent on the physically cross-linked networks maintained in the course of shape memorization. As shown in Fig. 2.7, the segmented polyurethanes with 15% $\leq$ HSC $\leq$ 40% show over 90% shape recovery. However, as HSC reaches 45 and 50%, shape recovery dramatically drops to 83 and 75%. This decrease of shape recovery should arise from the hard phase changing from isolated to interconnected states. Because the interconnected hard domains in PCL4000-B-45 and PCL4000-B-50 form a rigid framework, the hard phase is forced to undergo a large deformation. This results in the lower shape recovery of the two

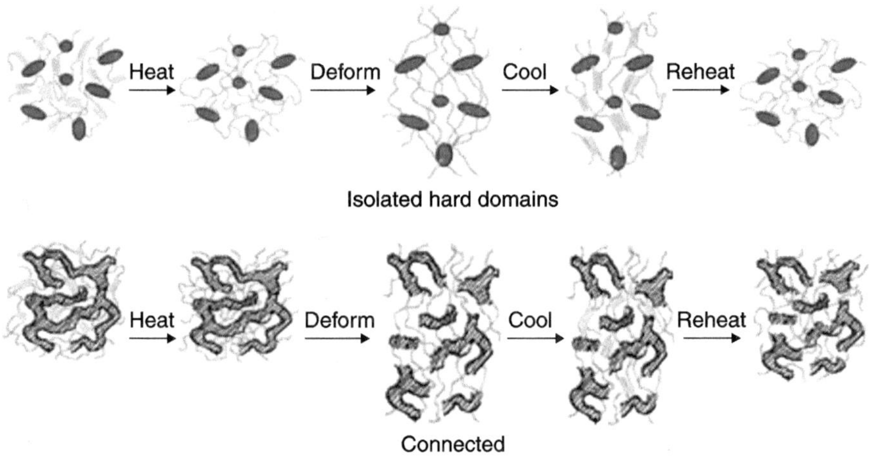

2.9 Demonstration of the effect of hard phase morphology on shape recovery.

polyurethanes at the recovery temperature $T_{rec} = T_m + 20°C$. The effect of the morphological change on shape recovery is demonstrated in Fig. 2.9.

In the evaluation of the SME, thermo-mechanical cyclic tensile tests were extensively employed. This method can provide highly reproducible experimental results. As described previously, the values of shape fixity are obtained in the unloading process. However, we found that if the segmented polyurethanes were continuously kept at T_{fix}, their shape fixity would decrease gradually after unloading. A similar phenomenon was observed in an investigation on SMPU foam (Tey *et al.*, 2001). In the present study, the shape fixity of the segmented polyurethanes changing with time after unloading is presented in Fig. 2.10. The

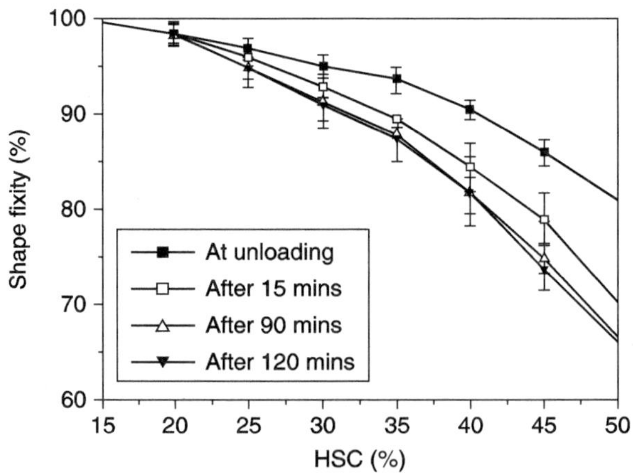

2.10 The shape fixity vs time after unloading of the T_m-SMPUs.

shape fixity decreases in the first 15 minutes after unloading. Furthermore, this tendency is more apparent for the segmented polyurethanes with higher HSC. The shape fixity curves tend towards constant values when the samples are maintained at T_{fix} for 120 minutes after unloading.

In the thermo-mechanical cyclic tensile tests, the shape recovery of the segmented polyurethanes occurs at $T_m + 20°C$. To examine the influence of T_{rec} on shape recovery, a series of shape recovery tests were performed. Figure 2.11 presents the plots of shape recovery as a function of recovery temperature of the segmented polyurethanes. The initial shape recovery of the segmented polyurethanes is roughly equal to the shape fixity obtained when they were kept at T_{fix} for 120 minutes after unloading. For example, PCL4000-B-50 remains at 65% of shape fixity in this case and the starting shape recovery in Fig. 2.9 is at 35% accordingly. It is found that all the segmented polyurethanes show over 90% shape recovery when heated to a high temperature T_∞, where no more shape recovery occurs with the increasing temperature. All the segmented polyurethanes show abrupt increases in shape recovery in a narrow temperature range around the T_m of the soft phase.

The shape recovery in this stage arises from the melting of the soft phase. After this stage the shape recovery appears to slow down and the segmented polyurethanes with HSC $\leq$ 40% almost recover to their original shape. In contrast, PCL4000-B-45 and PCL4000-B-50 show shape recovery after the melting transition of the soft phase. The shape recovery after melting transition is mainly ascribed to the recovery of the deformation of the hard phase. At $T_{rec} = T_m + 20°C$, the shape recovery of PCL4000-B-45 and PCL4000-B-50 are much lower than that of the other segmented polyurethanes. This is consistent with the experimental results of the previous thermo-mechanical cyclic tensile tests. The lower shape

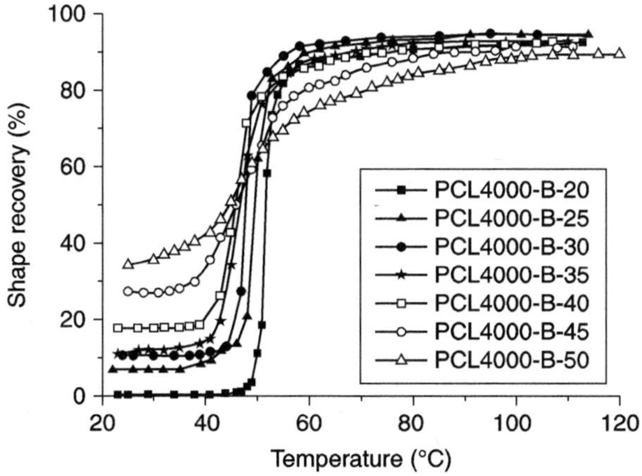

2.11 Shape recovery vs recovery temperature of the T_m-SMPUs.

recovery of the PCL4000-B-45 and PCL4000-B-50 at $T_{rec} = T_m + 20°C$ is mainly caused by the deformation of the hard phase.

To eliminate the effect of thermal expansion, the shape recovered segmented polyurethanes were cooled to T_{room} and the final strains were measured. As shown in Fig. 2.12, the final irreversible deformation strain of all the segmented polyurethanes is less than 5% and the ultimate shape recovery of the segmented polyurethanes is over 95%. The tests suggest that the deformation of the hard phase can be recovered if the segmented polyurethanes are heated to sufficiently high temperature. Therefore the deformation of the hard phase is basically viscoelastic rather than plastic. From the shape recovery tests it is inferred that the shape fixity of the segmented polyurethanes with lower HSC is mainly ascribed to the strain-induced crystallization of the soft phase, while that of the segmented polyurethanes with higher HSC, for example, PCL4000-B-45 and PCL4000-B-50, consists of both the strain-induced crystallization of the soft phase and the deformation of the hard phase.

The deformation of the polymers consists of elastic deformation, viscoelastic deformation and irreversible deformation (Anon, 2003). The total deformation ε of the segmented polyurethanes with higher HSC can be given by:

$$\varepsilon = \varepsilon_e + \varepsilon_{vis} + \varepsilon_{ir} \approx \varepsilon_{vis} + \varepsilon_{ir} \qquad [2.8]$$

where ε_e, ε_{vis} and ε_{ir} represent elastic deformation, viscoelastic deformation and irreversible deformation, respectively. Since the ε_e is usually negligible, the total deformation ε is approximately composed of ε_{vis} and ε_{ir}. The foregoing shape recovery tests proved that ε_{ir} is less than 5%. Hence the deformation of the segmented polyurethanes in this study is determined by the viscoelastic deformation ε_{vis}, which included the viscoelastic deformation of the soft phase

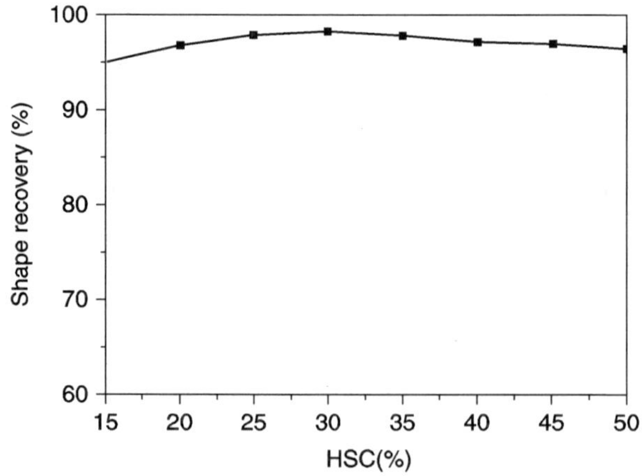

2.12 Final shape recovery of the T_m-SMPUs.

ε_{vis} (SS) and the hard phase εX_c (HS). The shape fixing and shape recovery are strongly related to ε_{vis} (SS) and ε_{vis} (HS). The distribution of viscoelastic deformation between soft and hard phases is dependent on the two-phase morphology such as phase separation, phase composition, domain sizes, domain concentration and connectivity, etc. Thus the shape memory properties are defined by the two-phase morphological structure of the segmented polyurethanes.

2.4 Thermo-mechanical conditions affecting T_m-type shape memory properties

2.4.1 Influences of stress relaxation

It is impossible to fully avoid stress relaxation in the course of shape memorization. For example, the shape memory polymers (SMPs) cannot always be quickly cooled down to a frozen state in practical applications, due to the limitation of cooling rate and heat transfer efficiency. In some cases, they may be constrained at a high temperature for a long time and stress relaxation can thus occur significantly. It is assumed that segmented polyurethanes with different structures show different dependencies on stress relaxation. In this study, the influences of stress relaxation on the SME of the segmented polyurethanes were investigated. In the thermo-mechanical cyclic tensile tests, the segmented polyurethanes were maintained in the constraint state at T_m + 20°C for 30 minutes prior to being cooled down to T_{fix}. The segmented polyurethanes could thus undergo more apparent stress relaxation in the thermo-mechanical tests. The effect of stress relaxation on the shape fixity of the segmented polyurethanes is shown in Fig. 2.13,

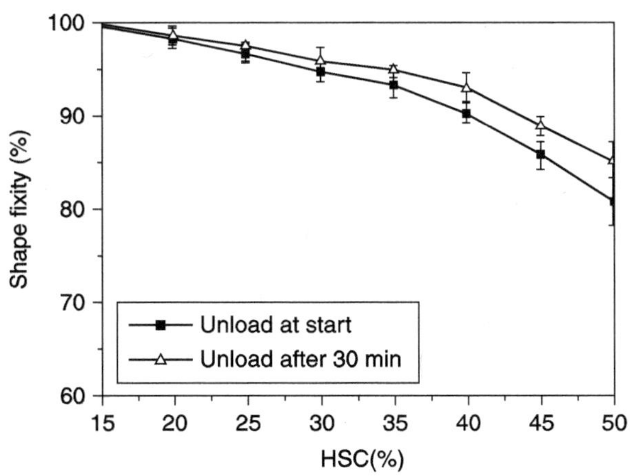

2.13 Effect of stress relaxation on shape fixity of the T_m-SMPUs.

where the stress relaxation results in an increase of shape fixity. Moreover, this tendency is increased with the increase of HSC.

As shown in Fig. 2.14, stress relaxation results in the decrease of shape recovery for all the segmented polyurethanes. Furthermore, the decrease of shape recovery becomes more noticeable when HSC is reduced to 20 and 15%. There may be two reasons accounting for this phenomenon. First, besides the hard domains, the molecular interactions such as the intermolecular hydrogen bonds, dipole–dipole interaction, induced dipole–dipole interaction and even molecular entanglements among the segmented polyurethane molecules, can induce weak physical cross-links which prevent molecular slippage in a short timescale. SAXS results showed that the domain sizes of the soft phase of PCL4000-B-15 and PCL4000-B-20 are much larger than those of the others. But the molecular interactions can help prevent molecular slippage on a short timescale, which should play a comparatively important role in PCL4000-B-15 and PCL4000-B-20. But the weak polymer network conjunctions arising from the molecular interactions have a short lifetime and can disappear upon application of an external force over the long timescale.

Second, in PCL4000-B-15 and PCL4000-B-20, some of the hard domains may not be strong enough and therefore can also break and reconstruct. The polymer networks thus cannot be held for a long time at high temperature. The lower strength of the hard domains results in PCL4000-B-15 and PCL4000-B-20 showing a sharper decrease in shape recovery. In contrast, the segmented polyurethanes with the higher HSC of 25 to 50% can better endure stress relaxation and exhibit comparatively less decrease in shape recovery due to the stronger hard domains as physical cross-links.

The shape recovery tests were also performed on PCL4000-B-15, PCL4000-B-30 and PCL4000-B-45 to study the influence of stress relaxation. The segmented

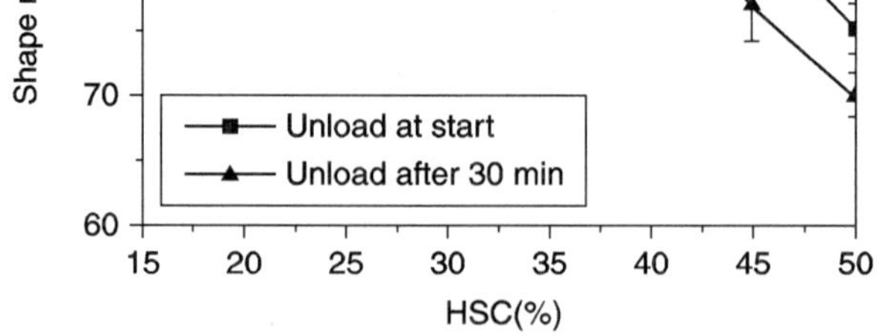

2.14 Effect of stress relaxation on shape recovery of the T_m-SMPUs.

polyurethane samples for these tests were prepared by maintaining the deformed polyurethanes at T_m + 20°C for 30 minutes and subsequently cooling them to T_{fix}. Hereafter the samples from the two processes are simply designated as the segmented polyurethanes without relaxation and with relaxation. Of course, the stress relaxation can occur even if the deformed polyurethanes are cooled to T_{fix} rapidly after deformation.

Figure 2.15 presents the effect of stress relaxation on the shape recovery as a function of temperature. The segmented polyurethanes with relaxation exhibit lower shape recovery at T_m + 20°C (~65°C) in comparison with those without relaxation, which is in agreement with the results obtained from the preceding thermo-mechanical cyclic tensile tests. With increasing temperature, the shape recovery of the segmented polyurethanes with relaxation gradually rises to over 90% and the final shape recovery is close to that of the segmented polyurethanes without relaxation. Therefore it was inferred that the stress relaxation gives rise to no more irreversible deformation, but only resulted in more viscoelastic deformation of the hard phase or more breakage and reconstruction of molecular interactions and weak hard domains.

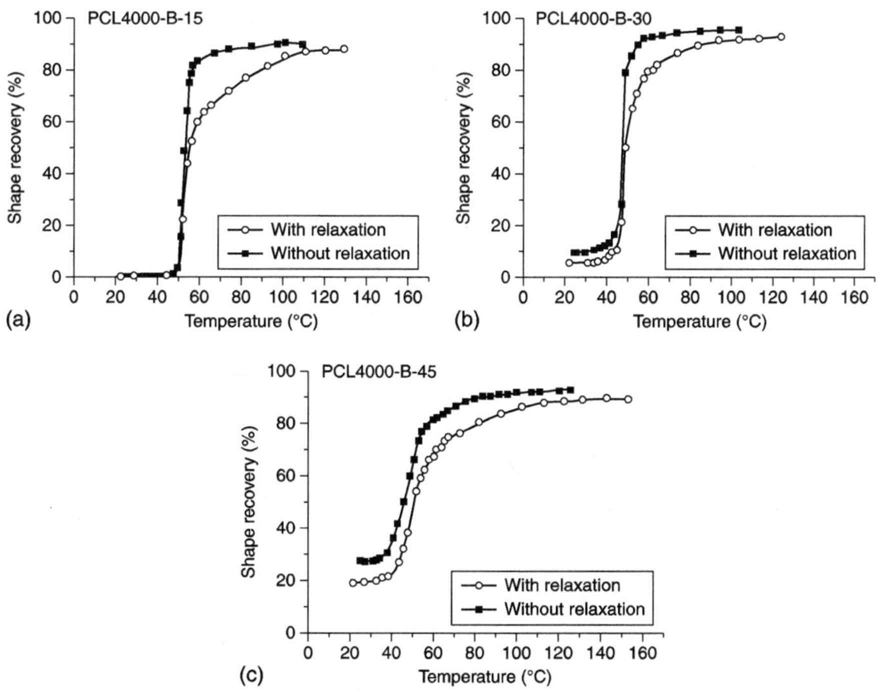

2.15 Effect of stress relaxation on shape recovery curves of the T_m-SMPUs.

2.4.2 Influences of deformation amplitude

Figure 2.16 presents the effect of deformation amplitude on shape fixity of the segmented polyurethane. It can be seen that the shape fixity rises with increase in deformation amplitude. For all the tests with varying deformation amplitude, the shape fixity decreases with an increase of HSC. This is because the crystallizability of the soft segments decreases with the increase of HSC. In the zone HSC ≤ 25%, the shape fixity obtained from all the tests is over 95%. This suggests that the shape fixity of segmented polyurethanes with low HSC is mainly determined by their high X_c.

For the segmented polyurethanes with 30% ≤ HSC ≤ 50%, the shape fixity decreases from 95 to 76% in the tests with 50% deformation amplitude. But the shape fixity is much lower than the corresponding result obtained in the tests with 100% deformation amplitude. This is because the soft segments do not fully orientate in the tests with 50% of deformation amplitude and very little strain-induced crystallization develops. This suggests that the segmented polyurethanes with 30% ≤ HSC ≤ 50% have to be extended to over 100% of elongation to obtain higher shape fixity. As deformation amplitude increased from 100 to 250%, the shape fixity shows no distinct change for the segmented polyurethanes with 30% ≤ HSC ≤ 35%. Whereas for the segmented polyurethanes with 40% ≤ HSC ≤ 50%, the shape fixity rises gradually as deformation amplitude increases from 100 to 250%.

Figure 2.17 demonstrates the effect of deformation amplitude on shape recovery of the segmented polyurethanes. For the segmented polyurethanes with 20% ≤ HSC ≤ 40%, the shape recovery decreases as the deformation amplitude increases from 50 to 200%. This tendency stops when the deformation amplitude rises to 250%. For the segmented polyurethanes with 45% ≤ HSC ≤ 50%, the

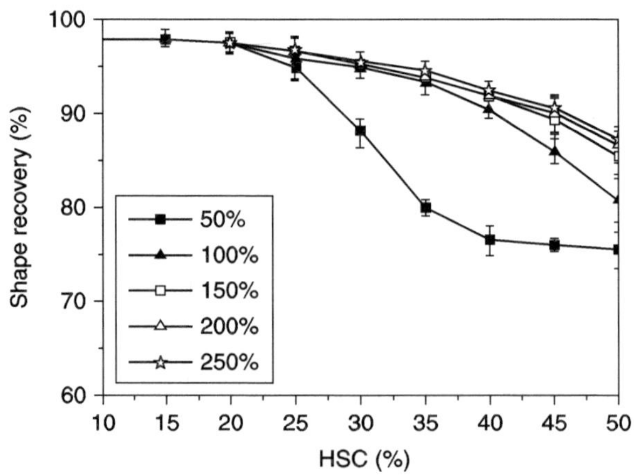

2.16 Effect of deformation amplitude on shape fixity of the T_m-SMPUs.

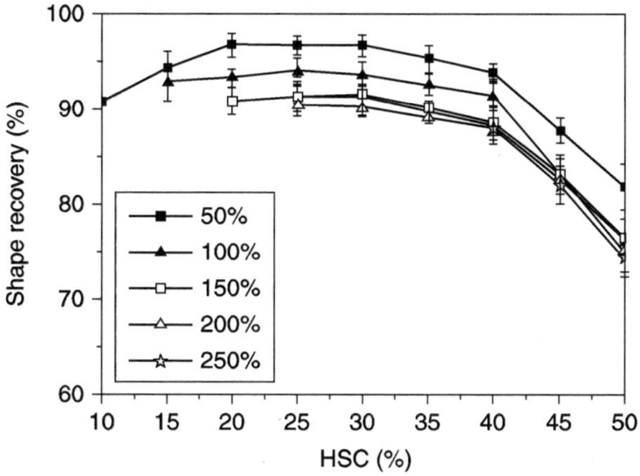

2.17 Effect of deformation amplitude on shape recovery of the T_m-SMPUs.

shape recovery decreases as the deformation amplitude increases from 50 to 100% and does not exhibit obvious change as the deformation amplitude increases from 100 to 250%. The segmented polyurethanes with intermediate HSC of 25 and 30% show the best shape recovery in all the tests with varying deformation amplitude. PCL4000-B-35 and PCL4000-B-40 show a slight decrease of shape recovery. The shape recovery drops when HSC $\geq$ 45%.

This results from larger deformation of the hard phase as hard domains change from an isolated to an interconnected state. In the tests with 50% deformation amplitude, the polyurethanes PCL4000-B-20, PCL4000-B-25 and PCL4000-B-30 show 97% of shape recovery. For the segmented polyurethanes with HSC $\leq$ 20%, shape recovery decreases with decrease of HSC. Notably, PCL4000-B-10, the polyurethane manifested to possess no hard domain, still shows over 90% of shape recovery. It could be the molecular interactions that prevent the molecular slippage in the deformation of PCL4000-B-10 and result in it memorizing its original shape. This suggests that besides hard domains, molecular interaction can also play the role of physical cross-links and enable polymers to show the SME under particular conditions.

Shape recovery tests were also performed on PCL4000-B-25 and PCL4000-B-45 to study the influence of the deformation amplitude. Figure 2.18 presents the effect of the deformation amplitude on shape recovery of the two segmented polyurethanes, PCL4000-B-25 and PCL4000-B-45. They exhibit the best shape recovery at T_m + 20°C when the deformation amplitude was 50%, which is in agreement with the results of the foregoing tests. With the increase of recovery temperature, the shape recovery of the segmented polyurethanes gradually increases to over 90%. Moreover, the segmented polyurethanes with varying deformation amplitude exhibit almost identical final shape recovery. Therefore

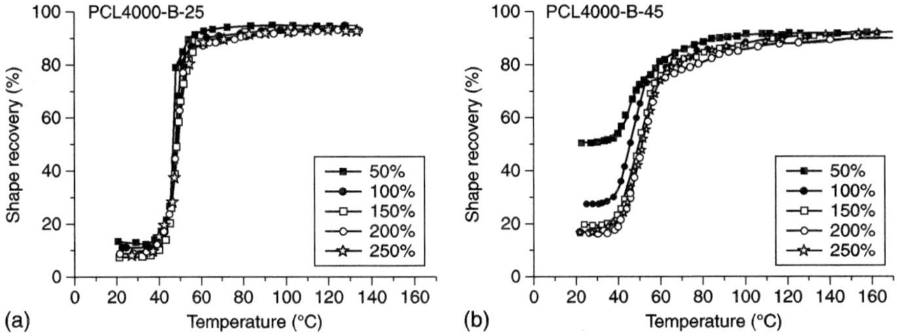

2.18 Effect of deformation amplitude on shape recovery on: (a) PCL4000-B-25; and (b) PCL4000-B-45.

the increase of deformation amplitude results in no more irreversible deformation, which indicates that the decrease of shape recovery at T_m + 20°C with the increase of amplitude arises from the viscoelastic deformation of the hard phase or the breakage and reconstruction of molecular interactions and unstable hard domains.

2.4.3 Influences of pre-deformation

As a type of smart material, precise control of the SME is essential for the application of SMPs. In the thermo-mechanical cyclic tensile tests of the segmented polyurethanes, the third loading curve is close to the second loading curve. In other words, the segmented polyurethanes turn completely elastic after the first deformation. This led to enhancing the shape recovery of the segmented polyurethanes by a pre-deformation treatment. In this study, the segmented polyurethanes were extended to 120% of strain on the tensile tester at T_m + 20°C. Afterwards, the specimens were removed from the tensile tester and kept at T_m + 20°C for 15 minutes for shape recovery. Then the recovered specimen was viewed as a new sample to perform the thermo-mechanical cyclic tensile tests with 100% of deformation amplitude.

As shown in Fig. 2.19, the shape fixity of the segmented polyurethanes is close to those without pre-extension. However, after the pre-deformation, the shape recovery of the segmented polyurethanes is nearly 100%. It seems the irreversible deformation and the viscoelastic deformation of the hard phase occurred in pre-deformation. In the following deformation, no more irreversible deformation and viscoelastic deformation takes place if the deformation amplitude does not exceed that of pre-deformation. Therefore the shape recovery is nearly 100%.

A series of segmented polyurethanes showing SME were synthesized with HSC varying from 10 to 50%. The segmented polyurethanes had a crystalline

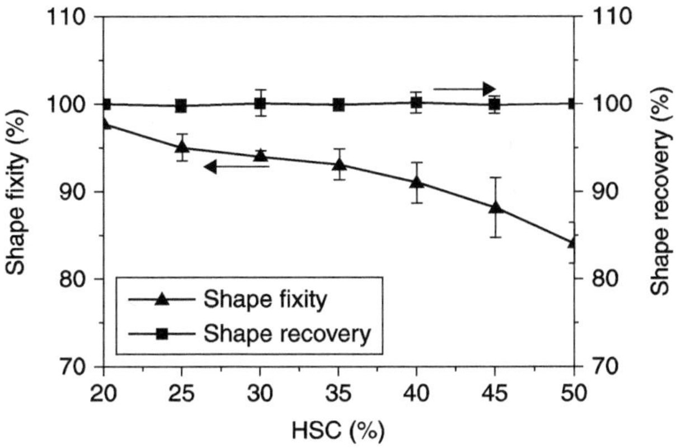

2.19 Effect of pre-deformation on shape memory effect of the T_m-SMPUs.

reversible phase for triggering the SME. The morphology and SME of the polyurethanes were investigated and the morphology of the segmented polyurethanes was characterized with a combination of multiple techniques including DSC, TMA, SAXS and tensile hysteresis tests. In the low HSC zone, the segmented polyurethane PCL4000-B-10 is basically homogeneous and has no hard domains. The segmented polyurethanes possess a more or less hard phase when HSC $\geq$ 15%. The hard domains change from an isolated state into an interconnected state as HSC increases from 40 to 45%. With the increase of HSC, the X_c of the soft phase decreases. The microdomain morphological structure of the segmented polyurethanes was investigated. The interdomain space decreases and the concentration and the sizes of the hard domains increase as HSC increases.

The shape fixity of the segmented polyurethanes decreases with the increase of HSC. The shape fixity is strongly related to the strain-induced crystallization of the soft segments. As for the segmented polyurethanes with lower HSC, deformation amplitude has little effect on shape fixity due to their high crystallizability. For the segmented polyurethanes with 30% $\leq$ HSC $\leq$ 50%, the shape fixity augments dramatically as the deformation amplitude increases from 50 to 100%, because the segmented polyurethanes cannot develop strain-induced crystallization under low deformation strain. It is found that the segmented polyurethanes should be extended to over 100% of strain to attain better shape fixity. In addition, stress relaxation can result in the increase of shape fixity and this tendency rises with increase of HSC.

The segmented polyurethane without hard domains shows over 90% of shape recovery in the thermo-mechanical cyclic tensile tests with 50% of elongation. This suggests that the molecular interactions can also play the role of physical cross-links and enable the segmented polyurethane to show the SME under some conditions. The segmented polyurethanes with isolated hard domains showed better shape recovery.

The shape recovery of segmented polyurethanes decreases dramatically when the hard domains change from the isolated into the interconnected states. When they are heated up to a sufficiently high temperature, the shape recovery of all the segmented polyurethanes can be raised to over 90%. In the segmented polyurethanes with higher HSC, deformation of the hard phase results in lower shape recovery at $T_m + 20°C$. Stress relaxation results in the decrease of shape recovery at $T_m + 20°C$, while the decrease of shape recovery is more apparent for the segmented polyurethanes with HSC = 15% and HSC = 25%. The shape recovery at $T_{rec} = T_m + 20°C$ decreases as the deformation amplitude increases from 50% to 200%, but the decrease of shape recovery ceases when the deformation amplitude increases to 250%, but the shape recovery of the segmented polyurethanes can be enhanced to nearly 100% by a pre-deformation treatment.

2.5 References

Anon. (2003), Polymer physics, in: Rubinstein, M., Colby, R. H., eds. New York, Oxford University Press, 199–202.

Blundell, D. J., Eeckhaut, G., Fuller, W., Mahendrasingam, A., Martin, C. (2002), Real-time SAXS/stress-strain studies of thermoplastic polyurethanes at large strains, *Polymer*, **43**, 5197–207.

Blundell, D. J., Eeckhaut, G., Fuller, W., Mahendrasingam, A., Martin, C. (2004), Time-resolved SAXS/stress-strain studies of thermoplastic polyurethanes during mechanical cycling at large strains, *J. Macromol. Sci. Phys.*, **43**, 125–42.

Bogart, V., John, W. C., Gibson, P. E., Cooper, S. L. (1983), Structure–property relationships in polycaprolactone-polyurethanes, *J. Polym. Sci., Polym. Phys. Ed.*, **21**, 65–95.

Born, L., Crone, J., Hespe, H., Wolf, K. H. (1982), Physical cross-linking of polyurethane of elastomers studied by X-ray investigation of model urethanes, *Colloid. Polym. Sci.*, **260**, 819.

Bras, W., Derbyshire, G. E., Bogg, D., Cooke, J., Elwell, M. J., *et al.* (1995), Simultaneous studies of reaction kinetics and structure development in polymer processing, *Science*, **267**, 996–9.

Chang, S. L., Yu, T. L., Huang, C. C., Chen, W. C., Linliu, K., Lin, T. L. (1998), Effect of polyester side-chains on the phase segregation of polyurethanes using small-angle X-ray scattering, *Polymer*, **39**, 3479–89.

Chu, B., Gao, T., Li, Y., Wang, J., Desper, C. R., Catherine, A. B. (1992), Microphase separation kinetics in segmented polyurethanes: Effects of soft segment length and structure, *Macromolecules*, **25**, 5724–9.

Elwell, M. J., Mortimer, S., Ryan, A. J. (1994), Synchrotron SAXS study of structure development kinetics during the reactive processing of flexible polyurethane foam, *Macromolecules*, **27**, 5428–39.

Garrett, J. T., Lin, J. S., Runt, J. (2002), Influence of preparation conditions on microdomain formation in poly(urethane urea) block copolymers, *Macromolecules*, **35**, 161–8.

Garrett, J. T., Runt, J., Lin, J. S. (2000), Microphase separation of segmented poly(urethane urea) block copolymers, *Macromolecules*, **33**, 6353–9.

Huang, W. M., Lee, C. W., Teo, H. P. (2006), Thermo-mechanical behavior of a polyurethane shape memory polymer foam, *J. Intell. Mater. Sys. Struc.*, **17**, 753.

Kim, B. K., Lee, S. Y., Xu, M. (1996), Polyurethanes having shape memory effects, *Polymer*, **37**, 5781.

Koberstein, J. T., Leung, L. M. (1992), Compression-molded polyurethane block copolymers. 2. Evaluation of microphase compositions, *Macromolecules*, **25**, 6205–13.

Koberstein, J. T., Galambos, A. F., Leung, L. M. (1992), Compression-molded polyurethane block copolymers. 1. Microdomain morphology and thermomechanical properties, *Macromolecules*, **25**, 6195–204.

Krakovsky, I., Urakawa, H., Kajiwara, K. (1997), Inhomogeneous structure of polyurethane networks based on poly(butadiene)diol: 2. Time-resolved SAXS study of the microphase separation, *Polymer*, **38**, 3645–53.

Leung, L., M., Koberstein, J. T. (1985), Small-angle scattering analysis of hard-microdomains structure and microphase mixing in polyurethane elastomers, *J. Polym. Sci., Polym. Phys. Ed.*, **23**, 1883–913.

Li, F., Zhang, X., Hou, J., Xu, M., Luo, X., *et al.* (1997), Studies on thermally stimulated shape memory effect of segmented polyurethanes, *J. Appl. Polym. Sci.*, **64**, 1511.

Li, W., Ryan, A. J., Meier, I. K. (2002), Morphology development via reaction-induced phase separation in flexible polyurethane foam, *Macromolecules*, **35**, 5034–42.

Li, Y., Gao, T., Chu, B. (1992a), Synchrotron SAXS studies of the phase-separation kinetics in a segmented polyurethane, *Macromolecules*, **25**, 1734–42.

Li, Y., Gao, T., Liu, J., Linliu, K., Desper, C. R., Chu, B. (1992b), Microphase separation kinetics in segmented polyurethanes: Effects of soft segment length and structure, *Macromolecules*, **25**, 7365–72.

Martin, C., Eeckhaut, G., Mahendrasingam, A., Blundell, D. J., Fuller, W., *et al.* (2000), Micro-SAXS and force/strain measurements during the tensile deformation of single struts of an elastomeric polyurethane foam, *J. Synchrotron Radia.*, **7**, 245–53.

O'Sickey, M. J., Lawrey, B. D., Wilkes, G. L. (2003), Structure–property relationships of poly(urethane-urea)s with ultralow monol content poly(propylene glycol) soft segments, Part III: Influence of mixed soft segments of ultralow monol poly(propylene glycol), poly(tetramethylene ether glycol), and tri(propylene glycol), *J. Appl. Polym. Sci.*, **89**, 3520–9.

Petrovic, Z. S., Javni, I., Schaefer, D. W., Justice, R. (2005), Soy-oil-based segmented polyurethanes, *J. Polym. Sci., Polym. Phys. Ed.*, **43**, 3178–90.

Ping, P., Wang, W., Chen, X., Jing, X. (2005), Poly(ε-caprolactone) polyurethane and its shape-memory property, *Biomacromolecules*, **6**, 587.

Ryan, A. J., Willkomm, W. R., Bergstrom, T. B., Macosko, C. W., Koberstein, J. T., *et al.* (1991), Dynamics of (micro)phase separation during fast, bulk copolymerization. Some synchrotron SAXS experiments, *Macromolecules*, **24**, 2883–9.

Ryan, A. J., Macosko, C. W., Bras, W. (1992), Order–disorder transition in a block copolyurethane, *Macromolecules*, **25**, 6277–83.

Ryan, A. J., Stanford, J. L., Tao, X. Q. (1993), Copolyureas formed by reaction injection moulding: Correlations between chemical structure, thermal properties and microphase separation, *Polymer*, **34**, 4020–31.

Strobl, G. R., Schnieider, M. J., Voigt-Martion, I. G. (1980), Model of partial crystallization and melting derived from small-angle X-ray scattering and electron microsopic studies on low-density polyethlene, *J. Polym. Sci., Polym. Phys. Ed.*, **18**, 1361–81.

Tey, S. J., Huang, W. M., Sokolowski, W. M. (2001), Influence of long-term storage in cold hibernation on strain recovery and recovery stress of polyurethane shape memory polymer foam, *Smart Mater. Struct.*, **10**, 321.

Valenkar, S., Cooper, S. L. (1998), Microphase separation and rheological properties of polyurethane melts, Part I: Effect of block length, *Macromolecules*, **31**, 9181–92.

Valenkar, S., Cooper, S. L. (2000), Microphase separation and rheological properties of polyurethane melts, Part II: Effect of block incompatibility on the microstructure, *Macromolecules*, **33**, 382–94.

3

T_g-type shape memory polymers

DOI: 10.1533/9780857098542.47

Abstract: This chapter introduces three sets of T_g-type shape memory polyurethanes. The thermal properties and morphological structure of polymers are shown to vary with chemical composition. The morphology of the T_g-SMPUs is thus correlated to shape memory properties. The influences of several important thermo-mechanical conditions including deformation temperature, deformation strain, and shape fixing temperature and time on shape memory properties are demonstrated.

Key words: T_g-type shape memory polymer, phase separation, phase morphology, shape memory property.

3.1 Introduction

Although attempts have been made to investigate the morphological structure of T_g-SMPUs, by studying phase composition and microdomain sizes (the latter defining the physically cross-linked polymer networks), the structure has not yet been demonstrated systematically (Takahashi *et al.*, 1996; Lin and Chen, 1998a,b; Lee *et al.*, 2001).

In this study, a series of segmented polyurethanes based on PCL400 diols and PPG400 and PPG725 glycols with varying hard segment content (HSC) were prepared. The formulations of the T_g-SMPUs are detailed in Table 3.1. Three series of segmented polyurethanes with varying HSC were synthesized with PCL400 diols and PPG400 and PPG725 polyols as the soft segments and with 4,4′-diphenylmethane diisocyanate (MDI) and BD as the hard segments. The polyester and polyether glycols were dried and dehydrated at 80°C under vacuum for six hours. The solvent dimethyl formamide (DMF) (Aldrich) was dried with 4Å molecular sieves prior to use.

The segmented polyurethanes were prepared by a two-step copolymerization with 2% excess MDI to ensure complete reaction. The isocynate-terminated prepolymers were first prepared by reacting the dried polyols with the mole excessive MDI (Acros) in the DMF solution at 65°C for three hours. The prepolymers were then chain extended with BD (International Laboratory, USA) at 80°C for another four hours. The molecular characteristics of the segmented polyurethanes are shown in Table 3.1. The series of polyurethanes based on PCL400 are designated as PCL400-xx, where 'xx' stands for the wt.% HSC. Likewise, the polyurethanes based on PPG400 and PPG725 are designated as PPG400-xx and PPG725-xx, respectively.

Table 3.1 The formulations of the T_g-SMPUs

Samples	Feed ratio (mole)		Theoretical HSC	
	Polyols	MDI	BDO	Wt.%
PCL400-60	1	1.78	0.78	60.0
PCL400-65	1	2.16	1.16	65.0
PCL400-70	1	2.68	1.68	70.0
PCL400-75	1	3.39	2.39	75.0
PCL400-80	1	4.47	3.47	80.0
PPG400-60	1	1.78	0.78	60.0
PPG400-65	1	2.16	1.16	65.0
PPG400-70	1	2.68	1.68	70.0
PPG400-75	1	3.39	2.39	75.0
PPG400-80	1	4.47	3.47	80.0
PPG725-65	1	3.94	2.94	65.0
PPG725-70	1	4.91	3.91	70.0
PPG725-75	1	6.26	5.26	75.0
PPG725-80	1	8.29	7.29	80.0

The morphology of the segmented polyurethanes was studied with DSC, DMA and SAXS. The shape fixity and shape recovery, as well as recovery stress of the segmented polyurethanes, were examined through a series of thermo-mechanical cyclic tensile tests. The segmented polyurethanes with similar T_g but different structure were compared in terms of shape memory behaviors. In addition, the influences of the deformation conditions including deformation temperature, pre-deformation, shape fixing temperature and shape fixing time, on shape memory behaviors were investigated. The insights into the relationship between morphology and shape memory effect (SME) derived from this study are anticipated to facilitate development and application of high performance T_g-SMPUs.

3.2 Structure and properties of T_g-type shape memory polymers

3.2.1 Thermal properties of T_g-type shape memory polymers

Figure 3.1 illustrates the DSC thermograms of the three series of segmented polyurethanes. Transition temperatures, heat capacity change (ΔC_p) and heat of fusion (ΔH_m) are tabulated in Table 3.2. All three series of segmented polyurethanes exhibit at least two thermal transitions.

As shown in Fig. 3.1(a) and (c), the PCL400- and PPG725-based polyurethanes show a second-order transition in the temperature range 20 to 50°C, which is

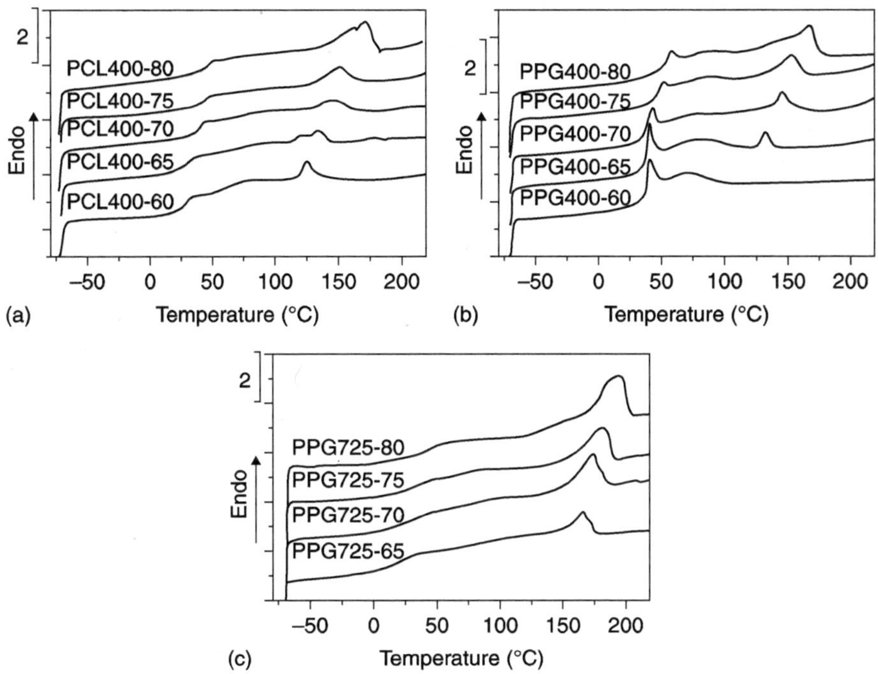

3.1 DSC traces of the T_g-SMPUs at a scanning rate of 10°C/min.

Table 3.2 Thermal properties of the T_g-SMPUs

	DSC experiments				DMA experiments	
Samples	T_g (δC)	ΔC_p (J/(g × °C))	T_m (°C)	ΔH_m (J/g)	T_g (°C)	E' at T_g + 20°C (MPa)
PCL400-60	28.1	0.42	125.2	4.2	36.5	157
PCL400-65	30.2	0.37	132.2	13.0	49.0	162
PCL400-70	37.9	0.33	144.5	14.3	61.0	320
PCL400-75	42.0	0.31	150.3	21.7	70.4	326
PCL400-80	45.9	0.29	168.3	30.3	81.8	478
PPG400-60	–	–	–	–	59.7	40
PPG400-65	–	–	133.5	2.8	61.9	44
PPG400-70	–	–	146.9	8.0	70.1	1077
PPG400-75	–	–	157.9	16.9	76.9	205
PPG400-80	–	–	169.0	27.0	86.2	255
PPG725-65	24.0	0.34	165.5	16.0	39.7	360
PPG725-70	26.5	0.29	172.5	20.7	51.6	474
PPG725-75	31.3	0.22	180.0	22.0	58.5	572
PPG725-80	36.2	0.18	193.7	26.2	71.0	752

ascribed to the glass transition of the soft phase. The microphase separation of segmented polyurethanes has been found to be incomplete due to the broad distribution of hard segment sequence lengths (Peebles, 1974, 1976). The shorter hard segments tend to reside within the soft matrix due to lack of driving force for microphase separation, which accounts for the phase mixing (Koberstein *et al.*, 1992).

Specifically, the segmented polyurethanes in this study should exhibit marked phase mixing because of the relatively short soft and hard segments. As shown in Table 3.2, the T_g of the soft phase in the PPG725-based polyurethanes spans a temperature range from 24 to 36°C, which is much higher than the T_g (−69°C) of the pure PPG725 polyols reported previously. Therefore the soft phase of the two series of segmented polyurethanes is inherently a mixed phase composed of soft segments incorporated with considerable hard segments. The mixed soft phase plays the role of the reversible phase of the T_g-SMPUs. For convenience, it is called the soft phase in this chapter.

It is found that the T_g of the two series of segmented polyurethanes increases with the increase of HSC, for two reasons. First, the fraction of hard domains rises with the growth of HSC, which enhances the restriction imposed on the soft segments and raises the T_g of the soft phase. Second, as HSC increases, a higher percentage of hard segments are dissolved in the soft matrix, resulting in an increase in T_g of the soft phase. It is also noted that ΔC_p is reduced progressively as HSC increases, suggesting a decreasing fraction of the soft phase.

As shown in Fig. 3.1(b), the PPG400-based polyurethanes exhibit a hysteresis peak in a similar temperature range. This is usually associated with some physical relaxation (Sperling, 1992). In the intermediate temperature range from 50 to 100°C, all three series of segmented polyurethanes show broad and obscure deflections, which are ascribed to the glass transition of their hard phase. According to previous studies, the glass transition of the hard phase is barely detectable in DSC thermograms (Leung and Koberstein, 1986). Moreover, the T_g of the hard phase in the segmented polyurethanes is usually lower than the T_g of MDI-BD homopolymers (~110°C), because of the residual soft segments dissolved in the hard phase.

In the higher temperature range of 130 to 200°C, the three sets of segmented polyurethanes exhibit a melting transition attributed to long-range ordering of the hard phase. As shown in Table 3.2, T_m and ΔH_m of the hard phase rise with the growth of HSC, suggesting the fraction and ordering degree of hard domains are raised. In addition, PPG400-60 exhibits no high temperature endotherm, indicating that it is basically homogeneous, i.e. has no hard domains. In comparison with PCL400-based polyurethanes with identical HSC, the PPG400 series of polyurethanes exhibit less phase separation.

Figure 3.2 shows the plots of dissipation factor tanδ vs temperature obtained from the DMA thermograms of the segmented polyurethanes. In the temperature range 0 to 150°C, all three series of segmented polyurethanes show a distinct

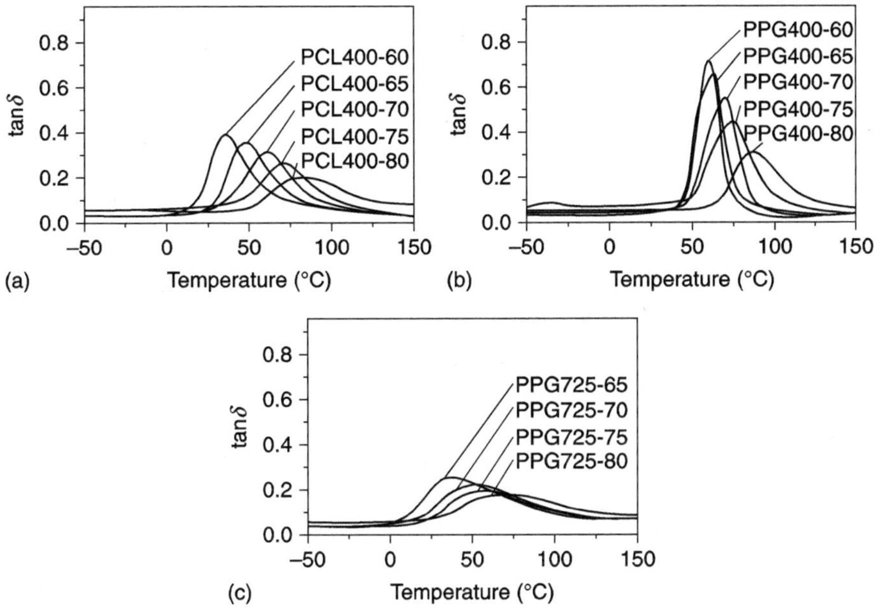

3.2 Tanδ vs temperature of the T_g-SMPUs.

dissipation peak. In a previous study of segmented polyurethanes with low-molecular-weight soft segments, Takahashi *et al.* (1996) attributed a similar transition observed in DMA tests to the glass transition of the soft phase. However, the transitions illustrated in that report ended at temperatures over 100°C, which were much higher than the T_g of the soft phase measured by the DSC tests (Takahashi *et al.*, 1996). Here the transition is ascribed to the glass transition of the soft phase superimposed on that of the hard phase.

Bogart *et al.* (1983) and Lin and Chen (1998a,b) also proposed a similar conclusion in their investigations. This may result from the fact that the glass transition of the soft phase is situated close to that of the hard phase in segmented polyurethanes with low-molecular-weight soft segments. The dissipation factor is associated to molecular mobility, namely the relaxation time, whose temperature dependence defines the SME of the segmented polyurethanes. Thereby the T_g extracted from the mamixa of tanδ curves should be T_{trans} for triggering the SME.

For all three series of polyurethanes, the tanδ maxima steadily shift to higher temperatures with the growth of HSC. There may be two aspects accounting for this tendency. The first is the raised T_g of the soft phase. The second is the fact that the fraction of the soft phase descends while that of the hard phase rises in comparison, resulting in the superimposed peaks shifting to higher temperatures. Furthermore, it is noted that the width of the glass transition is broadened with increasing HSC, suggesting the contribution of the hard phase is raised.

A rubbery modulus is believed to be a measure of physical cross-linking density of segmented polyurethanes (Kim *et al.*, 1996). Figure 3.3 presents the storage modulus (E′) as a function of temperature of the segmented polyurethanes. As can be seen, all the segmented polyurethanes exhibit apparent modulus drops in certain temperature regions, which correspond to the dissipation peaks on the tanδ plots. Moreover, the modulus drops shift to a higher temperature with an increase of HSC. The segmented polyurethanes with lower HSC give rise to narrow glass transition regions. In addition, for a given HSC, the glass transition region of the PPG400-based segmented polyurethane is the narrowest, the PCL400-based is the intermediate, and the PPG725 is the widest. It is also noted that the rubbery plateau of the segmented polyurethanes is elevated with the growth of HSC, indicating the increase of physical cross-linking density.

The values of the storage modulus at T_g + 20°C of the polyurethanes extracted from DMA tests are listed in Table 3.2. Comparing the segmented polyurethanes with identical HSC, we find that the PPG725-based segmented polyurethanes give rise to the highest rubbery modulus. In contrast, those that are PPG400-based exhibit much lower rubbery modulus than those of the PPG725 and PCL400-bases, suggesting a lower physical cross-linking density of the latter. Specifically, the segmented polyurethanes PPG400-60 and PPG400-65 show extremely low rubbery modulus, due to absence of a distinct phase separation.

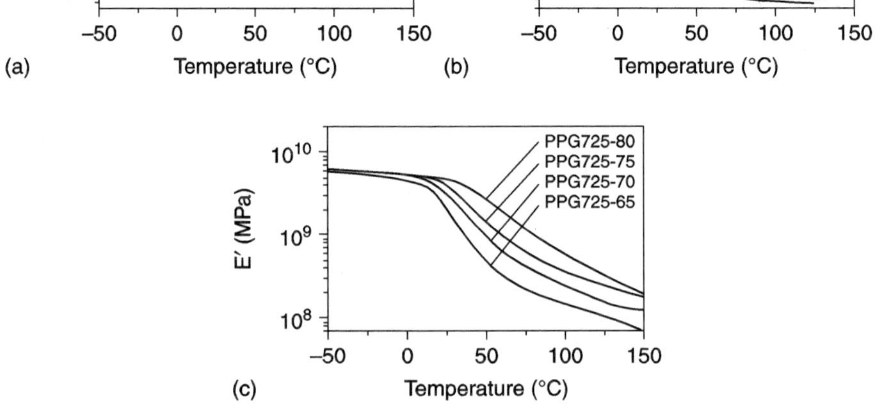

3.3 Storage modulus vs temperature of the T_g-SMPUs.

3.2.3 Microdomain morphology

Figure 3.4 shows the SAXS profiles of the segmented polyurethanes. The PCL400 and PPG725-based polyurethanes exhibit a single broad scattering peak, indicating a polydisperse two-phase morphology (Bogart *et al.*, 1983; Leung and Koberstein, 1985). As illustrated in Fig. 3.4(b), PPG400-60 is homogeneous while PPG400-65 is slightly phase separated, suggesting that the short hard segments are mostly dissolved in the PPG400 matrix in the two materials. This suggests that the PPG400 is more compatible with hard segments than PCL400 and PPG725. As HSC rises from 70 to 80%, a scattering peak appears in the SAXS profiles of PPG400-based segmented polyurethanes, despite possibly significant phase mixing. This is because the increased hard segment length raises the driving force for microphase separation.

From the positions of the maxima of the scattering profiles, the interdomain spaces, d, of the segmented polyurethanes can be evaluated with Bragg's law:

$$d = \frac{2\pi}{q_m} \quad [3.1]$$

where q_m is the scattering vector corresponding to the maximum of the scattering peak. The scattering vector, q, is defined by $q = (4\pi/\lambda)\sin(2\theta/2)$, wherein λ refers

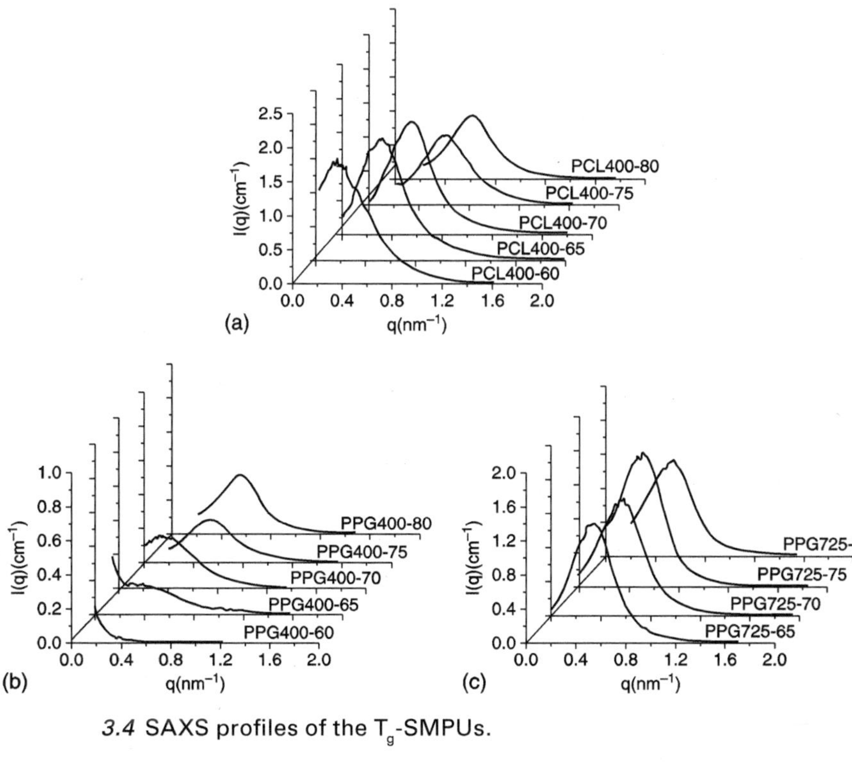

3.4 SAXS profiles of the T_g-SMPUs.

Table 3.3 Phase composition and microdomain morphology of the T_g-SMPUs

Samples	Interdomain spacing (nm)	$\overline{\Delta\rho^{2}}$ ((mol e/ cm^3)^2×10^{-3})	ϕ_1 (%)	ϕ_2 (%)	$\bar{l}_1$ (nm)	$\bar{l}_2$ (nm)
PCL400-60	18.3	1.025	23.4	76.6	3.0	9.8
PCL400-65	12.0	1.120	29.9	70.1	3.6	8.4
PCL400-70	11.4	1.101	35.7	64.3	3.9	7.0
PCL400-75	10.2	1.060	41.9	57.1	3.9	5.2
PCL400-80	10.3	0.856	48.1	51.9	4.5	4.6
PPG400-60	–	–	–	–	–	–
PPG400-65	–	–	–	–	–	–
PPG400-70	18.5	0.463	21.7	78.3	2.9	10.4
PPG400-75	11.7	0.536	31.1	68.9	3.1	6.8
PPG400-80	11.2	0.556	41.5	58.2	4.2	5.9
PPG725-65	11.0	1.112	33.5	66.5	3.7	7.4
PPG725-70	10.6	1.084	39.3	60.7	4.0	6.2
PPG725-75	12.2	1.076	47.4	52.6	5.1	5.6
PPG725-80	11.4	0.864	52.2	47.8	5.4	4.9

to the wavelength of the incident X-ray radiation and 2θ is the scattering angle. The interdomain spaces of the polyurethanes are summarized in Table 3.3. For all the PCL400- and PPG400-based-segmented polyurethanes, d is reduced with the increase of HSC, indicating that the concentration of the hard domains, namely the physical cross-linking density, augments. However, for the PPG725-based segmented polyurethanes, as HSC increases from 65 to 80%, d decreases at first and then rises at HSC = 75% and decreases again at HSC = 80%. This particular changing tendency of d may be associated to some morphological change with chemical composition variation.

Phase composition is crucial in determining the morphology of the polyurethanes and significant for governing their shape memory properties. Several models have been developed to estimate the phase compositions of the copolymers (Camberlin and Pascault, 1983, 1984; Couchman, 1980, 1982; Koberstein and Rs, 1983). Here we follow a method proposed by Koberstein *et al.* (1992) based on SAXS tests. At first it is assumed that the soft phase contains hard segments while the hard domains are pure, i.e. have no residual soft segments. The phase compositions can thus be estimated by comparing the experimental electron density variance with the theoretical prediction. The estimation is based on the analysis of Porod's law:

$$\lim_{q \to \infty} I(q) = \frac{K}{q^4} \qquad [3.2]$$

where K is the Porod's law constant. Taking into account the influence of the background, which is caused by the phase mixing within phases into the observed scattering intensity Iobs(q), Eq. [3.3] is revised as:

$$\lim_{q\to\infty} I_{obs}(q) = \lim_{q\to\infty} \frac{K}{q^4} + I_B \qquad [3.3]$$

The background corrected intensity is therefore equal to Iobs(q)-I_B. By considering the effect of the diffuse boundary between the hard and soft microdomains, the above equation is modified to:

$$\lim_{q\to\infty} I(q) = \lim_{q\to\infty} \frac{K}{q^4} \exp(-\sigma^2 q^2) \qquad [3.4]$$

where σ is a measure of the interfacial boundary thickness. By analyzing the plot of $I(q)q^4$ against q^4, and that of $\ln\{[I(q) - I_B(q)]q^4\}$ against q^2, the parameters I_B and σ are obtained. Thus the experimental electron density variance $\overline{\Delta\rho}^{2''}$ can be evaluated with Eq. [3.5]:

$$\overline{\Delta\rho}^{2''} = c\int_0^\infty \frac{[I(q) - I_B(q)]}{\exp(-\sigma^2 q^2)} q^2\, dq \qquad [3.5]$$

where c is a constant equal to 1.76 ρ 10^{-24} mol^2/cm^2. However, the electron density variance of the polyurethanes can be predicted theoretically:

$$\overline{\Delta\rho}^{2''} = \phi_1 \cdot (1 - \phi_1) \cdot (\rho_1 - \rho_2)^2 \qquad [3.6]$$

where ϕ_1 and $1 - \phi_1$ refer to the volume fraction of the soft phase and hard phases, respectively; and ρ_1 and ρ_2 stand for the electron density of the soft and hard phases, respectively. The values of the physical density of PCL400, PPG400 and PPG725 polyols at T_{room} were experimentally determined as 1.062, 1.016 and 1.010 g/cm^3, respectively. The physical density of the MDI/BD hard segments is taken as 1.354 (Leung and Koberstein, 1985). Assuming ΔW_h to be the weight fraction of hard segments dissolved in the soft matrix, the weight fractions of the soft and hard phases are expressed as $W_1 = W_s + \Delta W_h$ and $W_2 = W_h - \Delta W_h$, respectively. Here W_s and W_h represent the weight fraction of soft and hard segments, respectively. Then the volume fraction of soft segments in the soft phase $\phi_{1,s}$ is given by:

$$\phi_{1,s} = \frac{W_s / v_s}{(W_s / v_s) + (\Delta W_h / v_h)} \qquad [3.7]$$

where v_s and v_h refer to the specific volume of the polyols and the pure hard phase, respectively; and v_T stands for the specific volume of the polyurethanes by assuming additional volume. Then the volume fraction of soft phase ϕ_1 is calculated by Eq. [3.8]:

$$\phi_1 = \frac{W_1}{v_T\left(\dfrac{\phi_{1,s}}{v_s} + \dfrac{1-\phi_{1,s}}{v_h}\right)} \qquad [3.8]$$

The electron density of the PCL400 is 0.59 (Bogart *et al.*, 1983) and that of PPG400 and PPG725 is 0.604 (Leung and Koberstein, 1985). The electron density of the hard phase is taken as 0.716 (Leung and Koberstein, 1985). Therefore the electron density of soft phase can be obtained from the following equation:

$$\rho_1 = \phi_{1,s}\rho_s + (1-\phi_{1,s})\rho_h \qquad [3.9]$$

Combining Eqs [3.5] and [3.6], the volume fraction of soft phase ϕ_1, and hard phase $1 - \phi_1$ can be calculated. With the phase composition, the domain sizes can also be evaluated. The Porod's law constant K is related to the interfacial surface-to-volume ratio S/V:

$$\frac{S}{V} = \pi \cdot \phi_1 \cdot (1-\phi_1)\frac{c \cdot K}{\overline{\Delta\rho^{2}}} \qquad [3.10]$$

The average length of soft domains $\bar{l}_1$ and hard domains $\bar{l}_2$ can be calculated by:

$$\bar{l}_1 = 4\phi_1 \Big/ \left(\frac{S}{V}\right) \qquad [3.11]$$

$$\bar{l}_2 = 4(1-\phi_1) \Big/ \left(\frac{S}{V}\right) \qquad [3.12]$$

The phase compositions and domain sizes are summarized in Table 3.3. It can be seen that the volume fractions of the soft phase of the segmented polyurethanes are almost over 50 wt.%, although MDI and BD account for over 60 wt.% of the raw materials. This means that a significant amount of hard segments are dissolved in the soft matrix, which is in agreement with the DSC results.

Koberstein *et al.* (1992) have investigated a series of segmented polyurethanes based on MDI, BD and PPG (M_w = 2000) glycols and proposed that the hard segments shorter than a critical length, corresponding to a chain with 3 to 4 MDI residues, were assumed to dissolve into the soft matrix. As shown in Table 3.3, the average MDI numbers in hard segments of the PCL400 series and PPG400 series polyurethanes are less than 3.8. The short lengths of both soft and hard segments raise their compatibility, resulting in significant phase mixing. However, the soft and hard segments are not fully miscible as expected due to three possible reasons. First, the length of the hard segments in segmented polyurethanes is always dispersive as is well documented in Peebles (1974, 1976). Particularly, the sequence length distribution can be broadened in the case of short average hard segments (Peebles, 1974, 1976). The short and long hard segments co-exist in the two series of segmented polyurethanes, which lead to both significant phase mixing as well as phase separation.

Second, besides the critical length, the concentration of the hard segments should also be essential in governing phase separation. The HSC values of the polyurethanes are more than 60%, resulting in the formation of hard domains.

Third, the critical length of the hard segments should alter significantly due to the difference in the thermodynamic incompatibility of the soft and hard segments. The critical length with 3 to 4 MDI residues proposed by Koberstein and Rs (1993) is probably not suitable for the segmented polyurethanes in this study. For a certain HSC, the volume fraction of the hard phase of the segmented polyurethanes increases in the order PPG725-xx > PCL400-xx > PPG400-xx, indicating increasing physical cross-linking density.

It can thus be concluded that three types of polyols are ranked in the same order, according to their capability for driving phase separation. As would be expected, the volume fraction of the soft phase is reduced and accordingly the volume fraction of the hard phase is raised steadily as HSC increases. Likewise, the domain sizes of the soft and hard phases show similar tendencies. The PCL400 and PPG725 series of segmented polyurethanes with higher HSC are estimated to have a higher volume fraction of the hard phase (>40%). Hence they are expected to form a continuous hard phase morphology.

As usual, the T_g-SMPUs are deformed at a high temperature $T_{def} = T_g + 15°C$, where the deformation stress is significant for generating recovery stress. As shown in Fig. 3.5, the stresses of all three series of segmented polyurethanes rise with the growth of HSC, which is caused by the increasing physical cross-linking density. Comparing the polyurethanes with identical HSC, it is found that the PPG725-based segmented polyurethane yields the highest deformation stress, the PCL400-based one generates an intermediate deformation stress and the PPG400-based one yields the lowest deformation stress. According to the SAXS results, this is due to the decrease of physical cross-linking density in this order. In particular, PPG400-60- and PPG400-65-based segmented polyurethanes exhibit a low deformation stress, since they are proven to be homogeneous or only slightly phase separated.

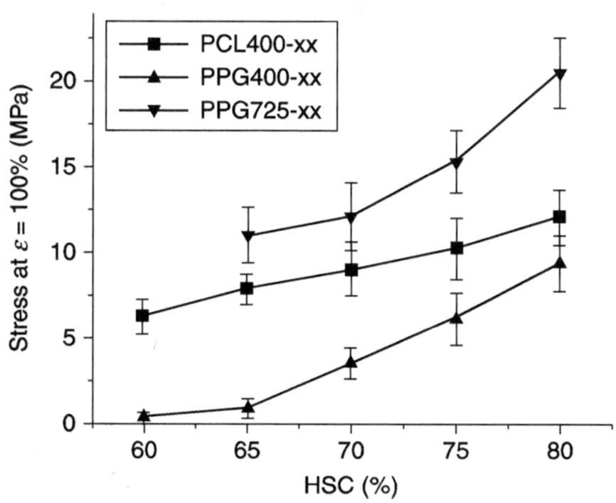

3.5 Stress at 100% elongation of the T_g-SMPUs in the case of $T_{def} = T_g + 15°C$.

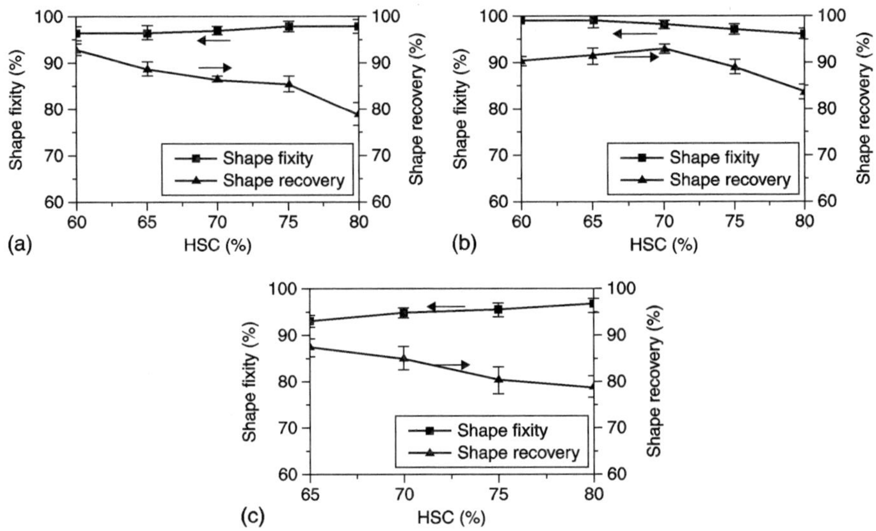

3.6 Influences of HSC on SME of the T_g-SMPUs.

Figure 3.6 presents the experimental results of shape fixity and shape recovery derived from the thermo-mechanical cyclic tensile tests. It is evident that the shape recovery of PCL400- and PPG725-based segmented polyurethanes is reduced with the growth of HSC. The reduction of shape recovery is caused by both the increasing faction of the hard phase and the broad glass transition region. A dramatic shape recovery drop occurs at higher HSC for the two series of segmented polyurethanes, which should arise from the continuous hard phase morphology. As far as the PPG400-based segmented polyurethanes are concerned, the shape recovery rises as HSC rises from 60 to 70% and then falls as HSC increases further.

On the lower HSC side, the lower physical cross-linking density of PPG400-60 and PPG400-65 results in a lower shape recovery. On the higher HSC side, the shape recovery depression primarily arises from the increasing volume fraction of the hard phase whose deformation cannot recover at T_g + 15°C. The shape fixity of the PCL400 and PPG725 series of segmented polyurethanes is reduced with the decrease of HSC, which arises from the decreasing T_g. In contrast, the PPG400 series of segmented polyurethanes exhibits the opposite tendency, which is primarily ascribed to the lower deformation stress of the segmented polyurethanes with lower HSC.

3.2.4 Shape fixing of T_g-type shape memory polymers

Figure 3.7(a) shows the shape fixity variation of the PCL400-based segmented polyurethanes 120 minutes after unloading at T_{room}. It is found that the time to

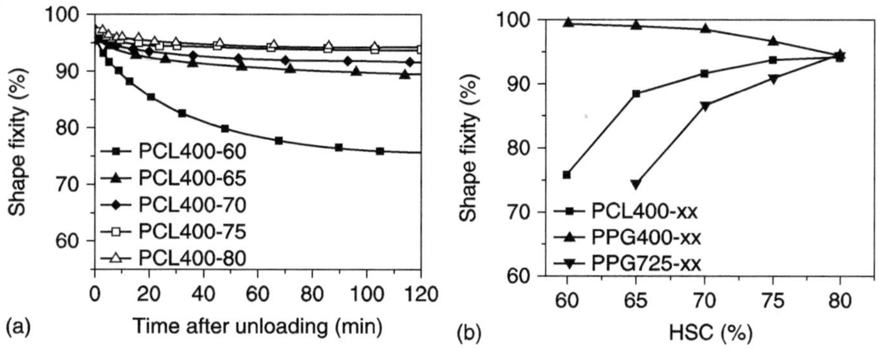

3.7 Shape fixing of the T_g-SMPUs at T_{room}.

shape fixity of the segmented polyurethanes is reduced, which can be explained as follows. The SME of T_g-SMPUs is ascribed to the molecular motions of soft segments below and above T_g (Takahashi *et al.*, 1996; Tobushi *et al.*, 1996). The polymer molecular mobility of main chain segments is characterized by a spectrum of characteristic times, which are inversely proportional to the temperature imposed on the materials (Perez, 1998). By being cooled down below T_g, the micro-Brownian motions of the soft segments are limited. Nevertheless, the molecular mobility in the glassy polymers is not fully suppressed, regardless of the long characteristic times. In particular, a part of the molecular segments can exist in the rubbery state if the shape fixing temperature is located in the glass transition regions of the segmented polyurethanes. Therefore the shape recovery is not fully stopped at a temperature below T_g, i.e. the shape fixity will decay slowly.

Because of the lower molecular mobility at T_{room}, the segmented polyurethanes with higher T_g preserve higher shape fixity. In contrast, the segmented polyurethanes with lower T_g show faster shape recovery and thus lower shape fixity because of their comparatively high molecular mobility. Though the shape recovery occurs constantly, the decrease rate of shape fixity is remarkably reduced 120 minutes after removing the external force where the shape fixity tends to certain values. Figure 3.7(b) presents the shape fixity 120 minutes after removing the external force. The shape fixity of the PCL400 and PPG725 series of segmented polyurethanes augments with the increase of HSC.

It can be envisaged that the temporary shape of the polymers with higher HSC can be readily preserved for a long time, whereas the shape fixity of those with lower HSC can only be kept in the short term at T_{room}, unless they are cooled down to even lower temperatures. It is also observed that the shape fixity of the PPG400-based polyurethanes decreases slightly with the increase of HSC. This may primarily arise from the lower stress of the segmented polyurethanes with lower HSC.

3.2.5 Shape recovery of T_g-type shape memory polymers

In order to study temperature-dependent shape recovery behaviors, the deformed T_g-SMPUs were heated at a rate of approximately 3°C/min from T_{room} to a high temperature where shape recovery stopped. Figure 3.8 presents the plots of the shape recovery as a function of shape recovery temperature. The initial shape recovery rates of the plots roughly correspond to the shape fixity values obtained 120 minutes after unloading. For example, the shape fixity of PCL400-60 is left as 75% and its initial shape recovery in Fig. 3.8(a) is 25%. It is found that the shape recovery of the segmented polyurethanes rises gradually with the increasing temperature. In a narrow temperature region around T_g, the shape recovery of the segmented polyurethanes abruptly increases. With the increase of HSC, the shape recovery curves constantly shift to higher temperature, which is caused by the increasing T_g of the segmented polyurethanes.

Exceptionally, the shape recovery curve of PPG400-70 arises in a lower temperature region in comparison to those of PPG400-60 and PPG400-65, as illustrated in Fig. 3.8(b). This is ascribed to the lower physical cross-linking density of the latter two segmented polyurethanes. As the temperature is raised to a sufficiently high level (often >100°C) the shape recovery of the segmented polyurethanes increases to over 90% and tends to constant values, suggesting the irreversible deformation strain of the polyurethanes is less than 10%. It is also noted that the segmented polyurethanes

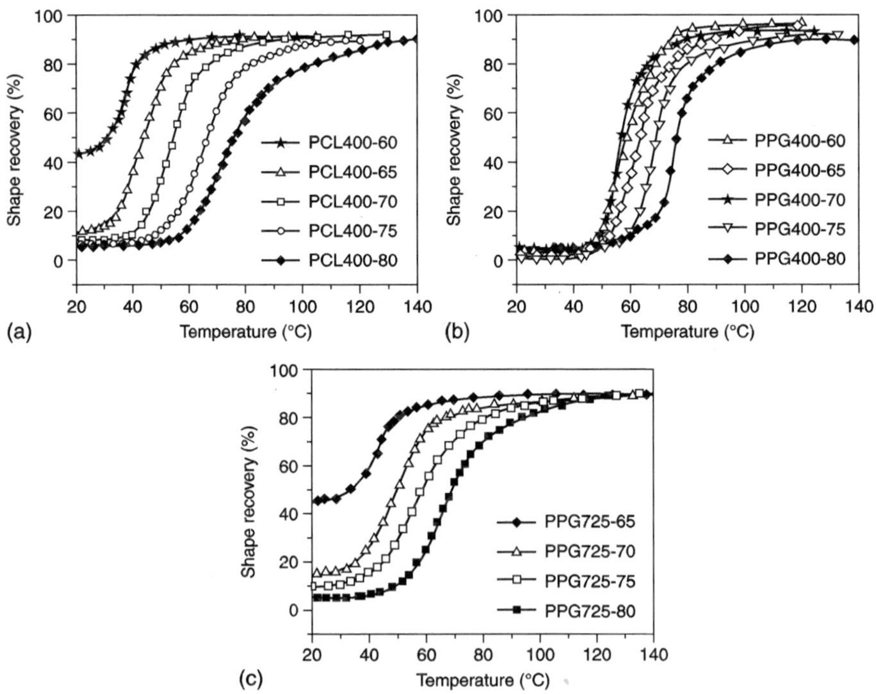

3.8 Shape recovery vs temperature of the T_g-SMPUs.

with higher HSC tend to exhibit broad shape recovery regions, which may arise from their wider glass transition regions. In addition, the segmented polyurethanes with higher HSC have relatively larger volume fraction of the hard phase, which results in larger deformation of the hard phase. The deformation of the hard phase cannot be restored at low temperature, leading to broad shape recovery regions. The broad shape recovery regions account for the lower shape recovery at $T_{rec} = T_g + 15°C$, as demonstrated by the thermo-mechanical cyclic tensile tests.

3.2.6 Recovery stress of T_g-type shape memory polymers

It is well established that the deformation stress of polyurethane elastomers consists of entropic and energetic parts (Petrovic and Ferguson, 1991). The recovery stress of the SMPUs should derive from the entropic part of their deformation stress. The preservation of entropic energy is dependent on the shape fixing, which is characterized by shape fixity. Hence the theoretical maximum recovery stress of a material is equal to the entropic part of its deformation stress excluding the loss caused by unfixed strain. However, it is inevitable that stress loss occurs during stress storage and release via stress relaxation. Furthermore, the modulus of the polymers decreases with increasing of temperature. Hence the observed recovery stress evolution in the experiments involves complex processes, including stress release, stress relaxation and modulus variation.

The plots of the recovery stress as a function of temperature of the segmented polyurethanes are shown in Fig. 3.9. In the course of heating, the

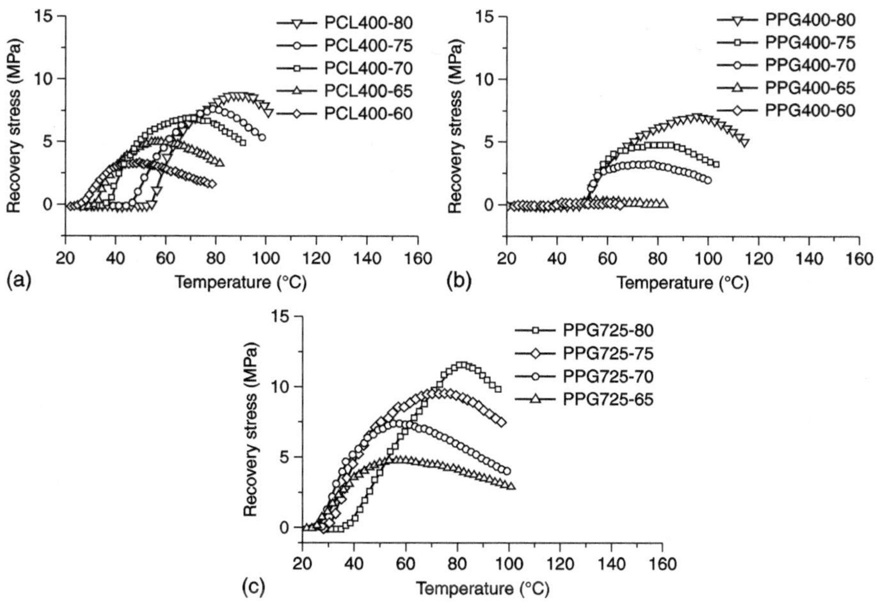

3.9 Recovery stress vs temperature of the T_g-SMPUs.

recovery stress of a polyurethane sample occurs at a certain temperature and is raised steadily up to a maximum where the stress stored is released entirely. Above the peak point the recovery stress starts to decrease as temperature increases, which arises from the stress relaxation and the descending modulus. The recovery stress curves shift to higher temperatures as HSC increases, which is consistent with the results of the shape recovery tests. Moreover, it is noted that as for a series of segmented polyurethanes, the maximum recovery stress is increased continuously with increasing HSC, due to increasing deformation stress yielding rising shape fixity and resistance to stress relaxation. It is noteworthy that the PPG400-60 and PPG400-65 exhibit low recovery stress but good shape recovery and shape fixity, as proven previously. Therefore it can be concluded that some polymer networks with weak cross-links and lower cross-linking density cannot exhibit good comprehensive shape memory properties, even though they may show higher shape fixity and shape recovery.

3.3 Segmented polyurethanes with similar T_g

By altering the species of soft segments and adjusting the chemical composition, the T_g, i.e. the T_{trans} for triggering the SME, the T_g-SMPUs can be readily located at a desired temperature, as demonstrated previously. T_g-SMPUs have been considered to be superior to T_m-SMPUs because of their higher flexibility for molecular design (Kusy and Whitley, 1994). In this study, the comparisons between three polyurethanes with similar T_g are conducted. Figure 3.10(a) presents the DMA thermograms of PPG400-70, PCL400-75 and PPG725-80. It is found that the peak maxima on the tanδ curves of the three segmented polyurethanes are all roughly located at 70°C, i.e. they have similar T_g. But the tanδ curves are distinguished from each other in terms of peak height and width because of the different morphological structures.

Figure 3.10(b) presents the thermo-mechanical cyclic tensile tests of the three segmented polyurethanes. Under similar thermo-mechanical conditions the residual strain of PPG400-70, PCL400-75 and PPG725-80 increases, suggesting that shape recovery is reduced in this order. In contrast, the deformation stresses of the three segmented polyurethanes show the opposite tendency. As illustrated in Fig. 3.10(c), from PPG400-70 to PCL400-75 and PPG725-80, the three polyurethanes exhibit increasingly broadened shape recovery regions, which accounts for the reduced shape recovery obtained from the thermo-mechanical cyclic tensile tests. Figure 3.10(d) reveals that the three polyurethanes are ranked in the order PPG400-70 < PCL400-75 < PPG725-80 in terms of recovery stress. Evidently, the segmented polyurethanes with higher fractions of the hard phase exhibit lower shape recovery at T_g + 15°C, wider shape recovery regions and higher recovery stress, although the three polymers have similar T_g. Thus, although T_g can be readily located at a particular temperature through molecular design, the

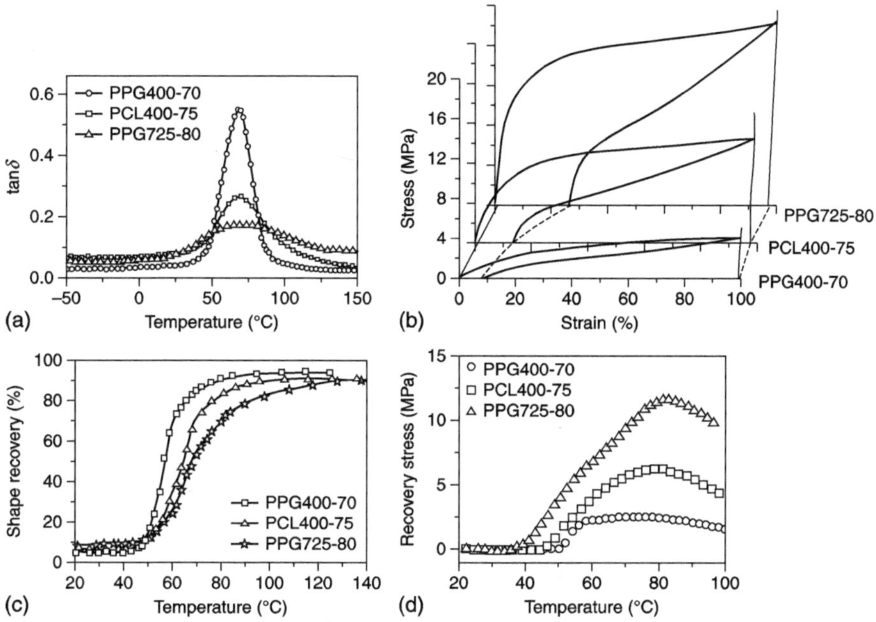

3.10 Comparisons of the polyurethanes with T_g = 70°C.

morphological structure should be controlled properly in order to derive optimal shape memory properties.

3.4 Thermo-mechanical conditions on T_g-type shape memory polymers

3.4.1 Influences of deformation temperature on shape memory polymers

In the previous investigations, T_g-SMPUs were usually deformed at a temperature above T_g, for example T_g + 15°C, to approach a temporary shape. However, it is believed that SMPUs could exhibit SMEs in a wider deformation temperature range (Kim *et al.*, 1996). Systematic investigation of the effect of deformation temperature on shape memory behavior has not yet taken place.

Figure 3.11(a) illustrates the influences of T_{def} on the SME of PCL400-75. It is evident that PCL400-75 exhibits pronounced shape fixity and shape recovery with T_{def} varying from T_g − 20°C to T_g + 30°C. It is also observed that the shape recovery curves shift to high temperature as T_{def} increases. This is a consequence of the dispersive characteristic times of the segmented polyurethanes around the glass transition region. The dispersive characteristic times correspond to a distribution of molecular fractions that exhibit different T_g (Perez, 1998).

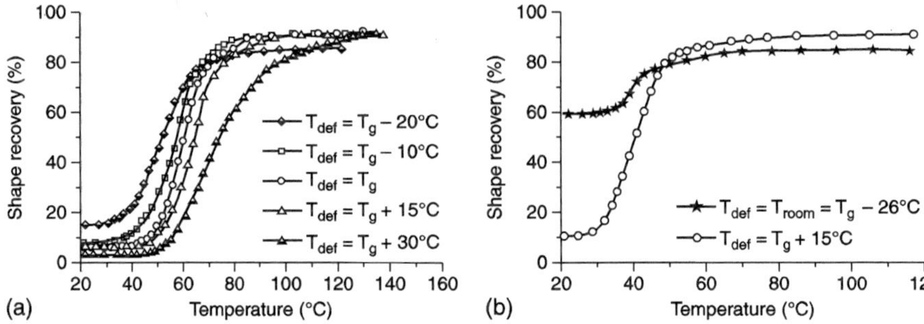

3.11 Influences of deformation temperature on shape recovery.

When the deformation occurs in the glass transition region, the molecular segments with T_g lower than T_{def} can readily deform and participate in the strain storage, whereas those with T_g higher than T_{def} hardly deform and so contribute less to strain storage. In other words, the shape recovery primarily arises from the structural recovery of the molecular segments with T_g lower than T_{def}. Therefore the shape recovery curves move to lower temperature with decreasing T_{def}. It is also noted that the shape recovery curve with $T_{def} = T_g + 30°C$ is broadened, suggesting that the strain storage is contributed to by molecular segments with a broad spectrum of T_g. It is also found that the shape recovery curve with $T_{def} = T_g - 20°C$ shows lower ultimate shape recovery, suggesting that a too low T_{def} may lead to significant structural destruction and thus result in a large irreversible deformation. Moreover the shape fixity, which corresponds to the starting point of a shape recovery curve, is lessened with decreasing T_{def}. This is because the strain storage is mainly contributed to by the molecular segments with lower T_g in the case of lower T_{def}. Hence the shape fixity tends to decrease via rapid relaxation. As indicated in Fig. 3.11(b), the effect of lower T_{def} on PCL400-65 was also examined in order to verify this phenomenon. Likewise the shape recovery and especially the shape fixity decrease markedly when T_{def} decreases from $T_g + 15°C$ to T_{room} (here $T_{room} = T_g - 26°C$).

3.4.2 Influences of pre-deformation on shape memory polymers

The segmented polyurethanes with higher HSC, especially those with a continuous hard phase, exhibited higher recovery stress but broader shape recovery region and lower shape recovery at $T_g + 15°C$. SMPUs, as a class of smart material, are expected to show good shape recoverability at a low temperature. In 1965, Hicks and co-worker improved the recoverability of Spandex via a stretching process followed by heat setting to reduce the interconnectivity of the hard domains (Aharoni, 1979). In order to enhance shape recovery at low recovery temperature, in a previous study (Zhu *et al.*, 2006) concerning T_m-SMPUs, a measure was proposed to increase shape recovery via pre-deformation. The study examined the validity of the measure for T_g-SMPUs and described its scope of application. As illustrated in Fig. 3.12(a), the segmented polyurethane with higher HSC PCL400-80 was subjected to two thermo-mechanical cycles with most conditions identical to the descriptions in the experimental sections except that the maximum strain, which is designated as ε_{m0}, is 150%. As marked in the second loading, the strain ε_{p0} (here ~ 45%) where the stress starts to be more than zero represents the residual strain in the first cycle. After the second unloading the polyurethane sample was reheated up to $T_{rec} = T_g + 15°C$ for 20 minutes.

A residual strain roughly equal to ε_{p0} applied. Then the slightly deformed sample was taken as a 'new' sample and subjected to a series of thermo-mechanical cyclic tensile tests to examine their SME. As demonstrated in Fig. 3.12(a), the fourth loading curve almost coincides with the third loading, indicating no more residual strain. This can be explained by the fact that in the preceding thermo-mechanical cycles some of the molecular segments with lower molecular mobility, especially in the interconnected hard domains, were forced to deform. This part of the deformation could not be restored at $T_{rec} = T_g + 15°C$ and was left as ε_{p0}. It is assumed that there would be no more residual strain generated in the subsequent cycles, provided the following relation is fulfilled:

$$\varepsilon_m \leq \varepsilon_{m0} - \varepsilon_{p0} \qquad [3.13]$$

where ε_m refers to the maximum deformation strain imposed on the 'new' sample. To test the validity of the assumption that no more strain would be generated, a series of SME tests with varying deformation strains were conducted. As illustrated in Fig. 3.12(b), the shape recovery obtained from the shape memory tests with different deformation strains is close to 100% when ε_m is less than 105% and starts to descend as ε_m continues to increase, which agrees well with the assumption that no more residual strain would be generated in subsequent cycles. However, ε_{p0} is temperature-dependent and can recover as temperature gradually increases. Therefore the recovery temperature was elevated to $T_{def} + 10°C$ to examine the influence of T_{rec} on shape recovery.

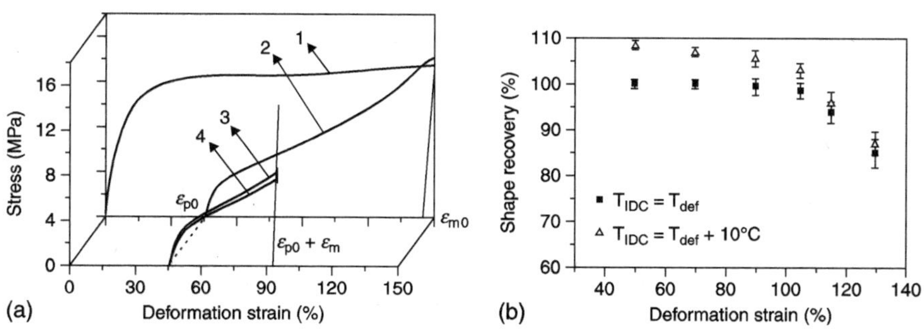

3.12 Influence of pre-deformation on the shape memory effect.

As shown in Fig. 3.12(b), the resultant shape recovery is increased to more than 100% but less than 110%, suggesting that more recovery of ε_{p0} occurs at higher shape recovery temperature. Therefore 100% of shape recovery can be approached only if the T_{rec} in the SME tests is consistent with the T_{rec} used in the pre-deformation program. It is significant that experiments corresponding to Fig. 3.12(b) were finished with numerous samples made of a uniform PCL400-80 thin film. The samples were proven to be highly consistent with each other in terms of physical properties. It can thus be safely concluded that the shape recovery of these T_g-SMPUs can be raised to about 100% via appropriate pre-deformation.

3.4.3 Influences of shape fixing temperature on shape memory polymers

The molecular mobility of the segmented polyurethanes is dependent on the cooling temperature. Figure 3.13 illustrates the influences of shape fixing temperature T_{fix} on the shape fixing behaviors of PPG400-70 and PCL400-75. As would be expected, the shape fixity of each segmented polyurethane augments with decreasing T_{fix}. As for PPG400-70, the shape fixing curve with $T_{fix} = T_g - 38°C$ is essentially identical to that with $T_{fix} = T_{room}$ (~$T_g - 48°C$), suggesting $T_{fix} = T_g - 38°C$ is low enough to obtain a high shape fixity. Likewise, PCL400-75 exhibits similar shape fixing behaviors under $T_{fix} = T_g - 42°C$ and $T_{fix} = T_{room}$. $T_g - 38°C$ and $T_g - 42°C$ almost correspond to the starting temperature of the glass transition regions of PPG400-70 and PCL400-75, as indicated by the DMA tests. This seems to suggest that the segmented polyurethanes should be at least cooled down to a temperature below the starting point of their glass transition region in order to obtain good shape fixing.

The shape fixity of the segmented polyurethanes is dependent on not only their T_g and the T_{fix} but also the morphology structure. Figure 3.14 shows the shape

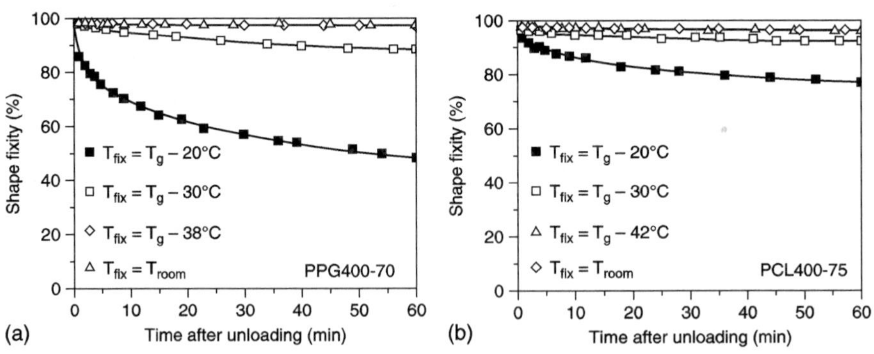

3.13 Influences of fixing temperature on shape fixity.

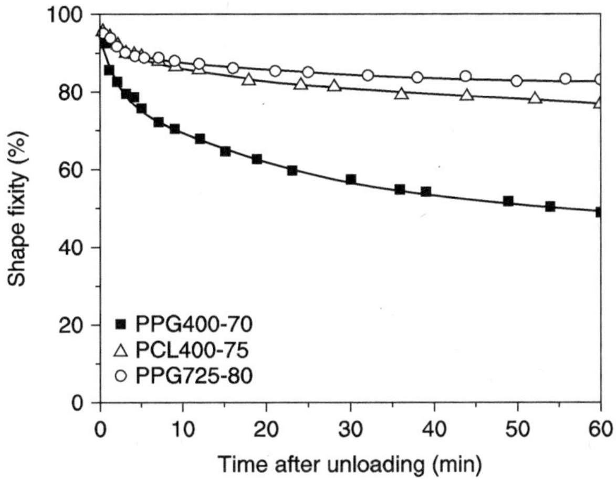

3.14 Shape fixing of the T_g-SMPUs having similar T_g at $T_{fix} = T_g - 20°C$

fixity changing with time of PPG400-70, PCL400-75 and PPG725-80, at $T_{fix} = T_g - 20°C$. Although they have identical T_g, as proved by the DMA tests, they exhibit different shape fixing ability at the same T_{fix}. With the elapse of time, the shape fixity of PPG400-70 descends rapidly, while that of PPG400-75 and PPG725-80 decreases slowly.

The volume fractions of the hard phase of PPG400-75 and PPG725-80 are higher than PPG400-70. Higher fraction of the hard phase can result in larger deformation of the hard phase, which in turn results in lower shape recovery at $T_{rec} = T_{room}$ and thus higher shape fixity at $T_{fix} = T_g - 20°C$. Furthermore, in PPG400-75 and PPG725-80, the hard phase glass transition accounts for larger parts in the overall glass transition as compared to PPG400-70. In other words, PPG400-75 and PPG725-80 have more molecular segments with higher T_g participating in deformation. Hence they exhibit higher shape fixity at $T_g - 20°C$ than PPG400-70. This result suggests that the segmented polyurethanes with higher fraction of hard phase tend to exhibit higher shape fixing ability.

3.4.4 Influences of shape fixing time on shape memory polymers

When the deformed T_g-SMPUs are cooled from T_{def} to T_{fix} for shape fixing, the free volume in the amorphous region is squeezed out progressively, which is also time-dependent, similar to the physical aging in glassy polymers. Hence the shape fixing time (t_{fix}), the time for shape fixing in which the samples are constrained at T_{fix}, can influence the shape fixity of the T_g-SMPUs.

Figure 5.15 presents the influence of shape fixing time (t_{fix}) on shape fixity of PPG725-80. Evidently, the shape fixity augments with the growth of t_{fix}. The

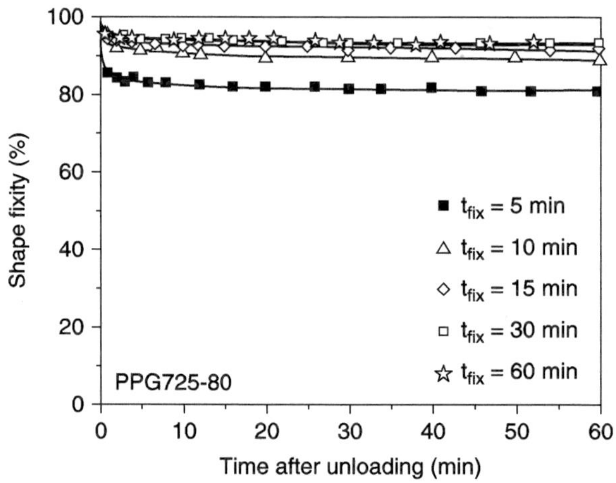

3.15 Influence of fixing time on shape fixity of PPG725-80.

shape fixing curves with t_{fix} = 30 min and t_{fix} = 60 min are almost the same, indicating that most of the free volume in the material is reduced to a minimum value and the shape fixity tends to a constant level at t_{fix} = 30 min.

3.5 Conclusions

Three sets of T_g-SMPUs were prepared from three kinds of low-molecular-weight polyols. The relationship between morphological structure and shape memory properties of the T_g-SMPUs were investigated:

1. DSC, DMA and SAXS were employed to explore the thermal properties and morphology of the T_g-SMPUs. Most of the T_g-SMPUs exhibit distinct soft and hard phase separation, indicating the presence of physical cross-linking networks. It is found that the three polyols are ranked in the order PPG400 < PCL400 < PPG725 for phase separation. DSC tests reveal that the T_g of the mixed soft phase varies in the temperature range 20 to 50°C. But the DMTA tests prove that the T_g of the T_g-SMPUs covers a wide temperature range of 30 to 90°C, which is assumed to be a result of the superposition of soft and hard phase glass transitions. The complex glass transition is considered as the molecular switch for triggering the SME of the T_g-SMPUs. Based on a two-phase model, the phase compositions of the polyurethanes are estimated via the SAXS analysis. Thereby the microdomain morphology of the T_g-SMPUs is evaluated.
2. The shape fixity of the PCL400- and PPG725-based polyurethanes is raised with increasing HSC because of the increased T_g, while that of the PPG400-based polyurethanes exhibits the opposite tendency, because the

PPG400-based segmented polyurethanes with lower HSC exhibit less deformation stress. The T_g-SMPUs with higher HSC exhibit broader shape recovery regions, resulting in lower shape recovery under $T_{rec} = T_g + 15°C$. The recovery stress of the T_g-SMPUs is elevated with increasing HSC, because both deformation stress and shape fixity are increased. Comparison between the T_g-SMPUs with similar T_g show that although T_g can be readily located at a specific temperature via altering types of soft and hard segments and their chemical composition, the resultant polymers have different shape memory properties due to their different morphological structure. Higher fraction of the hard phase results in higher recovery stress but a broader shape recovery region.

3. The T_g-SMPUs can exhibit pronounced shape recovery and shape fixity when T_{def} is in the vicinity of T_g. The shape recovery curves shift to lower temperature with decreasing T_{def}. The shape recovery and shape fixity are dramatically decreased when T_{def} is reduced to below $T_g - 20°C$. After a pre-deformation process, the shape recovery of the polyurethane with higher HSC can be improved to nearly 100% upon appropriate programming.
4. The shape fixity of T_g-SMPUs is mainly dependent on their T_g and the T_{fix} employed for shape fixing. Where $T_{fix} = T_{room}$, the T_g-SMPUs with higher T_g exhibit better shape fixing. The T_g-SMPUs should be cooled down to a temperature lower than the starting point of the glass transition to obtain stable shape fixing. In addition, the T_g-SMPUs with a greater fraction of the hard phase tend to exhibit better shape fixing. With the increase of shape fixing time, the T_g-SMPUs show higher shape fixity.

3.6 References

Aharoni, S. M. (1979), Hysteresis behavior in polyether poly(urethaneureas) based on 2,4-toluene diisocyanate, ethylenediamine, and poly(tetramethylene oxide), *Macromolecules*, **12**, 539–41.

Bogart, V., John, W. C., Gibson, P. E., Cooper, S. L. (1983), Structure–property relationships in polycaprolactone-polyurethanes, *J. Polym. Sci., Polym. Phys. Ed.*, **21**, 65–95.

Camberlin, Y., Pascault, J. P. (1983), Quantitative DSC evaluation of phase segregation in linear segmented polyurethanes and polyurethaneureas, *J. Polym. Sci., Polym. Chem.*, **21**, 415–23.

Camberlin, Y., Pascault, J. P. (1984), Phase segregation kinetics in segmented linear polyurethanes—relations between equilibrium time and chain mobility and between equilibrium degree of segregation and interaction parameter, *J. Polym. Sci., Polym. Phys. Ed.*, **22**, 1835–44.

Couchman, P. R. (1980), Prediction of glass-transition temperatures for compatible blends formed from homopolymers of arbitrary degree of polymerization compositional variation of glass-transition temperatures, *Macromolecules*, **13**, 1272–6.

Couchman, P. R. (1982), Composition-dependent glass-transition temperatures and copolymers, *Nature*, **298**, 729–30.

Kim, B. K., Lee, S. Y., Xu, M. (1996), Polyurethanes having shape memory effects, *Polymer*, **37**, 5781.

Koberstein, J. T., Rs, S. (1983), Small-angle X-ray scattering measurements of diffuse phase-boundary thickness in segmented polyurethane elastomers, *J. Polym. Sci., Polym. Phys. Ed.*, **21**, 2181–200.

Koberstein, J. T., Leung, L. M. (1992), Compression-molded polyurethane block copolymers. 2. Evaluation of microphase compositions, *Macromolecules*, **25**, 6205–13.

Koberstein, J. T., Galambos, A. F., Leung, L. M. (1992), Compression-molded polyurethane block copolymers, Part I: Microdomain morphology and thermo-mechanical properties, *Macromolecules*, **25**, 6195–204.

Kusy, R. P., Whitley, J. Q. (1994), Thermal characterization of shape memory polymer blends for biomedical implantations, *Thermochimi. Acta*, **243**, 253.

Lee, B. S., Chun, B. C., Chung, Y. C., Sul, K. I., Cho, J. W. (2001), Structure and thermomechanical properties of polyurethane block copolymers with shape memory effect, *Macromolecules*, **34**, 6431.

Leung, L. M., Koberstein, J. T. (1985), Small-angle scattering analysis of hard-microdomains structure and microphase mixing in polyurethane elastomers, *J. Polym. Sci., Polym. Phys. Ed.*, **23**, 1883–913.

Leung, L. M., Koberstein, J. T. (1986), DSC annealing study of microphase separation and multiple endothermic behavior in polyether-based polyurethane block copolymers, *Macromolecules*, **19**, 706–13.

Lin, J. R., Chen, L. W. (1998b), Study on shape-memory behaviors of polyether-based polyurethanes, Part I: Influence of the hard-segement content, *J. Appl. Polym. Sci.*, **69**, 1563–74.

Peebles, L. H. (1974), Sequence length distribution in segmented block copolymers, *Macromolecules*, **7**, 872–82.

Peebles, L. H. (1976), Hard block length distribution in segmented block copolymers, *Macromolecules*, **9**, 58–61.

Perez, J. (1998), *Physics and Mechanics of Amorphous Polymers*, Rotterdam, Netherlands: A.A. Balkema, 253.

Petrovic, Z. S., Ferguson, J. (1991), Polyurethanes elastomers, *Prog. Polym. Sci.*, **16**, 695–836.

Sperling, L. H. (1992), *Introduction to Physical Polymer Science*, New York, John Wiley & Sons.

Takahashi, T., Hayashi, N., Hayashi, S. (1996), Structures and properties of shape-memory polyurethane block copolymers, *J. Appl. Polym. Sci.*, **60**, 1061.

Tobushi, H., Hara, H., Yamada, E., Hayashi, S. (1996), Thermo-mechanical properties in a thin film of shape memory polymer of polyurethane series, *Smart Mater. Struct.*, **5**, 483.

Zhu, Y., Hu, J. L., Yeung, L. Y., Liu, Y., Ji, F. L., Yeung, K. W. (2006), Development of shape memory polyurethane fiber with complete shape recoverability. *Smart Materials and Structures*, **15**, p. 1385–94.

4
High performance type shape memory polymers prepared by modified two-step polymerization

DOI: 10.1533/9780857098542.71

Abstract: The properties of SMPUs can be improved significantly by improving the phase separation and the stability of the hard segment. This chapter introduces three strategies to improve the properties of SMPUs. The first is achieved by employing 4,4′-methylene-bis(2-chloroaniline) (MOCA) as the chain extender to form a stable hard segment phase. The second strategy is to employ cycloalphatic diisocyanates (isophorone diisocyanates (IPDI)) in the hard segment instead of aromatic diisocyanates, which can improve the stability of the polyurethane to ultraviolet radiation. The third strategy is achieved by employing a modified two-step copolymerization method to significantly improve the phase separation of the polyurethane.

Key words: high performance shape memory polymer, MOCA, thermal property, shape memory property, isophorone diisocyanate.

4.1 Introduction

Although both T_m-SMPUs and T_g-SMPUs have attracted increasing attention, especially in respect of their shape fixing and recovery behaviors, there are some difficulties in application. These are:

- low shape recovery force;
- instability to ultraviolet (UV) radiation.

The proper shape recovery stress is essential for high performance application of both T_m-SMPUs and T_g-SMPUs. Beside the shape recovery stress, the long-term instability to UV light will also have a negative impact on the performance of shape memory polyurethane (SMPU), as UV irradiation will cause both a deeper color and a decrease in mechanical properties. To solve these problems and to increase the performance of SMPUs, 4,4′-methylene-bis(2-chloroaniline) (MOCA) has been employed in T_m-SMPU as the chain extender to form a stable hard segment phase. By employing cycloalphatic diisocyanates (isophorone diisocyanates (IPDI)) in the hard segment of T_m-SMPU instead of aromatic diisocyanates, superior stability of the polyurethane to UV radiation is achieved yielding superior color stability and good mechanical property retention. At the same time, a modified two-step copolymerization method has also been employed in the synthesis of T_g-SMPU, which can significantly improve the phase separation degree of the polyurethane.

There have been a number of investigations into T_m-SMPUs (Hu *et al.*, 2005a; Kim *et al.*, 1996, 1998; Li *et al.*, 1997), primarily concerned with the shape fixing and shape recovery behaviors of SMPUs, whereas recovery stress has been less studied. Recovery stress is essential for most of the applications for SMPUs, since in practical applications they are often required to realize shape memory effects (SMEs) under external hindrance. Nevertheless, one of the disadvantages of SMPUs is their low recovery stress (Wei *et al.*, 1998). Therefore, SMPUs have been considered only for applications involving lower recovery force or free recovery.

Physical cross-links play a central role in governing the shape memory properties of SMPUs. The SME of T_m-SMPUs displaying interconnected, isolated and no hard domains has been demonstrated. It is found that the T_m-SMPUs with 25% ≤ HSC ≤ 40% show better shape recovery. The shape fixity of T_m-SMPUs arises from the crystallization of the soft phase. Physical cross-links can reduce the crystallization rate and the crystallizability of the soft phase (Bogart *et al.*, 1983; Kim *et al.*, 1996). Recovery stress is derived from deformation stress and thus leads to higher physical cross-linking density. However, it is also dependent on shape fixity, which usually indicates lower physical cross-linking density. Because of the dilution effect, higher hard segment content (HSC) can result in lower crystallinity (X_c) of the soft phase. Therefore raising physical cross-linking density without employing too high a HSC is essential in exploring the stress storage potential of T_m-SMPUs.

Segmented polyurethane-ureas chain extended with aromatic diamines are well phase separated and show higher deformation stress and shape recovery because of their strong tendency to hard segment aggregation (Garrett *et al.*, 2000, 2002, 2003; Luo *et al.*, 1996, 1997). Currently, few efforts have been devoted to studying the SME of polyurethane-ureas (SMPUUs). Rabani *et al.* (2006) prepared a series of T_m-SMPUUs with short hard segments. However, probably because of a too low HSC, the T_m-SMPUUs did not exhibit good shape memory properties. In the research reviewed in this chapter, a series of T_m-SMPUUs with higher HSC were developed. The influences of physical cross-links in T_m-SMPUs and T_m-SMPUUs on the shape memory properties, including shape recovery, shape fixity and recovery stress, are compared.

4.2 High performance shape memory polymers

We synthesized a series of T_m-SMPUUs with PCL4000 diols as soft segments, with MDI and MOCA as hard segments. The formulations of the MOCA-based T_m-SMPUs are detailed in Table 4.1. A set of segmented polyurethane-ureas were synthesized with polycaprolactone (PCL) diols as soft segments and MDI and MOCA as hard segments. The PCL4000 was dried and dehydrated at 80°C under vacuum for six hours prior to use. The solvent dimethyl formamide (DMF) was dried with 4Å molecular sieves before copolymerization. The

Table 4.1 The formulations of the MOCA-based T_m-SMPUUs

Samples	Feed ratio (mole)			HSC (wt.%)
	PCL4000	MDI	MOCA	
PCL4000-M-30	1.0	3.74	2.74	30
PCL4000-M-35	1.0	4.47	3.47	35
PCL4000-M-40	1.0	5.57	4.57	40

isocyanate-terminated prepolymers were first prepared by reacting the dried PCL4000 with the mole excessive MDI to ensure complete reaction in the DMF solution at 65°C for three hours. Then the prepolymers were chain extended with MOCA at 80°C and the reaction system was kept at this temperature for another four hours. The T_m-SMPUUs are designated as PCL4000-M-xx where the character 'M' means that segmented polyurethane-ureas are chain extended with MOCA, and 'xx' stands for the wt.% HSC.

We compared the MOCA-based T_m-SMPUUs with the T_m-SMPUs prepared from PCL4000, MDI and BD. Figure 4.1 illustrates the chemical structure of the hard segments of the (a) T_m-SMPUs and (b) MOCA-based T_m-SMPUUs. The T_m-SMPUUs contain more benzene rings, which are assumed to raise the rigidity of the hard segments and prompt phase separation between the soft and hard segments. More importantly, the urea hydrogen bonds were considered to be stronger than the urethane hydrogen bonds.

It is assumed that the shape memory properties of the T_m-SMPUs and T_m-SMPUUs are dependent on their chemical compositions. However, too low an HSC can yield lower deformation stress and thus lower recovery stress, though

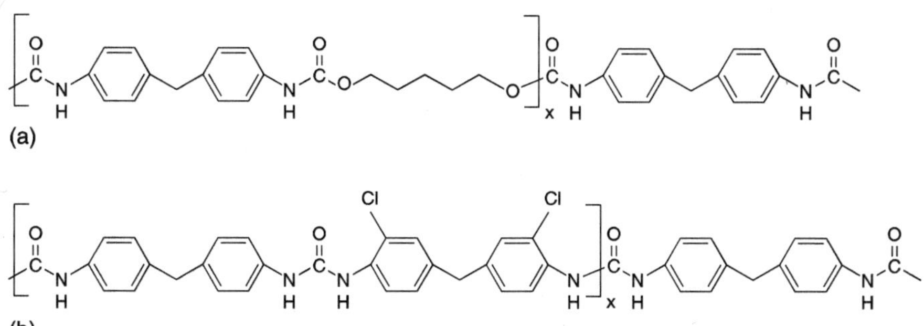

4.1 Illustrations of hard segments of T_m-SMPUs and MOCA-based T_m-SMPUUs.

it can give rise to high shape fixity and shape recovery under particular conditions (Kim *et al.*, 1996; Li *et al.*, 1997). However, too high an HSC can severely limit crystallization of the soft phase and gives rise to lower shape recovery, shape fixity and recovery stress. Thus the HSC of the T_m-SMPUs and the T_m-SMPUUs in this chapter covers an intermediate HSC range from 30 to 40%. The influences of HSC on the shape memory properties are discussed systematically. It is found that the T_m-SMPUUs show much better shape recovery and yield larger recovery stress as compared with the T_m-SMPUs with identical HSC.

4.2.1 Molecular structures

FT-IR tests were conducted on the T_m-SMPUUs and T_m-SMPUs in order to characterize their hydrogen bonding. The FT-IR spectra were normalized using the area of the 1412 cm^{-1} peak, which is assigned to the C–C stretching mode of the aromatic ring (Garrett *et al.*, 2003). The spectra were adjusted so that the relative peak areas of the absorbance at 1412 cm^{-1} were the same as the calculated ratio of aromatic rings in the copolymers. Figure 4.2(a) presents the carbonyl stretching region of the FT-IR spectra of the T_m-SMPUs PCL4000-B-30 and PCL4000-B-40. The absorbance peaks at 1724 and 1700 cm^{-1} are related to the stretching vibration of free carbonyl (C=O) groups and hydrogen bonded C=O groups, respectively (Yilgör *et al.* 2002a,b, 2006). The FT-IR spectra of the T_m-SMPUUs PCL4000-M-30 and PCL4000-M-40 in the same region are shown in Fig. 4.1(b). The absorbance at 1635 cm^{-1} represents the C=O groups in the form of planar bifurcated hydrogen bonds, where one C=O group is hydrogen bonded to two N-H groups of a nearby urea moiety (Garrett *et al.*, 2003; Luo *et al.*, 1996, 1997; Yen *et al.*, 1999). The length of the urea bonds is shorter than those of the urethane bonds, so the urea hard domains should be stronger than the urethane hard domains (Garrett *et al.*, 2003).

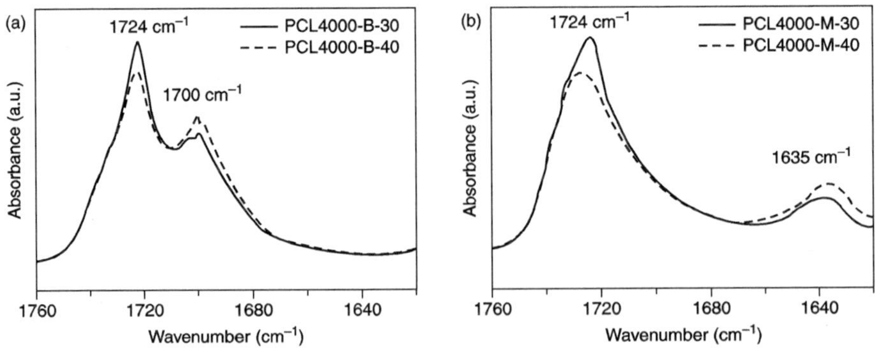

4.2 FT-IR spectra of T_m-SMPUs and T_m-SMPUUs in C=O stretching region.

As a measure of phase separation, the increase of degree of C=O hydrogen bonding indicates the presence of more aggregates of hard segments. In the T_m-SMPUs and T_m-SMPUUs, the C=O groups in the PCL soft segments might also hydrogen bond to the N-H groups in the hard segments. However, as shown in the FT-IR spectra, with increasing HSC the normalized intensity of the hydrogen bonded C=O peak increases, whereas that of the free C=O peak decreases for both the T_m-SMPUs and T_m-SMPUUs. This indicates that the hydrogen bonding degree is raised with increasing HSC in the T_m-SMPUs and T_m-SMPUUs.

4.2.2 Thermal properties

The DSC traces of the T_m-SMPUUs are presented in Fig. 4.3. All the T_m-SMPUUs exhibit an endothermic behavior at about 40°C, which is ascribed to the melting of the crystalline soft phase. T_m and ΔH_m of the soft phase decrease with increasing HSC, suggesting the depression of X_c. This is caused by both the reduced fraction of soft segments and the increased limitation of the hard domains to the soft segments. The T_m, ΔH_m and X_c of the T_m-SMPUUs are summarized in Table 4.2. It can be seen that T_m, ΔH_m and X_c of the soft phase of the T_m-SMPUUs are lower than those of the T_m-SMPUs. This can be attributed to two reasons.

First, urea hard segments are superior to urethane ones in packing into hard domains, resulting in the T_m-SMPUUs having higher physical cross-linking density as compared with the T_m-SMPUs with identical HSC. Second, urea groups contain more proton donors (N–H groups) compared with urethane groups. Therefore C=O groups in PCL segments might more readily hydrogen bond to the N–H groups in urea hard segments than to those in the urethane ones, which leads

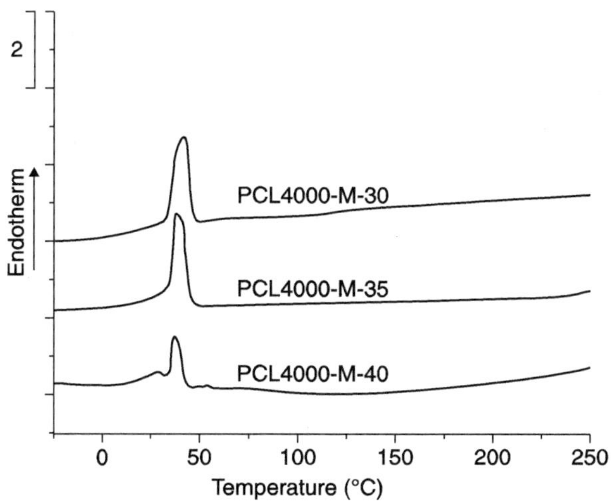

4.3 DSC traces of the MOCA-based T_m-SMPUUs.

Table 4.2 The thermal properties of the T_m-SMPUs and the T_m-SMPUUs

	DSC			DMA		
	T_m (°C)	ΔH_m (J/g)	X_c (%)	E'_{25} † (MPa)	E'_{65} ‡ (MPa)	E'_{25}/E'_{65} –
PCL4000-B-30	46.5	32.8	23.4	664.6	21.5	30.9
PCL4000-B-35	44.7	31.2	22.3	654.6	28.7	22.8
PCL4000-B-40	42.5	18.7	13.4	628.9	60.5	10.4
PCL4000-M-30	41.3	21.3	15.2	668.3	95.6	7.0
PCL4000-M-35	39.9	17.6	12.5	641.2	156.9	4.1
PCL4000-M-40	38.3	9.2	6.5	632.8	250.6	2.5

to lower X_c of the soft phase. The T_m-SMPUUs do not show the endotherms pertaining to the hard phase. A previous investigation showed that the melting point of a model compound composed of MDI and 3,5-diethyltoluenediamine (DETDA), another kind of aromatic diamine polyurethane chain extender, was about 347°C (Luo *et al.*, 1997). However, the PCL soft segments would apparently thermally decompose at such a high temperature. Hence it is difficult to manifest the presence of a MDI/MOCA hard phase in these T_m-SMPUUs via DSC tests.

Figure 4.4 illustrates the storage modulus as a function of temperature derived from the DMA thermograms of the T_m-SMPUs and T_m-SMPUUs. All of the polymers exhibit a substantial modulus drop at approximately 40°C, indicating a melting transition of the soft phase. Above the melting transition, a rubbery plateau region is observed for all the polymers, which indicates the presence of the physically cross-linked polymer networks defined by the hard domains. The rubbery modulus can serve as a measure of physical cross-linking density. As illustrated in Fig. 4.4, the plateau modulus is increased with increasing

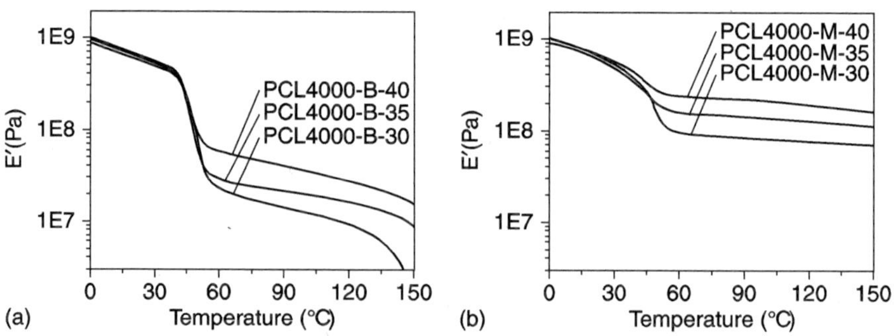

4.4 Storage modulus vs temperature of the MOCA-based T_m-SMPUUs.

HSC, suggesting an increase of physical cross-linking density. A higher ratio of glassy modulus to rubbery modulus, preferably of two orders of magnitude, was considered favorable for shape fixing and shape recovery (Kim *et al.*, 1996). The storage modulus at 25°C (E'_{25}) and the storage modulus at 65°C (E'_{65}), as well as the modulus ratio E'_{25}/E'_{65}, are tabulated in Table 4.2. The T_m-SMPUUs exhibit a much higher rubbery modulus compared with the T_m-SMPUs, indicating that the T_m-SMPUUs possess higher physical cross-linking density due to the strong aggregation tendency of the urethane-urea hard segments. The modulus ratios of the T_m-SMPUs are no more than 31, whereas those of the SMPUUs are lower than 10. All the polymers show good shape memory properties.

4.2.3 Shape memory properties

The shape fixity and shape recovery of the T_m-SMPUs and T_m-SMPUUs were determined via a series of thermo-mechanical cyclic tensile tests. Figure 4.5 demonstrates the thermo-mechanical cyclic tensile tests of PCL4000-B-30 and PCL4000-M-30. The strains marked by the arrows are the fixed strain ε_u and residual strain ε_p. The ε_p of PCL4000-M-30 is apparently less than that of PCL4000-B-30, suggesting that PCL4000-M-30 exhibits higher shape recovery, whereas the ε_{um} of PCL4000-M-30 is lower than that of PCL4000-B-30. In

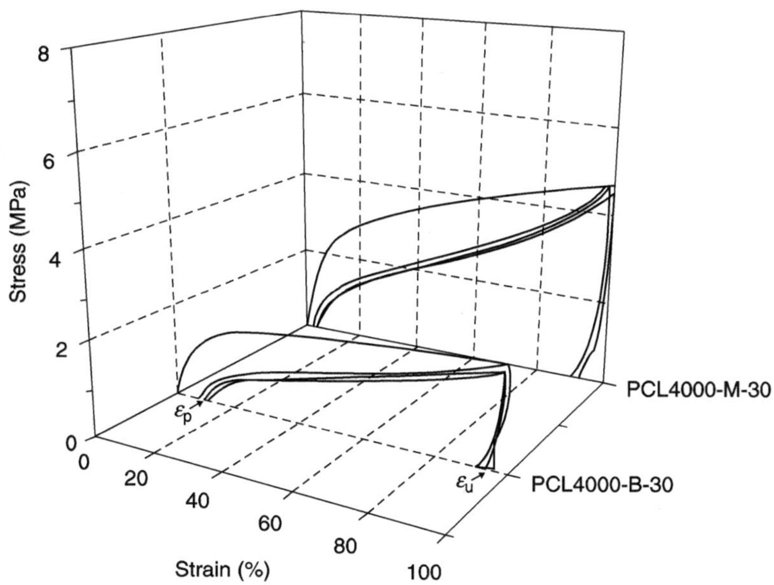

4.5 Thermo-mechanical cyclic tensile tests of the polymers with HSC = 30%.

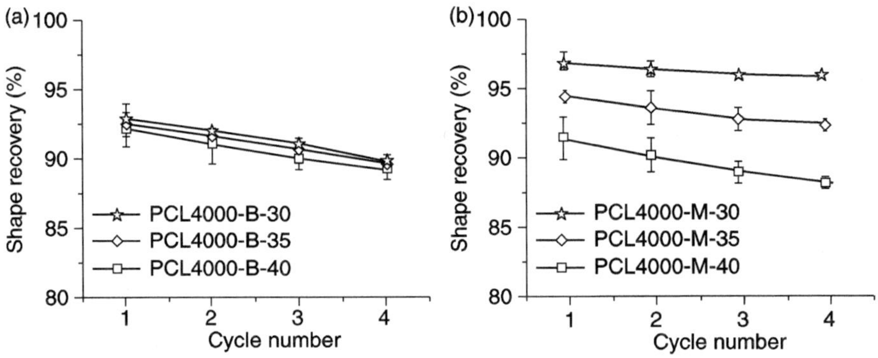

4.6 Shape recovery as a function of cycle numbers.

addition, the deformation stress at 100% of elongation $\sigma100$ in the first loading of PCL4000-M-30 is nearly twice that of PCL4000-B-30.

The results of shape recovery derived from the thermo-mechanical cyclic tensile tests of all the polymers are summarized in Fig. 4.6. As demonstrated in Fig. 4.6(a), all the T_m-SMPUs show over 90% shape recovery. Shape recovery is reduced slightly with increasing HSC. As for the T_m-SMPUUs, an obvious depression of shape recovery with increasing HSC is illustrated in Fig. 4.6(b). This appears to be due to the volume fraction of the hard phase rising with increasing HSC in the T_m-SMPUUs, which results in increased deformation of the hard phase. It is also found that PCL4000-M-30 and PCL4000-M-35 show higher shape recovery than PCL4000-B-30 and PCL4000-B-35, respectively, indicating that the two T_m-SMPUUs form more complete and stable physically cross-linked networks. These results suggest that the T_m-SMPUUs are superior to the T_m-SMPUs in yielding good shape recovery. However, the shape recovery of PCL4000-M-40 is similar to that of PCL4000-B-40, which is ascribed to the increasing fraction of hard phase in the T_m-SMPUUs. With increasing cycle numbers, the shape recovery of the T_m-SMPUs and T_m-SMPUUs reduce. Moreover, PCL4000-M-30 shows shape recovery depression because of its stronger hard domains and lower volume fraction of the hard phase. Therefore it can be concluded that for the T_m-SMPUs and T_m-SMPUUs, the shape recovery is proportional to the strength of the hard domains and inversely proportional to the fraction of the hard phase.

4.2.4 Deformation stress, long-term shape fixing and shape memory properties

Deformation stress is significant in giving rise to recovery stress. Figure 4.7 presents the deformation stress at 100% of elongation of the T_m-SMPUs and T_m-SMPUUs. It can be seen that the stresses of the T_m-SMPUUs are almost twice

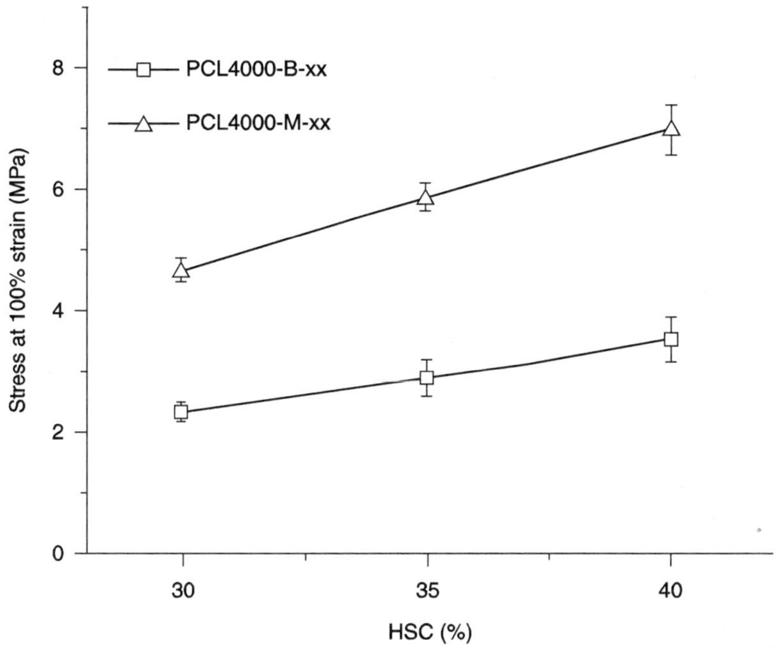

4.7 Deformation stress at 100% of elongation.

those of the T_m-SMPUs, due to the higher physical cross-linking density in the former.

The shape fixing of T_m-SMPUs is ascribed to the strain-induced crystallization of the soft segments. A number of investigations have reported strain-induced crystallization of the soft phase in the deformation of polyurethanes and polyurethane-ureas (Curgul *et al.*, 2004; Fu *et al.*, 2001; Yeh *et al.*, 2003). The extra X_c caused by strain-induced crystallization probably enables the polymers with high physical cross-linking density to show higher shape fixity. This makes it possible to enhance the recovery stress by employing appropriately higher physical cross-linking density. Because of the limitation of physical cross-links, the crystallization rate of the soft segments is reduced. Therefore extending t_{fix} is necessary to explore the maximum shape fixing ability of the T_m-SMPUs and T_m-SMPUUs.

With t_{fix} varying from 15 to 240 minutes, the long-term shape fixing behaviors after unloading, i.e. after removing the external force, of the T_m-SMPUs and T_m-SMPUUs are investigated. As illustrated in Fig. 4.8, the shape fixity of PCL4000-B-40 rises as t_{fix} increases. For a test with a given t_{fix}, the shape fixity decreases in the first 10 minutes after unloading. The unfixed strain, namely the shape recovery at T_{fix}, should be caused by the molecular motion of amorphous soft segments. After 120 minutes, the shape fixity curves of polyurethane tend to

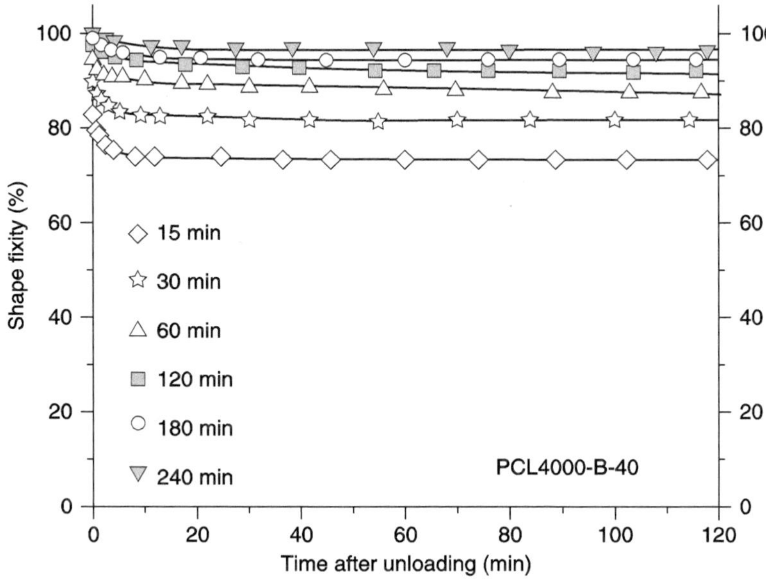

4.8 Long-term shape fixity after unloading of PCL4000-B-40.

constant values for all the tests with varying t_{fix}. The solid lines in Fig. 4.8 are the results of curve fitting based on a second-order exponential decay model. The shape fixity at 120 minutes derived from the fitted results is considered to be the steady-state shape fixity for a particular test.

The plots of the steady-state shape fixity as a function of t_{fix} are presented in Fig. 4.9. The solid lines in the figures represent the curve fitting results based on a second-order exponential growth model. In the first 60 minutes, the shape fixity of all the T_m-SMPUs and the T_m-SMPUUs increases, suggesting the X_c increases

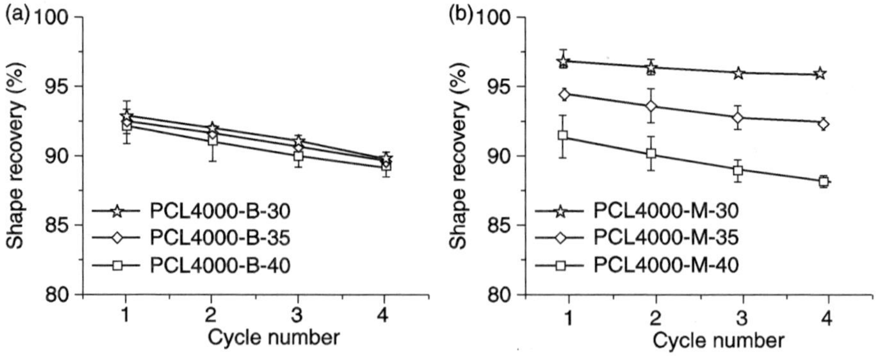

4.9 Effect of t_{fix} on shape fixity.

rapidly. With respect to the T_m-SMPUUs, the shape fixity curves tend to constant values when t_{fix} is extended to 120 minutes, suggesting that the strain-induced crystallization of the soft phase is completed and the maximum shape fixity is approached at that point. In contrast, the T_m-SMPUs exhibit a different shape fixity dependency on t_{fix}. The shape fixity of PCL4000-B-30 reaches a maximum at t_{fix} = 120 minutes, whereas that of PCL4000-B-40 augments continuously as t_{fix} increases from 15 to 240 minutes. This can be explained as follows.

After crystallization of the soft phase, the urethane hard domains experience continuous viscoelastic deformation because they are still subjected to external force in the course of shape fixing. PCL4000-B-40 has a higher percentage of hard phase and this effect is more apparent than PCL4000-B-35 and PCL4000-B-30. In contrast, the urethane-urea hard domains are so strong that they resist further deformation more effectively than urethane hard domains. The T_m-SMPUUs thus do not show continuous augmentation of shape fixity with increasing t_{fix}. In this study, t_{fix} = 120 minutes is considered a better condition for obtaining optimal comprehensive shape memory properties. It is found that the shape fixity at t_{fix} = 120 minutes of the T_m-SMPUUs decreases constantly with the growth of HSC. This suggests that with the increase of physical cross-linking density, the fraction of amorphous soft segments increases. In contrast, the T_m-SMPUs exhibits better shape fixity. However, all the T_m-SMPUs and T_m-SMPUUs exhibit over 86% shape fixity.

4.2.5 Recovery stress and shape memory properties

Shape memorization is essentially a manner of strain or stress storage and release. The recovery stress of the T_m-SMPUs and T_m-SMPUUs should be equal to the deformation stress excluding its energetic part, the losses caused by unfixed strain, stress relaxation and decreasing modulus.

Figure 4.10(a) and (b) present the plots of recovery stress as a function of heating temperature of the T_m-SMPUs and T_m-SMPUUs, respectively. It can be seen that the recovery stress of a polymer appears at a certain temperature and rises progressively. After a peak point the recovery stress decays continuously. In contrast, the recovery stress of the T_m-SMPUUs appears at the start of heating, due to its imperfect crystallites of the soft phase, which broadens the melting transition. The recovery stress augments with the increase of HSC for both the T_m-SMPUs and T_m-SMPUUs. The recovery stress of a polyurethane-urea is nearly twice as high as that of the polyurethane with identical HSC. As marked in Fig. 4.10(a) and (b), the maximum recovery stresses of PCL4000-B-30 and PCL4000-M-30 are 2.07 and 4.16 MPa, respectively. This means that urethane-urea hard domains can indeed enhance the recovery stress of SMPUs.

In order to understand the stress storage and release behavior of the different polymers, a parameter called the stress conversion ratio, i.e. the ratio of maximum recovery stress to the deformation stress at maximum strain, is defined.

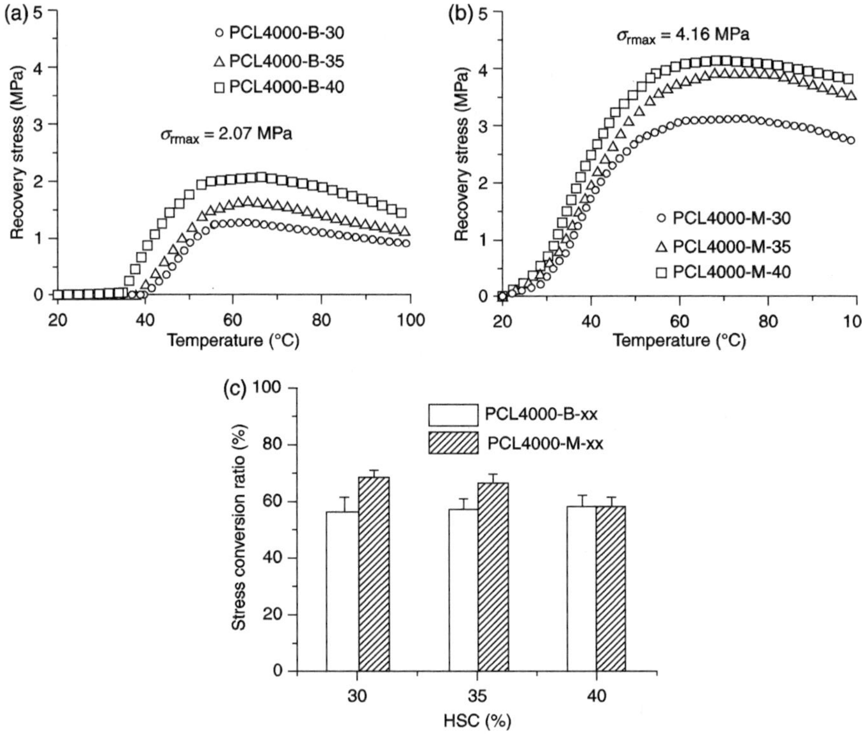

4.10 Recovery stress of the T_m-SMPUs and the T_m-SMPUUs.

Figure 4.10(c) demonstrates the stress conversion ratios of the polymers. It is found that the stress conversion ratios of the T_m-SMPUs and T_m-SMPUUs are in the range of approximately 55 to 70%. The stress conversion ratio of the T_m-SMPUUs is reduced with increasing HSC, which is mainly caused by the apparent decrease of shape fixity. In contrast, the stress conversion ratios of the T_m-SMPUs increase with increasing HSC. This is because the urethane hard domains in the T_m-SMPUs with lower HSC are comparatively weak and cannot effectively endure stress relaxation, resulting in more stress loss in stress storage and release.

4.2.6 Stress relaxation

In investigations, it is found that shape recovery of the T_m-SMPUs can be reduced by stress relaxation (Hu *et al.*, 2005b). This study compares the T_m-SMPUs and T_m-SMPUUs in terms of the endurance to stress relaxation. As demonstrated in Fig. 4.11(a), stress relaxation results in the depression of shape recovery for both PCL4000-B-35 and PCL4000-M-35. The decrease of shape recovery caused by stress relaxation is summarized in Fig. 4.11(b). The

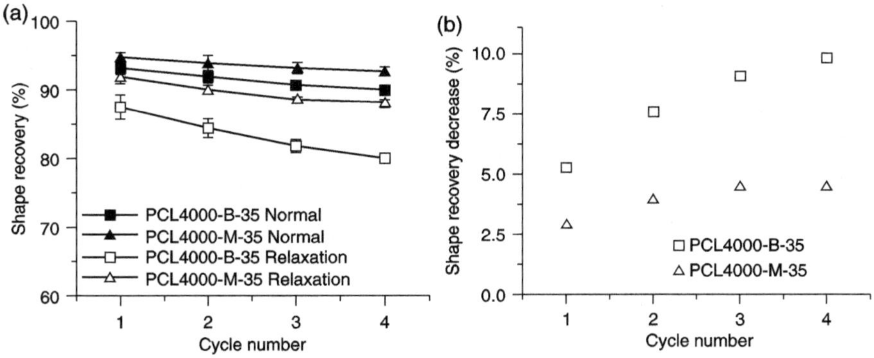

4.11 Effects of stress relaxation on shape recovery.

shape recovery decrease of PCL4000-M-35 is less than 5%, whereas that of PCL4000-B-35 is approximately 5 to 10%. Moreover, with the increase of cycle number, the shape recovery decrease of PCL4000-B-35 is raised, whereas that of PCL4000-M-35 is slightly reduced. These results suggest the urethane-urea hard domains are superior to urethane hard domains for enduring stress relaxation and retaining good shape recovery.

4.3 High performance isophorone diisocyanate (IDPI) T_m- type shape memory polymers

In order to ensure good shape recoverability of SMPUs, the previous investigators used rigid aromatic hard segments. Moreover, aromatic chain extenders containing mesogenic moiety (Jeong *et al.*, 2000a), planar shaped hard segments (Yang *et al.*, 2003), short aramid hard segments (Rabani *et al.*, 2006), and even chemical cross-links (Hu *et al.*, 2005a; Kim *et al.*, 2001), were used to strengthen the hard domains. Exposed to UV radiation in the presence of oxygen, such as incidental solar radiation, aromatic polyurethanes initiate auto-oxidation degradation, which results in a deepening color from colorless to yellow to amber and even to brown, together with loss of mechanical properties on long-term exposure (Hu *et al.*, 2005a; Kim *et al.*, 2001; Petrovic and Ferguson, 1991; Szycher, 1999). This is due to the fact that the aromatic ring, under such conditions, can be auto-oxidized to a chromophore such as a quinone-imine structure. Aliphatic and cycloalphatic diisocyanates cannot undergo such quinoid formation. The non-aromatic polyurethanes thus show a superior stability to UV radiation and thus superior color stability and good mechanical property retention (Huh *et al.*, 2008; Rogulska *et al.*, 2006). Compared with their aromatic counterparts, the segmented polyurethanes with SME containing non-aromatic hard segments have been less investigated (Meng *et al.*, 2008; Ping *et al.*, 2007). However, the non-aromatic urethane hard segments usually yield low phase separation and thus render weak physical cross-links and resultant low strength (Petrovic and Ferguson, 1991). Polyurethane-ureas chain extended with diamines are well phase separated because of stronger hydrogen bonds and higher hydrogen bond density (Sheth *et al.*, 2005; Yilgör *et al.*, 2004).

In this study, a series of T_m-SMPUUs based on IPDI were prepared with PCL4000 and PCL10000 diols as soft segments and with Jeffamine polyetheramine (D-230) and isophorone diisocyanates (IPDA) as chain extenders. Figure 4.12 illustrates the chemical structure of the T_m-SMPUUs chain extended with (a) D-230 and (b) IPDA. The formulations of the IPDI-based T_m-SMPUUs are detailed in Table 4.3. A series of T_m-SMPUUs were synthesized with IPDI and with PCL4000 and PCL10000 diols as soft segments. The PCL polyols were dried and dehydrated at 80°C under vacuum for six hours prior to usage. The solvent DMF was dried with 4Å molecular sieves prior to use. IPDI, D-230 IPDA and Dibutyltin Dilaurate (DBTL) were used as received. With DMF as solvent, the isocyanate-terminated prepolymers were first prepared by reacting the dried PCL diols with the mole excessive IPDI at 90°C for three hours. To prompt the complete reaction, a drop of DBTL was added into the reaction system as a catalyst. Then the prepolymers were

Table 4.3 Formulations of the IPDI-based T_m-SMPUUs

Samples	Feed ratio (mole)					HSC
	PCL4000	PCL10000	IPDI	D-230	IPDA	(wt%)
IPCL4000-D-30	1.0	–	4.30	3.30	–	30
IPCL4000-D-35	1.0	–	5.27	4.27	–	35
IPCL10000-D-30	–	1.0	9.99	8.99	–	30
IPCL10000-D-35	–	1.0	12.42	11.42	–	35
IPCL10000-D-40	–	1.0	15.26	14.26	–	40
IPCL10000-I-40	–	1.0	17.44	–	16.44	40

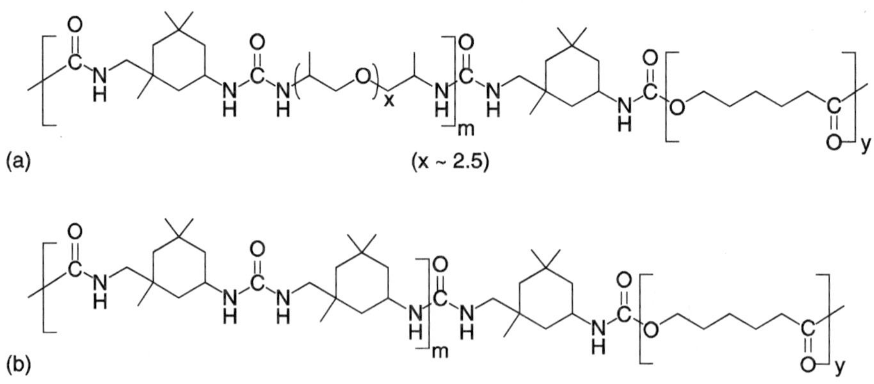

4.12 Illustration of the IPDI-based T_m-SMPUUs.

stoichiometrically chain extended with D-230 or IPDA, followed by being held at 90°C for another four hours. The formulations of the T_m-SMPUUs are shown in Table 4.3, wherein the SMPUUs are designated as IPCL4000-D-xx, IPCL10000-D-xx and IPCL10000-I-xx, respectively. Here the character 'I' before PCL means IPDI-based SMPUUs; 'D' and 'I' between two dashes refer to the chain extenders D-230 and IPDA, respectively; and 'xx' represents the wt.% HSC. HSC of a segmented polyurethane-urea is equal to its total wt.% of IPDI and chain extenders.

The structure and thermal properties of the T_m-SMPUUs were investigated by means of FT-IR, DSC and DMA. The shape memory properties including shape recovery and shape fixity were investigated via a series of thermo-mechanical cyclic tensile tests. The influence of the length of soft segments and HSC on the structure, thermal properties, shape fixity and shape recovery of the T_m-SMPUUs are discussed systematically. In addition, the recovery stress of the T_m-SMPUUs with hard segment content 40% was characterized. This study suggested that via appropriate molecular design, the cyclo-aliphatic segmented polyurethane-ureas can exhibit good shape memory properties similar to the aromatic T_m-SMPUs and T_m-SMPUUs.

4.3.1 Molecular structures

The physical cross-links of the IPDI-based T_m-SMPUUs in this study are predominantly determined by the hydrogen bonds between hard segments. Infrared spectroscopy is a powerful technique in characterizing the hydrogen bonding of polyurethanes and polyurethane-ureas (Coleman *et al.*, 1986; Luo *et al.*, 1996). Here FT-IR was used to study the hydrogen bonds of the IPDI-based T_m-SMPUUs. Each spectrum was first normalized using the area of the 1465 cm^{-1} peak, assigned to the CH_2 bending mode (Allara and Swalen, 1982; Wang *et al.*, 2006). Figure 4.13(a) shows the carbonyl (C=O) group stretching region of the FT-IR spectra of IPCL4000-D-30 and IPCL4000-D-35. The absorbance at 1724 cm^{-1} is ascribed to the stretching of the free C=O groups in either hard segments or PCL segments (Coleman *et al.*, 1986; Luo *et al.*, 1996).

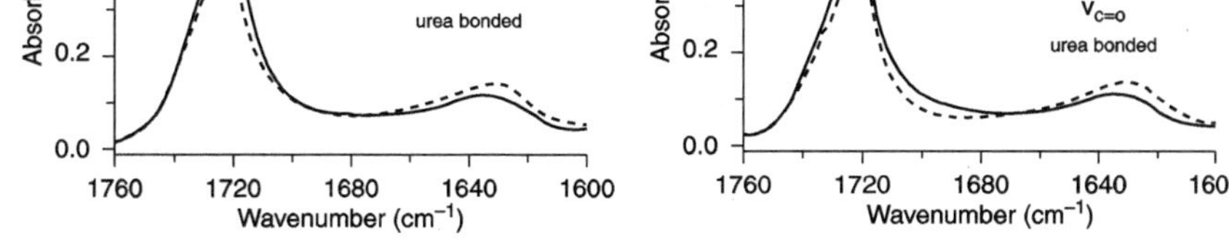

4.13 FT-IR spectra of the IPDI-based T_m-SMPUUs.

The absorbance centered at about 1637 cm^{-1} represents the C=O groups associated to the planar bifurcated hydrogen bonds, where one C=O group is hydrogen bonded to two N-H groups of a nearby urea moiety (Jena *et al.*, 2007). It is noted that with increasing HSC, the normalized intensity of hydrogen bonded C=O groups is raised, while that of free C=O groups is reduced, indicating the increasing degree of hydrogen bonding. In addition, the absorption of hydrogen bonded C=O groups shifts to lower frequency as HSC increases, suggesting that the strength of hydrogen bonding is raised. In the T_m-SMPUUs, the C=O groups in the PCL soft segments can also be hydrogen bonded to N–H groups in the hard segments. However, it can be envisaged that the apparent variation of the absorption peaks of free and hydrogen bonded C=O groups indicates the formation of harder segment aggregates with the increase of HSC. Figure 4.13(b) presents the comparison of the IR spectra of IPCL4000-D-30 and IPCL10000-D-30. It is observed that as the molecular weight of soft segments increases from 4000 to 10 000, the absorption peak of hydrogen bonded C=O groups is enhanced while that of free C=O is weakened. This is because the length of hard segments in IPCL10000-D-30 is over twice that of IPCL4000-D-30, as indicated in Table 4.3, which prompts the aggregation of hard segments. Likewise the absorption peak of hydrogen bonded C=O shifts to lower frequency, suggesting the strength of the hydrogen bonds of IPCL10000-D-30 is higher than that of IPCL4000-D-30.

4.3.2 Thermal properties

DSC traces of all the T_m-SMPUUs are presented in Fig. 4.14. In the temperature range 40 to 55°C, all the T_m-SMPUUs exhibited an endothermic behavior which is assigned to melting of the crystalline soft phase. The X_c of the PCL soft segments of the T_m-SMPUUs was calculated according to the melting peak area of the DSC trace, assuming the perfect PCL crystal has a melting enthalpy of 0.14 kJ/g (Ping *et al.*, 2005). The T_m, ΔH_m and X_c of the soft phase are summarized in Table 4.4.

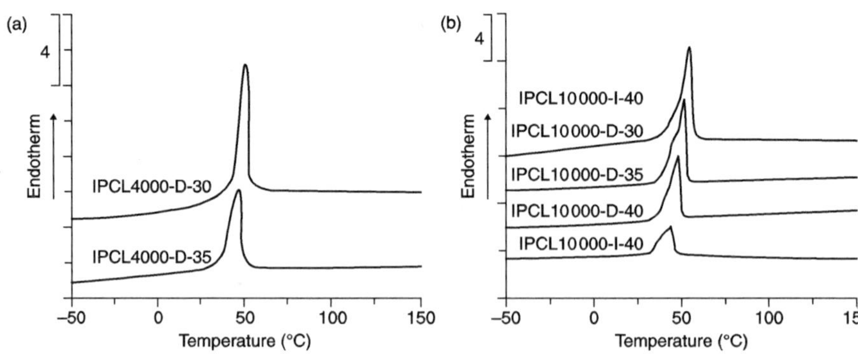

4.14 DSC thermograms of the IPDI-based T_m-SMPUUs.

Table 4.4 Thermal properties of IPDI-based T_m-SMPUUs

Samples	DSC			DMA		
	T_m (°C)	ΔH_m (J/g)	X_c (%)	E'_{25} † (MPa)	E'_{65} ‡ (MPa)	E'_{25}/E'_{65} –
IPCL4000-D-30	49.1	42.7	30.5	890.9	8.3	107.1
IPCL4000-D-35	44.4	15.8	11.3	843.4	23.5	35.9
IPCL10000-D-30	52.8	41.7	29.8	958.4	0.2	4563.9
IPCL10000-D-35	51.3	39.2	28.0	1015.7	13.0	77.9
IPCL10000-D-40	50.6	36.2	25.9	894.5	42.1	21.3
IPCL10000-I-40	43.5	17.8	12.7	835.8	57.6	14.5

†,‡ E'_{25} and E'_{65} denote the storage modulus at 25°C and 65°C, respectively.

As would be expected, T_m, ΔH_m and X_c are all reduced with increasing HSC (Bogdanov *et al.*, 1999). This is caused by both the reduced fraction of soft segments and the increased limitation of the hard domains to the soft segments. In addition, the N–H groups in the hard segments can hydrogen bond to the C=O groups in PCL segments. Thus the increasing HSC will intensify the limitation to soft segments and reduce their X_c. Moreover the ΔH_m and X_c values of PCL10000-based T_m-SMPUUs are correspondingly lower than those of the PCL4000-based ones with identical HSC. Similar results were reported in a previous investigation concerning T_m-SMPUs (Kim *et al.*, 1996). The X_c of IPCL10000-D-40 is lower than that of IPCL10000-I-40, suggesting that the hard segments in the latter impose more restriction on the PCL soft segments.

4.3.3 Dynamic mechanical properties

Figure 4.15 illustrates the storage modulus as a function of temperature derived from the DMA thermograms of the IPDI-based T_m-SMPUUs. All the polymers exhibit a substantial modulus drop at approximately 40°C, indicating a melting transition of the soft phase. Above T_m, a rubbery plateau region is viewed for all T_m-SMPUUs, which indicates the presence of physically cross-linked polymer networks. The storage modulus E′ is a measure of the stiffness and can render information relating to the cross-linking density of segmented polyurethanes (Asif *et al.*, 2005; Kim *et al.*, 1996). It can be seen that the plateau modulus of the IPDI-based T_m-SMPUUs is elevated with increasing HSC, which is caused by the rise of the fraction of the hard phase. The rubbery plateau of IPCL10000-I-40 is slightly higher than that of IPCL10000-D-40, indicating that the physical cross-linking density of the former is higher than that of the latter, which is consistent with the DSC results. This can be ascribed to the following two facts.

First, the concentration of urea groups of IPCL10000-I-40 is higher than that of IPCL10000-D-40, as indicated in Table 4.4, which favors the aggregation of hard

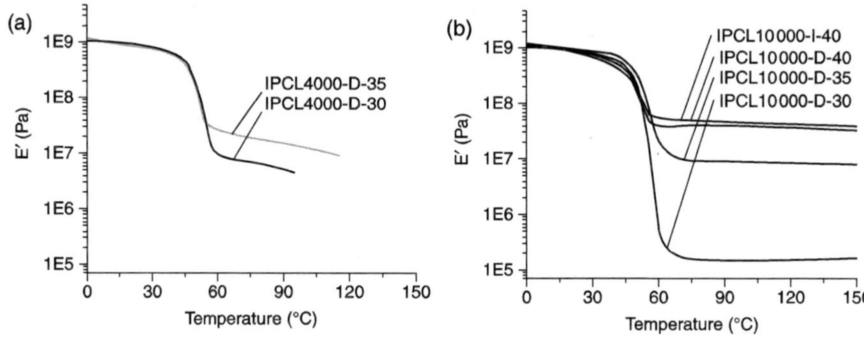

4.15 E′ vs temperature of the IPDI-based T_m-SMPUUs.

segments. Second, the chain extender IPDA is comparatively rigid, which prompts phase separation between soft and hard segments. It was considered that a higher ratio of glassy modulus to rubbery modulus, preferably of two orders of magnitude, would be favorable for shape fixity and shape recovery. Here the E′ at 25°C (E'_{25}) and at 65°C (E'_{65}), as well as the modulus ratio E'_{25}/E'_{65}, are tabulated in Table 4.2. It is shown that the modulus ratios of the PCL4000 series of T_m-SMPUUs are more than 30. However, the following experiments prove that they show lower shape fixity and shape recovery. On the contrary, IPCL10000-I-40 exhibit both excellent shape fixity and shape recovery, though its modulus ratio is only 14.49. Apparently, a higher modulus ratio does not necessarily lead to a higher SME because shape memory behaviors are shown in a series of thermo-mechanical processes with large deformation amplitude, while the dynamic mechanical characteristics are attained in cases of extremely low deformation amplitude.

4.3.4 Shape memory properties

Figure 4.16 illustrates the thermo-mechanical cyclic tensile tests of the IPDI-based T_m-SMPUUs from which the shape fixity and shape recovery of the polymers can be calculated via Eqs [3.1] and [3.2]. As shown in Fig. 4.16(a), ε_u and ε_p are the fixed strain and the residual strain for the first cycle of IPCL4000-D-35. The deformation stresses at 100% of elongation σ_{100} of the T_m-SMPUUs are given in Fig. 4.16. It can be seen that for the two series of T_m-SMPUUs σ_{100} is raised with increasing HSC, as a result of the increasing physical cross-linking density manifested by the DMA tests. In addition, σ_{100} of IPCL10000-I-40 is more than twice of that of IPCL10000-D-40, which results from the stronger hard domains and higher fraction of hard phase of IPCL10000-I-40.

In addition, the fixed strains of the PCL4000-based T_m-SMPUUs are apparently lower than those of PCL10000-based T_m-SMPUUs. However, the DSC tests reveal

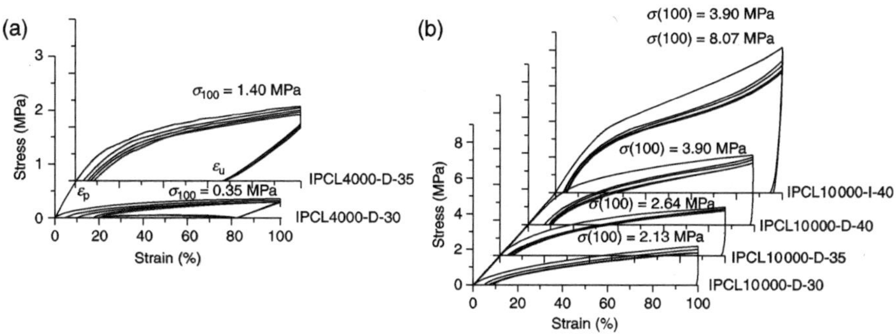

4.16 Thermo-mechanical cyclic tensile tests of the IPDI-based T_m-SMPUUs.

that the PCL4000-based T_m-SMPUUs exhibit higher X_c than their PCL10000-based analogs. The higher shape fixity of PCL10000-based SMPUUs should be ascribed to the strain-induced crystallization of the soft phase. The soft segments in the PCL10000-based T_m-SMPUUs possess higher molecular mobility, enabling them to easily fold and pack into crystallites because of the longer PCL polyols between adjacent cross-linking conjunctions. In contrast, the soft segments in the PCL4000-based T_m-SMPUUs tend to be more firmly restricted by hard segments due to the shorter PCL polyols, resulting in lower crystallization rate and X_c. The values of shape fixity derived from the thermo-mechanical cyclic tensile tests of the T_m-SMPUUs are summarized in Fig. 4.17. The shape fixity of the PCL10000-based T_m-SMPUUs is more than 95% and decrease slightly with the increase of HSC. Whereas the shape fixity of those that are PCL4000-based is no more than 80% and is reduced significantly with the increase of HSC. Hence it can be envisaged that the molecular weight of the polyols for the T_m-SMPUUs should

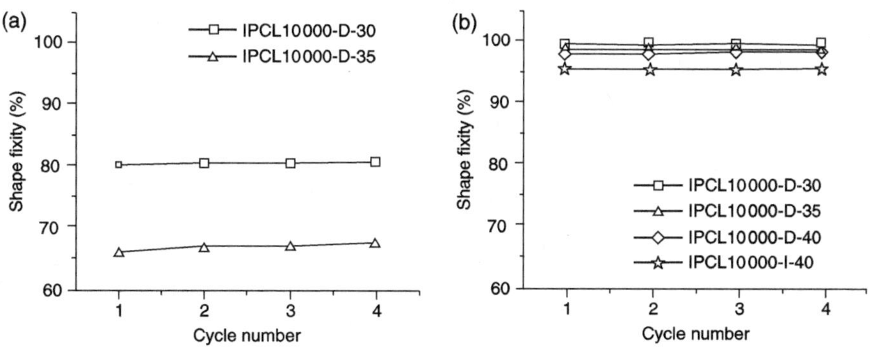

4.17 Shape fixity vs cycle number of the IPDI-based T_m-SMPUUs.

be higher than some critical value in order to attain satisfactory shape fixing. It is also observed that the shape fixity of the T_m-SMPUUs rarely changes as the cycle number increases.

Figure 4.18 presents the shape recovery as a function of the cycle number of the T_m-SMPUUs. As demonstrated in Fig. 4.18(a), the shape recovery of the PCL4000-based T_m-SMPUUs is raised with increasing HSC, indicating that the physical cross-linked polymer networks become more stable. It is also found that with the increase of cycle number the shape recovery is reduced steadily, suggesting that the physically cross-linked networks are gradually destroyed in repeated thermo-mechanical processes. IPCL4000-D-30 shows more apparent decrease tendency, suggesting that the physical cross-links in IPCL4000-D-30 are weak and cannot endure repeated deformations.

As illustrated in Fig. 4.18(b), the shape recovery of IPCL10000-D-35 is correspondingly higher than that of IPCL10000-D-30 and IPCL10000-D-40. This might be attributed to two opposite influencing factors. At lower HSC, the strength of the physical cross-links dominates the shape recovery of the polymers. Increasing HSC can prompt the formation of hard domains and thus favor the memorization of original shape. But as HSC increases further the fraction of hard phase is raised, resulting in the increase of the deformation of the hard phase, which leads to a reduction of shape recovery accordingly. It can also be seen that IPCL10000-I-40 exhibits the best shape recovery, indicating the formation of perfect physically cross-linked polymer networks which enable good shape memorization. In addition, the strong hard domains can resist deformation effectively. The PCL10000-based T_m-SMPUUs exhibit much better shape recovery than the PCL4000-based ones because of their stronger hard domains. It can also be seen that the PCL10000-based T_m-SMPUUs exhibit less decrease of shape recovery upon repeated deformations than those that are PCL4000-based.

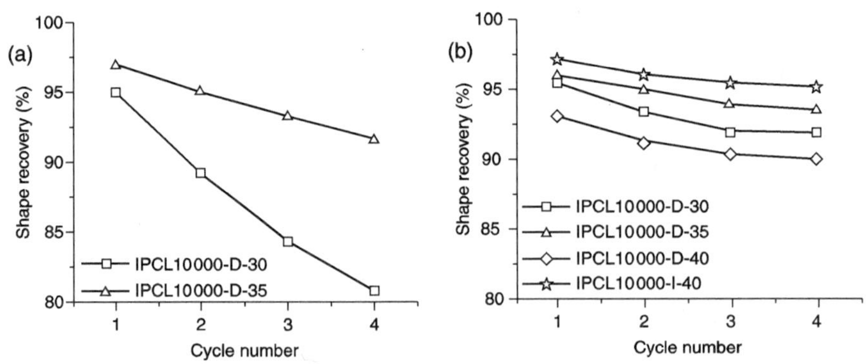

4.18 Shape recovery vs cycle number of the IPDI-based T_m-SMPUUs.

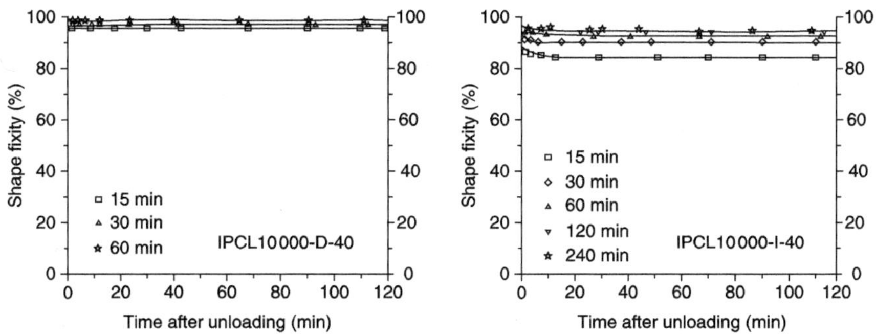

4.19 Influences of t_{fix} on shape fixity of IPCL10000-D-40 and IPCL10000-I-40.

4.3.5 Shape memory properties after long time fixing

Because of the limitation of physical cross-links, the crystallization rate of the soft segments can be slowed down. Therefore extending shape fixing time, t_{fix}, is necessary to explore the maximum shape fixing ability of the T_m-SMPUUs. Figure 4.19 presents the influences of t_{fix} on shape fixity of the IPCL10000-D-40 and IPCL10000-I-40. In general, the shape fixity curves descend in the first 10 minutes after removing the external force and tend to a constant value after 120 minutes. This suggests that some shape recovery occurs after unloading because of the molecular motion of amorphous soft segments. The solid lines in Fig. 4.19 are the curve fitting results based on a second-order exponential decay model. The shape fixity at 120 minutes derived from the fitted results is considered as the steady shape fixity for a particular test. It is also found that the shape fixity of the two T_m-SMPUUs is raised with increasing t_{fix}, suggesting that the X_c of the soft phase can be augmented by extending t_{fix}. As t_{fix} increases from 15 to 60 minutes, the shape fixity of IPCL10000-D-40 approaches 100%. In contrast, the shape fixity of IPCL10000-I-40 is raised gradually with the extension of t_{fix}. Hence increasing t_{fix} favors strain and stress storage.

4.3.6 Recovery stress

Figure 4.20 presents the plots of recovery stress as a function of heating temperature of IPCL10000-D-40 and IPCL10000-I-40. It can be seen that the recovery stress of a polyurethane-urea appears at a certain temperature above 30°C and rises steadily with increasing temperature up to a peak maximum, indicating complete melting of the crystalline soft phase. After a peak maximum, it descends constantly as temperature increases continuously, which is caused by the stress relaxation and decreasing modulus. It is noted that the maximum recovery stress of IPCL10000-I-40 is reached at a higher temperature in

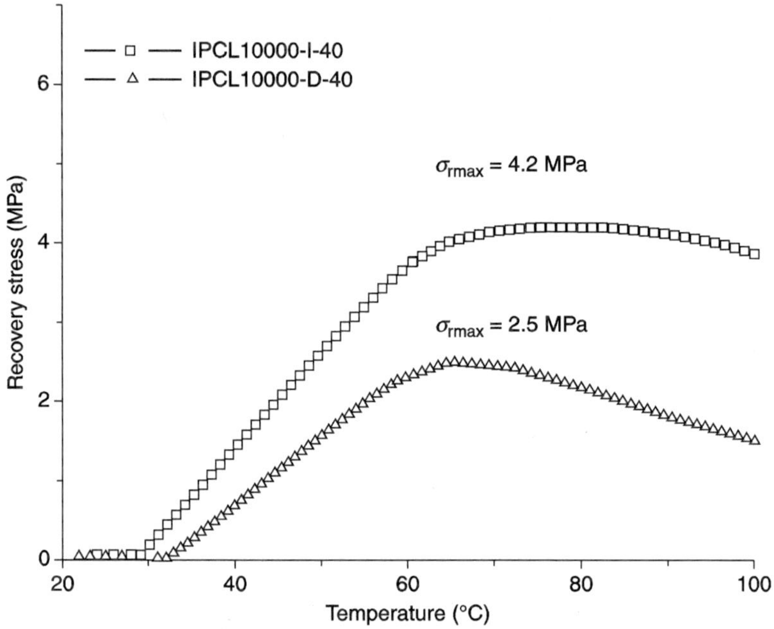

4.20 Recovery stress of IPCL10000-I-40 and IPCL10000-D-40.

comparison with that of IPCL10000-D-40. There may be two possible reasons to account for this phenomenon. First, the peak maxima of the recovery stress curve arise from the addition of the evolutions of stress release, stress relaxation and modulus decrease with increasing temperature. IPCL10000-I-40 is superior to IPCL10000-D-40 in enduring stress relaxation, resulting in the maximum recovery stress occurring at a higher temperature.

Second, in the T_m-SMPUUs with 40% of HSC, the deformation of the hard phase may play a significant role in their shape memory behaviors. The molecular segments in the deformed IPDI/IPDA hard phase should be activated at higher temperatures as compared with those of the IPDI/D-230 hard phase, resulting in the maximum recovery stress of IPCL10000-I-40 appearing at higher temperature. The maximum recovery stresses of IPCL10000-D-40 and IPCL10000-I-40 are 2.07 and 4.16 MPa, respectively. This should be ascribed to the higher physical cross-linking density as well as the stronger hard domains of the latter.

4.3.7 Resistance to stress relaxation

We have shown that if T_m-SMPUs were kept in the constraint state at high temperature for 30 minutes prior to being cooled down to the frozen state, their memory of original shape could be reduced significantly. The decrease of shape recovery is

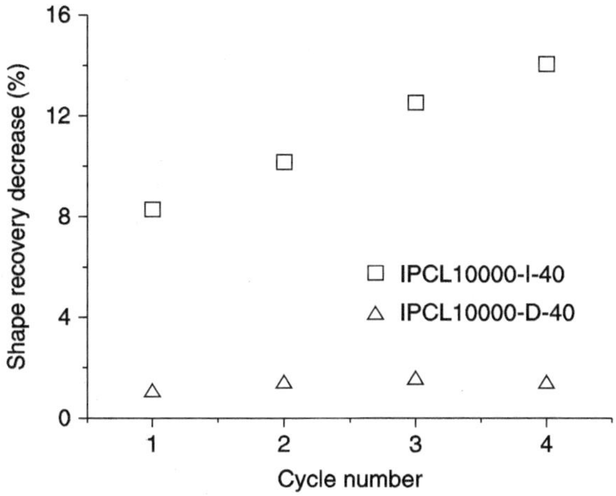

4.21 Effects of stress relaxation at high temperature.

mainly caused by the stress relaxation occurring to the T_m-SMPUs. Figure 4.21 shows the shape recovery decrease as a function of cycle number of the two T_m-SMPUUs. It is apparent that after the process of stress relaxation the shape recovery of IPCL10000-D-40 is reduced more strongly with increase of cycle number. This suggests that the IPDI/IPDA hard domains exhibit better resistance to stress relaxation as compared with the IPDI/IPDA ones and thus enable IPCL10000-I-40 to better memorize its original shape.

4.4 High performance T_g-type shape memory polyurethane (SMPU) prepared by modified two-step polymerization

Several attempts have been made to investigate T_g-SMPUs (Jeong *et al.*, 2000b; Kim *et al.*, 2001; Lee *et al.*, 2001; Lin and Chen, 1998; Takahashi *et al.*, 1996; Tobushi *et al.*, 1996). The properties, including T_g, shape recovery, shape fixity and recovery stress of the T_g-SMPUs are defined by the characteristics of the phases and the morphological structure such as phase composition and domain sizes, which are dependent on the chemical structure of the monomers, their compositions, and the sequence length distribution of the hard and soft segments. As compared with conventional segmented polyurethane elastomers the SMPUs have a much higher T_g, often close to or above T_{room} in order to attain shape fixing readily. T_g of segment polyurethanes can be raised either by employing short polyols, which results in the prominent phase mixing between soft and hard segments, or by increasing HSC, which enhances the physical cross-linking density. Hence, the T_g-SMPUs are usually fabricated with polyester or polyether diols having a number average molecular weight of 300 to 1000 as soft segments via a two-step copolymerization where polyols are first end-capped with isocyanate (NCO) groups followed by chain extension with short diols. Changing the species of the monomers and the chemical composition the T_g can be readily located at a desired temperature. However, in order to synthesize T_g-SMPUs with various T_g with optimal comprehensive shape memory properties, a series of polyols with various lengths and chemical structure are required.

This chapter presents a modified two-step copolymerization route by employing urethane chains made of short polyols and 4,4′-diphenylmethane diisocyanate (MDI) as soft segments, which is anticipated to expand the flexibility for molecular design of T_g-SMPUs. First, the PPG400 diols were used to alternatively copolymerize with MDI to produce N–C–O group terminated urethane prepolymers. Subsequently the prepolymers reacted with extra MDI and chain extenders to produce the T_g-SMPUs. The 'lone' MDI between PPG400 diols are considered miscible with the polyols (Garrett *et al.*, 2000).

A series of segmented polyurethanes with urethane chains composed of PPG400 glycols and MDI as soft segments were synthesized. The PPG400 polyols were dried at 80°C under vacuum for six hours prior to use. MDI, hydroquinone bis(2-hydroxyethyl)ether (HQEE) and MOCA were used as received. DMF and BD were dried with 4Å molecular sieves in advance. The segmented polyurethanes were synthesized in a modified prepolymer copolymerization (Fig. 4.22).

The isocyanate-terminated prepolymers were first prepared by reacting the dried PPG400 glycols with the mole excessive MDI in DMF solution at 65°C for three hours. Then the prepolymers were reacted with MDI and BD, HQEE or MOAC in particular stoichoimetric ratios in the temperature range 80 to 100°C for another four hours. Through this small change, the modified copolymerization changed the structure and properties of the resultant polyurethanes significantly. The detailed formulations of the segmented polyurethanes are shown in Table 4.5.

The series of segmented polyurethanes with urethane chains as soft segments are designated as iPPG400-b-xx or iiPPG400-b-xx, where 'b' refers to BD as chain extenders, and 'xx' the wt.% HSC. Likewise, iPPG400-h-xx and iPPG400-m-xx

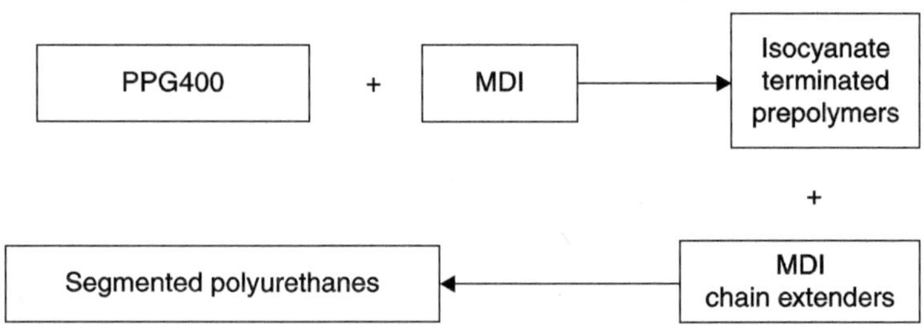

4.22 Scheme of the modified prepolymer copolymerization.

Table 4.5 Formulations of the T_g-SMPUs based on urethane chains

Samples	1st (mole)		2nd (mole)				HSC_1	HSC_2
	PPG400	MDI	MDI	BDO	HQEE	MOCA	(wt.%)	(wt.%)
iPPG400-b-35	4	5	1.98	2.98	–	–	55.72	35.00
iPPG400-b-40	4	5	2.87	3.87	–	–	59.14	40.00
iPPG400-b-45	4	5	3.92	4.92	–	–	62.55	45.00
iiPPG400-b-35	9	10	7.13	8.13	–	–	58.20	35.00
iiPPG400-b-45	9	10	11.00	12.00	–	–	63.74	45.00
iPPG400-h-35	4	5	1.27	–	2.27	–	55.76	35.00
iPPG400-h-40	4	5	1.94	–	2.94	–	59.15	40.00
iPPG400-m-35	4	5	0.96	–	–	1.96	55.71	35.00
iPPG400-m-40	4	5	1.55	–	–	2.55	59.16	40.00
PPG400-60	1	–	1.78	0.78	–	–	60.00	–

denote the copolymers chain extended by HQEE and MOCA, respectively. The characters 'i' and 'ii' are employed to distinguish the urethane prepolymers (MDI-PPG400)4-MDI and (MDI-PPG400)9-MDI. In Table 4.5, HSC_1 denotes the overall wt.% of MDI and chain extenders in the segmented polyurethanes; HSC_2 stands for the wt.% of the hard segments, which refers to the segments composed of MDI and chain extenders. Hereafter, HSC of the segmented polyurethanes produced via the modified two-step copolymerization refers to HSC_2 (Table 4.5). As a control, the sample PPG400-60 was prepared in the typical prepolymer copolymerization and is compared with the other segmented polyurethanes in terms of structure and properties.

The alternative copolymer of PPG400 and MDI was made, which was proven to be homogeneous and exhibited only a glass transition at approximately 17°C via the DSC test. In the T_g-SMPUs, the urethane chains are anticipated to reside in the soft phase and thus are viewed as soft segments. To strengthen the hard domains, the rigid chain extenders (HQEE) and MOCA were used for producing the T_g-SMPUs. Because of the presence of an aromatic ring in HQEE, the MDI/HQEE hard domains should be stronger than the MDI/BD ones. The MDI/MOCA hard segments should form the strongest domains because of the presence of aromatic rings and stronger urea hydrogen bonds.

However, introducing the rigid chain extenders via the typical two-step copolymerization can result in either too high a T_g because of phase mixing between polyols and rigid hard segments or too large a fraction of hard phase. Using the urethane chains as soft segments, the rigid chain extenders could be incorporated into the hard segments but only affect T_g slightly. The influences of the length of the urethane chains, HSC and type of chain extenders on the structure and properties such as T_g, shape recovery, shape fixity and recovery stress of the segmented polyurethanes, were investigated by DSC, DMA, SAXS and thermo-mechanical cyclic tensile tests. Increasing the length of the urethane chains or using rigid chain extenders are proven to increase the physical cross-linking density and raise deformation stress as well as recovery stress. This study is expected to shed some light on the fundamentals for developing high-performance T_g-SMPUs.

4.4.1 Thermal properties of T_g-type high performance shape memory polyurethane

Figure 4.23 presents the DSC thermograms of the segmented polyurethanes. The thermal properties including T_g, ΔC_p of the soft phase and T_m and ΔH_m of the hard

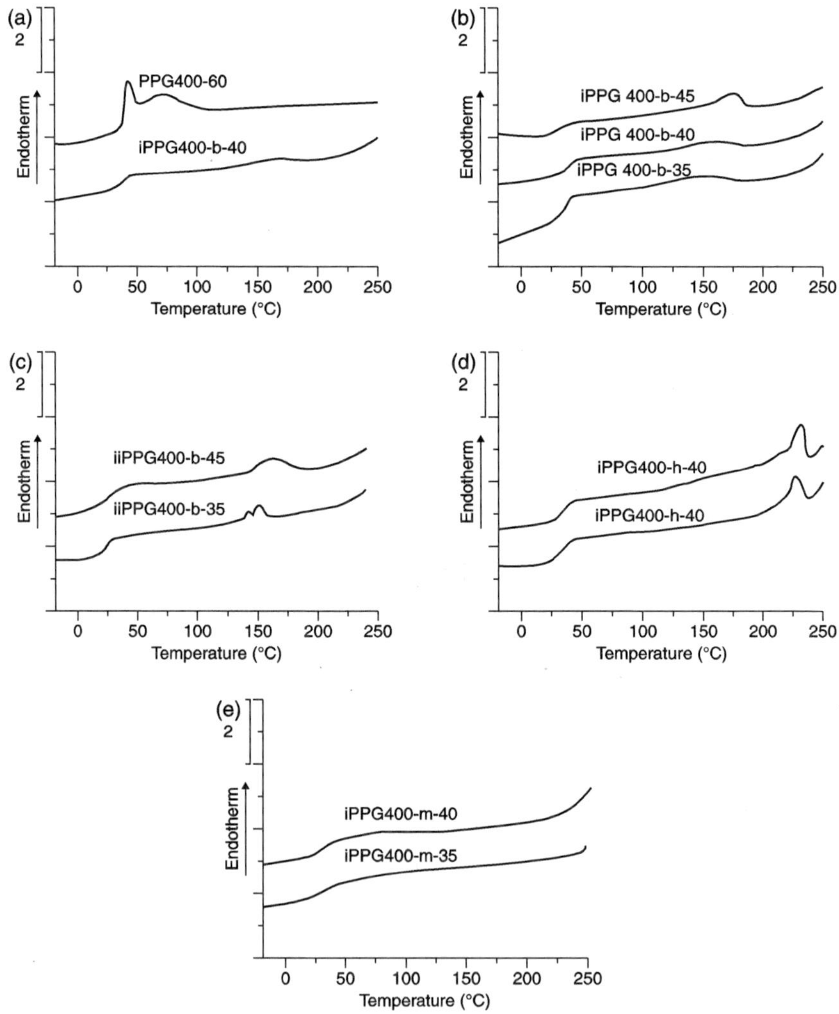

4.23 DSC traces of T_g-SMPUs having urethane chains as soft segments.

phase are summarized in Table 4.6. The PPG400-60 was synthesized through the typical two-step copolymerization, while the iPPG400-b-40 was synthesized via a modified one; however, they have similar chemical compositions, i.e. the same HSC_1 (Table 4.5). A deflection appears at approximately 34°C on the thermogram of the iPPG400-b-40, which is attributed to the glass transition of the soft phase (Fig. 4.23(a)). A broad and weak endothermic behavior is observed at approximately 156°C, suggesting the presence of a hard phase.

In contrast, PPG400-60 exhibits no endothermic behavior above 100°C and shows only a glass transition at approximately 38°C where a hysteresis peak appears. The hysteresis peak is associated with some physical relaxation. This suggests that the modified copolymerization enhances the phase separation and prompts the formation of hard domains because of the increased lengths of soft and hard segments. As would be expected, T_m and ΔH_m rise gradually with the growth of HSC (Fig. 4.23(b) and (c) and Table 4.6), resulting from the increasing fraction of hard phase. Simultaneously, T_g is raised and ΔC_p is reduced gradually, which primarily arises from the increased limitation imposed on the soft segments by hard domains and the decreased fraction of soft segments, respectively. Via the modified two-step copolymerization, the T_g values of the resultant copolymers are close to those chain extended by BD (Fig. 4.23(d) and (e) and Table 4.6). The iPPG400-h-xx polyurethanes exhibit endothermic behavior in the range 220 to 230°C, indicating the presence of a hard phase. In contrast, no endothermic peak appears in the DSC traces of the iPPG400-m-xx segmented polyurethanes because the thermal transition of MDI/MOCA hard domains should be situated at a higher temperature.

Table 4.6 Thermal properties and interdomain spacings of T_g-SMPUs with urethane chains as soft segments

Sample	DSC				DMA		SAXS
	T_g (°C)	ΔC_P (J/(g×°C))	T_m (°C)	ΔH_m (J/g)	T_g (°C)	E'_{+15}* (MPa)	d (nm)
iPPG400-b-35	34.1	0.62	156.9	2.6	49.6	34.2	16.7
iPPG400-b-40	35.2	0.48	166.7	3.5	52.9	58.2	15.8
iPPG400-b-45	37.4	0.43	173.8	6.7	59.8	187.2	13.9
iiPPG400-b-35	32.2	0.58	159.9	3.3	48.6	56.1	16.0
iiPPG400-b-45	36.1	0.44	174.9	8.7	58.1	231.8	13.5
iPPG400-h-35	32.9	0.53	221.9	9.2	48.2	87.9	15.5
iPPG400-h-40	34.6	0.42	228.6	8.1	52.3	231.8	13.9
iPPG400-m-35	30.6	0.43	–	–	53.3	109.8	–
iPPG400-m-40	32.4	0.40	–	–	55.5	245.8	14.1
PPG400-60	–	–	–	–	58.7	2.2	–

*E'_{+15} refers to the storage modulus at T_g + 15°C extracted from the DMA thermograms of the segmented polyurethanes

4.4.2 Dynamic mechanical properties of T_g-type high performance shape memory polyurethane

Figure 4.24(a) illustrates tanδ against temperature obtained from the DMA tests of PPG400-60 and iPPG400-b-40. According to the peak maximum of the tanδ plot, the T_g of PPG400-60 is centered at approximately 58.71°C, while that of iPPG400-b-40 is at 52.86°C (Table 4.6). This is in accordance to the DSC results, suggesting that employing the urethane chains as soft segments decreases phase mixing and results in a lower T_g. After the primary peak, the tanδ plot of PPG400-60 rises abruptly at approximately 80°C, whereas that of iPPG400-b-40 increases dramatically above 120°C (Fig. 4.24(a)). Some researchers attributed the rise of the tanδ plot appearing above T_g of the soft phase to the glass transition of the hard

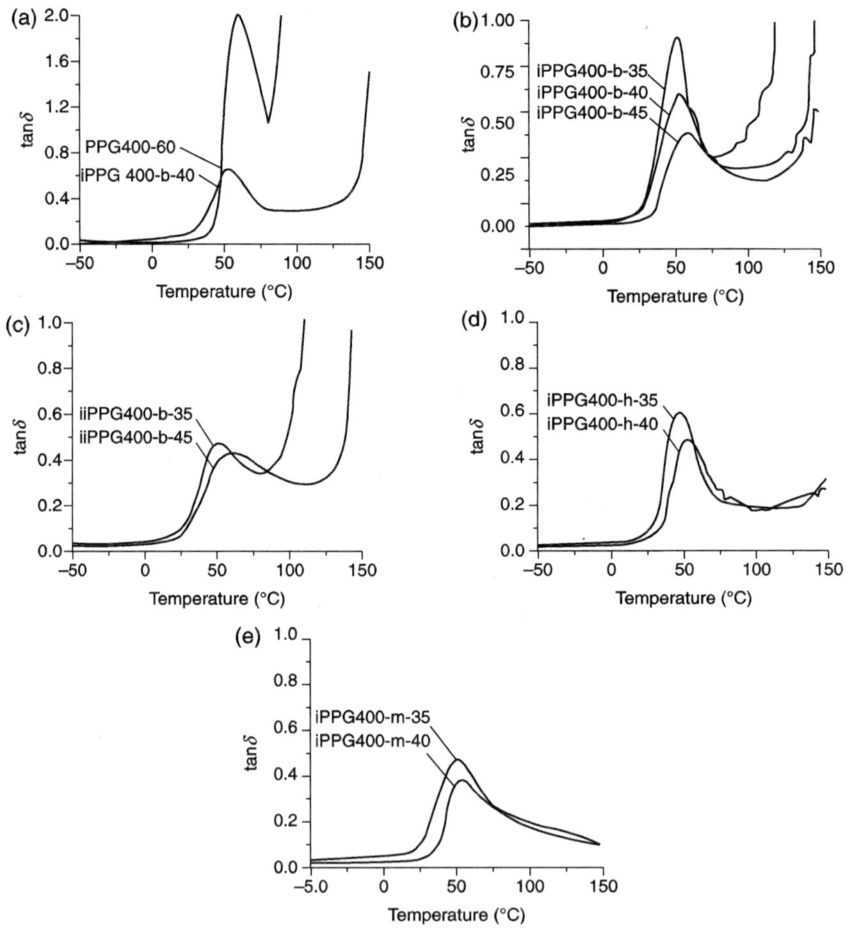

4.24 Tanδ vs temperature of T_g-SMPUs with urethane chains as soft segments.

phase (Kim *et al.*, 1996). Hence it can be assumed that the tanδ rise of iPPG400-b-40 appearing above 120°C is related to the thermal transition of the hard phase, whereas the tanδ rise of PPG400-60 should be associated to the deconstruction of weak physical interactions such as hydrogen bonds because it is homogeneous and has no hard phase, as shown in the following section. With increasing HSC, the peak maxima shift to higher temperatures (Fig. 4.24(b) and (c)), which is in agreement with the DSC results. The iPPG400-h-xx and iPPG400-m-xx segmented polyurethanes exhibit the same tendency (Fig. 4.24(d) and (e)). Increasing length of the soft segments results in little change to T_g for the segmented polyurethanes with identical HSC (Table 4.6). Chain extender HQEE brings no distinct change to T_g, whereas MOCA slightly raises T_g by 2 to 3°C because of the elevated restriction imposed on the soft segments by the strong MDI/MOCA hard domains.

Figure 4.25 illustrates storage modulus vs temperature of the segmented polyurethanes from which the storage modulus at T_g + 15°C (E'_{+15}) are extracted

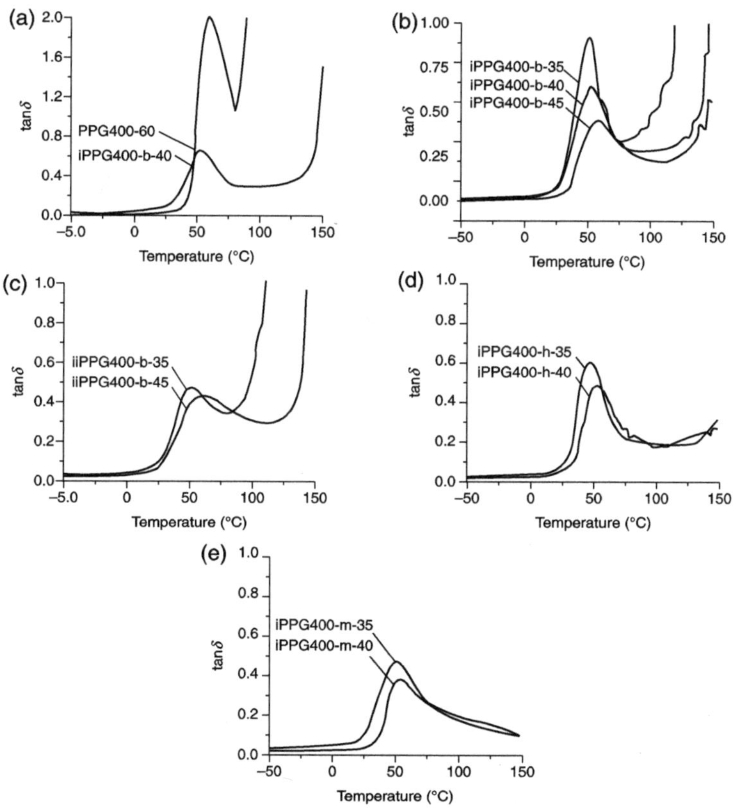

4.25 E′ vs temperature of T_g-SMPUs with urethane chains as soft segments.

and listed in Table 4.6 The rubbery modulus of segmented polyurethanes is associated to the physical cross-linking density. The rubbery modulus plateau of iPPG400-b-40 is both higher and longer than that of PPG400-60 (Fig. 4.25(a)), resulting from the presence of the hard phase. The rubbery plateau of the segmented polyurethanes is elevated and shifted toward higher temperature with increasing HSC (Figs 4.25(b) to (e)), arising from the increasing physical cross-linking density. Moreover, the E'_{+15} values are elevated by increasing the length of soft and hard segments and by employing rigid hard segments (Table 4.6), also due to the increasing physical cross-linking density.

4.4.3 Phase separation of T_g-type high performance shape memory polyurethane

Figure 4.26 illustrates the SAXS profiles of the segmented polyurethanes. A distinct scattering peak appears in the curve of iPPG400-b-40, indicating the presence of a two-phase morphology (Fig. 4.26(a)), whereas PPG400-60 yields no such a peak, suggesting that hard segments are dissolved in the amorphous soft matrix. This is consistent with the DSC and the DMA tests and indicates that employing the urethane chains as soft segments prompts phase separation. The other segmented polyurethanes exhibit more or less phase separation (Fig. 4.26(b) to (e)). With increasing HSC, the scattering peak shifts towards a higher angle, suggesting that the long period, d, of the segmented polyurethanes decreases, which arises from the increased concentration of hard domains, i.e., physical cross-linking density. The iiPPG400-b-xx segmented polyurethanes exhibit smaller d in comparison with the iPPG400-b-xx ones (Table 4.6), indicating that longer soft and hard segments enhance phase separation and thus prompt the formation of hard domains. The d values of iPPG400-h-xx and iPPG400-m-40 are correspondingly smaller than those of the iPPG400-b-xx ones with identical HSC, since HQEE and MOCA can prompt phase separation. Notably iPPG400-m-35 exhibits only a shoulder peak, suggesting its lower degree of phase separation. This may be caused by its comparatively low content of MOCA chain extenders.

4.4.4 Shape memory properties of T_g-type high performance shape memory polyurethane: thermo-mechanical cyclic tensile tests

According to the polymer viscoelasticity, the total deformation strain ε of the segmented polyurethanes consists of ε_e, the elastic deformation strain, which can be recovered instantly after the removal of the external stress; ε_{vis}, the viscoelastic deformation strain, which can be restored gradually by raising the temperature; and ε_{ir}, the irreversible deformation strain, which can no longer be eliminated. ε_e usually accounts for a very small part in the large deformation and is thus

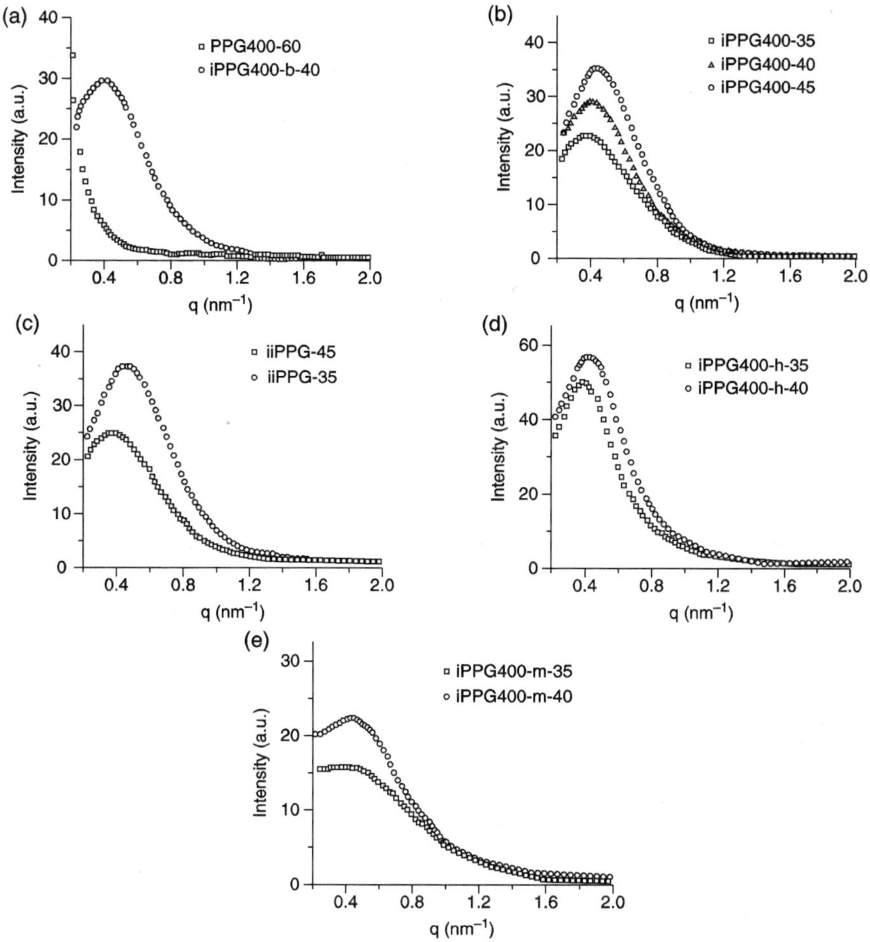

4.26 SAXS profiles of T_g-SMPUs with urethane chains as soft segments.

negligible. ε_{ir} is no more than 10%. Hence the shape memory behaviors of the segmented polyurethanes are primarily the generation, fixing and recovery of ε_{vis}.

As physical cross-links, the hard domains are not point cross-links and occupy a large volume and are deformable (Petrovic and Ferguson, 1991). Thus ε_{vis} arises from both the viscoelastic deformation strain of the soft phase ε_{vis}(SS) and that of the hard phase ε_{vis}(HS). Generally, ε_{vis}(SS) can be recovered above T_g, whereas ε_{vis}(HS) needs to be recovered at a higher temperature because of the lower molecular mobility of the hard phase. As a result, the shape recovery region can be broadened if ε_{vis}(HS) accounts for a high percentage. The phase composition and the moduli of soft and hard phases underline the ratios of ε_{vis}(SS) and ε_{vis}(HS) and ultimately define the shape memory behaviors.

Figure 4.27 demonstrates the thermo-mechanical cyclic tensile tests of the segmented polyurethanes from whence the shape memory parameters including shape recovery $R_r(1)$, $R_r(4)$ and shape fixity $R_f(1)$ and are listed in Table 4.6. Figure 4.27(a) shows the residual strain ε_p and fixed strain ε_u of iPPG400-m-35 for the first cycle. The $R_r(1)$ and $R_f(1)$ values of all the copolymers are more than 90% and nearly 100%, respectively (Table 4.6). The shape recovery of the materials is reduced by 2 to 4% after four thermo-mechanical cycles. The iiPPG400-b-xx segmented polyurethanes exhibit lower $R_r(1)$ compared with the iPPG400-b-xx ones with identical HSC (Fig. 4.27(a) and (c) and Table 4.6), because the higher fraction of hard phase results in larger ε_{vis}(HS), which cannot be recovered at T_g + 15°C. Likewise, because of their higher fraction of hard phase, the iPPG400-h-xx and the iPPG400-m-xx segmented polyurethanes show lower $R_r(1)$ in contrast with the iPPG400-b-xx ones (Fig. 4.27 (a) and (b) and Table 4.6).

Moreover, the $R_r(1)$ values decrease with increasing HSC, also due to the increasing fraction of the hard phase (Table 4.6). However, the $R_r(1)$ values of iPPG400-m-xx are higher than those of the iPPG400-h-xx values because of the stronger MDI/MOCA hard domains which give rise to lower ε_{vis}(HS). The deformation stress at 100% of elongation strain σ_{100} of the T_g-SMPUs is shown on the thermo-mechanical graphs. σ_{100} is monotonically increased with the increase of fraction of hard phase and strength of the hard domains. For example, the

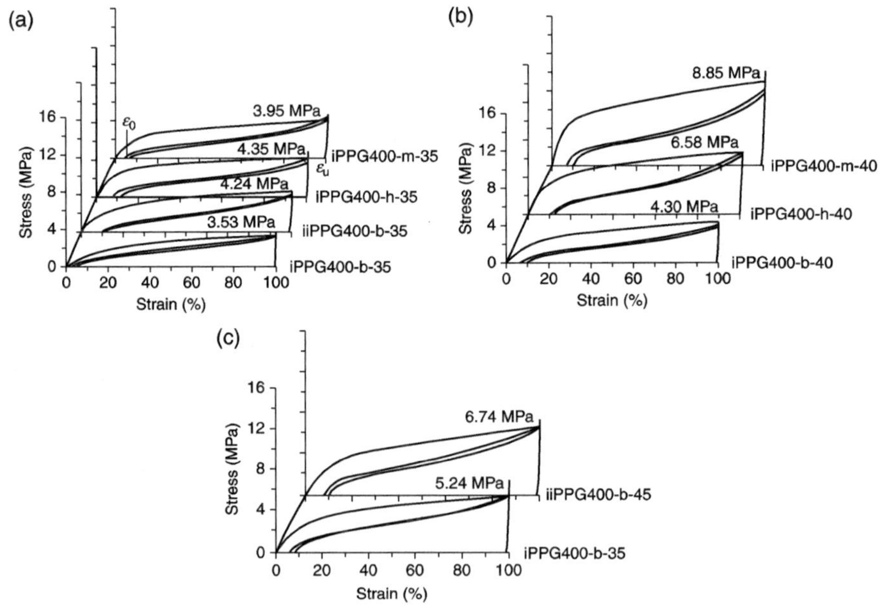

4.27 Thermo-mechanical cyclic tensile tests of T_g-SMPUs with urethane chains as soft segments.

iiPPG400-b-xx, iPPG400-h-xx and the iPPG400-m-xx copolymers exhibit higher σ_{100} values compared with those having identical HSC (Fig. 4.27(a) to (c)). Moreover, σ_{100} is raised with increasing HSC. It can be seen that the rigid chain extenders can rise to σ_{100}, which is more prominent as HSC increases from 35 to 40% (Fig. 4.27(a) and (b)), indicating that the rigid hard segments should be long enough to have a pronounced effect for reinforcing the segmented polyurethanes.

4.4.5 Shape memory properties of T_g-type high performance shape memory polyurethane: shape fixing

The shape fixity refers to the instant values attained from the thermo-mechanical cyclic tensile tests. It is reported that a SMPU foam did not hold shape fixity and recovered its original shape gradually. Figure 4.28 shows the shape fixity evolution with time after unloading at T_{room} of the iPPG400-b-xx segmented polyurethanes. With the elapse of time, the shape fixity of the segmented polyurethanes is reduced steadily, which can be explained as follows. In being cooled down below T_g, the micro-Brownian motions of the soft segments are mostly limited. Nevertheless the molecular mobility is not entirely suppressed regardless of their long characteristic times. In addition, some soft segments may be in a rubbery state if the shape fixing temperature is not sufficiently low, which may result in a rapid decrease of shape fixity. It is also found that the

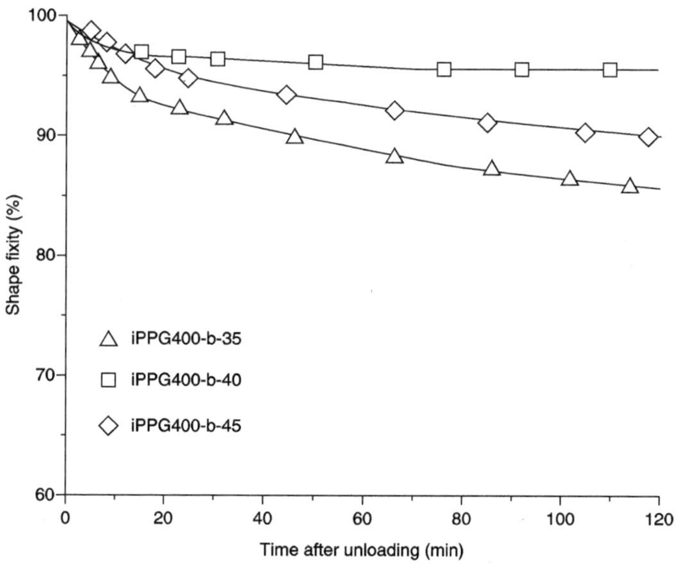

4.28 Long-term shape fixing of iPPG400-b-xx segmented polyurethanes.

Table 4.7 Shape memory properties of T_g-SMPUs with urethane chains as soft segments

Samples	$R_r(1)^{\dagger}$	$R_r(4)^{\ddagger}$	$R_r(1) - R_r(4)$	$R_f(1)$	R_f120	ε_{ir}	σ_{rmax}	$\sigma_{rmax}/\sigma_{100}$*
	(%)	(%)	(%)	(%)	(%)	(%)	(MPa)	(%)
iPPG400-b-35	97.3	95.0	2.3	99.0	85.4	3.2	2.2	61.2
iPPG400-b-40	94.5	90.8	3.7	99.1	89.2	5.5	2.7	63.5
iPPG400-b-45	93.6	91.3	2.3	98.8	95.5	7.0	3.7	70.5
iiPPG400-b-35	90.5	88.8	1.7	99.1	92.2	5.4	2.5	65.1
iiPPG400-b-45	92.7	89.2	3.5	98.5	96.6	7.0	4.9	72.1
iPPG400-h-35	93.1	89.1	4.0	99.2	88.4	5.5	2.9	66.9
iPPG400-h-40	93.1	89.1	4.0	98.8	91.4	5.5	4.4	68.9
iPPG400-m-35	96.1	93.8	2.3	98.7	93.5	7.1	2.4	61.2
iPPG400-m-40	93.0	90.0	3.0	98.8	95.0	7.6	5.9	66.3

†,‡ The notations (1) and (4) refer to the cycle number of the cyclic tests.

* σ_{100} denotes the deformation stress at 100% of elongation.

shape fixity of the iPPG400-b-xx copolymers descends more slowly with increasing time. Assuming that the shape fixity depression meets a two-order exponential decay model, we attained the fitting results at 120 minutes, R_f120 (Table 4.7). Because of higher T_g, the segmented polyurethanes with higher HSC exhibit higher R_f120. For a given HSC, the segmented polyurethanes with longer soft and hard segments, and those with rigid hard segments, exhibit higher R_f120 owing to higher fraction of the hard phase or higher T_g. Thus, 120 minutes after unloading, the segmented polyurethanes preserve more than 85% of shape fixity.

4.4.6 Shape memory properties of T_g-type high performance shape memory polyurethane: shape recovery

Figure 4.29 illustrates the plots of shape recovery vs heating temperature, where the starting shape recovery values of the curves are roughly equal to 100%-R_f120. With increasing temperature, the shape recovery of the polymers increase gradually at the beginning and rise abruptly in a narrow temperature range. At a certain temperature above 100°C, the shape recovery increases to above 90% and stops increasing at a certain temperature, from whence ε_{ir} is attained and is presented in Table 4.7. With increasing HSC, the shape recovery plots of the segmented polyurethanes shift toward higher temperature, which is caused by the increased T_g of the segmented polyurethanes (Fig. 4.28(a)). In addition, as compared with those of the iPPG400-b-xx segmented polyurethanes, the shape recovery processes of the iiPPG400-b-xx, iPPG400-h-xx and iPPG400-m-xx segmented polyurethanes appear at higher temperatures because of either larger fraction of the hard phase or higher T_g (Fig. 4.29(b) to (d)).

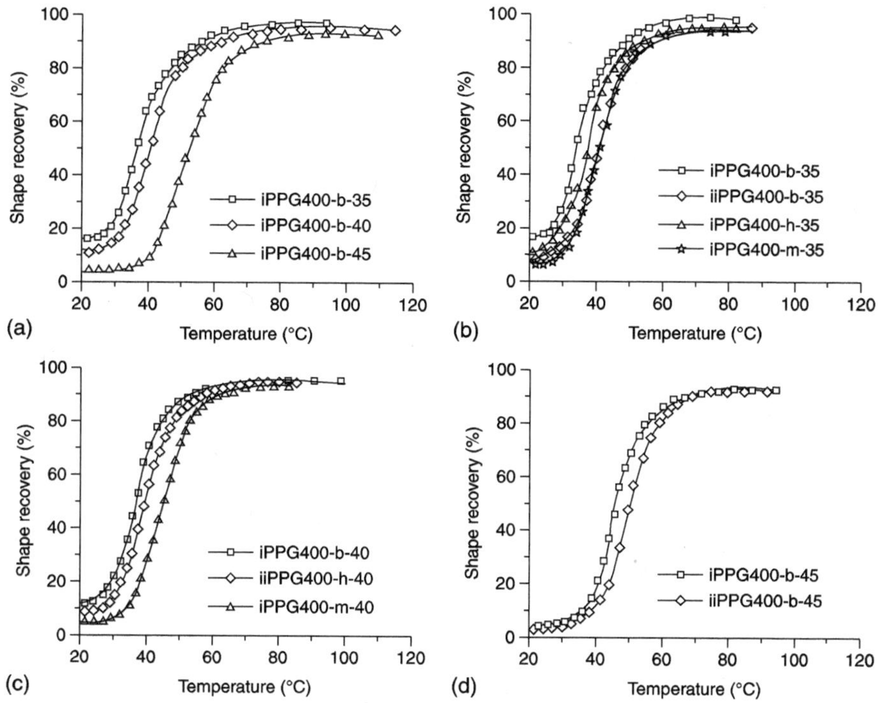

4.29 Shape recovery vs temperature of T_g-SMPUs with urethane chains as soft segments.

4.4.7 Shape memory properties of T_g-type high performance shape memory polyurethane: recovery stress

Figure 4.30 demonstrates the plots of recovery stress against temperature of the segmented polyurethanes. The recovery stress of a polymer appears at a certain temperature and rises steadily with increasing temperature up to a peak point where a maximum recovery stress (σ_{rmax}) is attained. After the peak maximum it goes down gradually as the temperature increases further, which is caused by the stress relaxation and the descending modulus. Compared with the iPPG400-b-xx segmented polyurethanes, the iiPPG400-b-xx ones exhibit higher σ_{rmax}, primarily resulting from their higher deformation stress (Fig. 4.30(a) and (c)). Likewise, iPPG400-h-xx and iPPG400-m-xx segmented polyurethanes show higher σ_{rmax} compared with iPPG400-b-xx ones (Fig. 4.30(a) and (b)). It is also noted that σ_{rmax} is elevated with increasing HSC (Table 4.7). The σ_{rmax} values of the segmented polyurethanes vary in the range 2 to 6 MPa with the changing lengths of the soft segments, HSC, and types of chain extenders. The parameter stress conversion ratio ($\sigma_{rmax}/\sigma_{100}$) is proposed to reflect the ratio of deformation stress

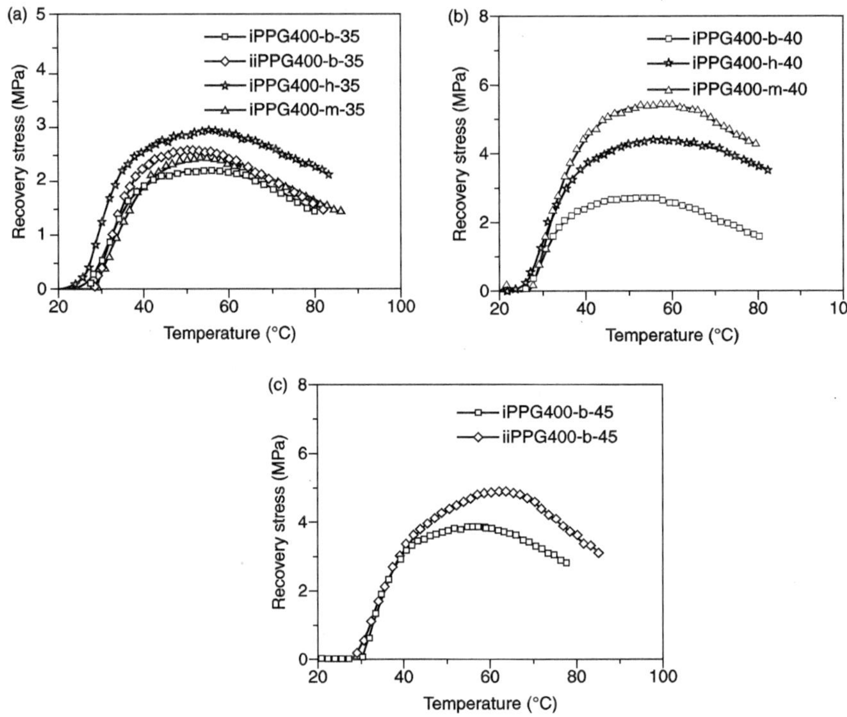

4.30 Recovery stress vs temperature of T_g-SMPUs with urethane chains as soft segments.

converting into recovery stress. It can be seen that the $\sigma_{rmax}/\sigma_{100}$ values cover the range 61 to 72%. It is found that the segmented polyurethanes with higher HSC and longer and more rigid hard segments give rise to higher stress conversion.

4.5 Conclusions

A series of T_m-SMPUUs chain extended with MOCA were prepared. They are compared with a set of T_m-SMPUs with the same soft segments and MDI. The structure and thermal properties of the T_m-SMPUs and the T_m-SMPUUs are investigated with FT-IR, DSC and DMA. Compared with the T_m-SMPUs, the T_m-SMPUUs with identical HSC exhibit apparently lower X_c of the soft phase as well as higher rubbery modulus and larger deformation stress. It is found that all the T_m-SMPUs and T_m-SMPUUs show over 90% shape recovery. Urethane-urea hard domains are superior to urethane hard domains for keeping good shape recovery. The T_m-SMPUUs exhibit lower shape fixity as compared with their polyurethane analogs. The recovery stress of the T_m-SMPUs and T_m-SMPUUs augments with increasing HSC at least in the HSC range 30 to 40%. The T_m-SMPUs are proved to store deformation strain well but do not yield high deformation stress and cannot effectively endure stress relaxation. In contrast, the T_m-SMPUUs can render higher deformation stress and endure stress relaxation more effectively and thus give rise to higher recovery stress provided adequate shape fixity is retained. The recovery stresses of the T_m-SMPUUs are twice as high as those of the T_m-SMPUs with identical HSC. For example, the maximum recovery stress of PCL4000-M-40 is 4.16 MPa, whereas that of PCL4000-B-40 is only 2.07 MPa.

At the same time, a series of non-aromatic T_m-SMPUUs with varying soft segment length and HSC were synthesized with PCL4000 and PCL10000 diols as soft segments and with IPDI and chain extenders D-230 or IPDA as hard segments. The structure, thermal and shape memory properties of the T_m-SMPUUs were investigated with FT-IR, DSC, DMA and a series of thermo-mechanical tests. Increasing the HSC prompts the formation of hard domains and thus raises the physical cross-linking density. In comparison with IPCL4000-D-30, IPCL10000-D-30 exhibits a higher hydrogen bonding degree, suggesting that the longer hard segments favor the formation of hard domains. The longer soft segments favor the strain-induced crystallization of the soft phase and thus can raise the shape fixity of the IPDI-based T_m-SMPUUs. Because of the presence of stronger physical cross-links, the PCL10000-based T_m-SMPUUs show much higher shape recovery compared with the corresponding PCL4000-based ones.

Increasing the HSC has two opposite effects: raising shape recovery due to increased strength of the hard domains and reducing shape recovery because of the increased fraction of the hard phase. As a chain extender, IPDA is superior to D-230 for raising the rigidity of the hard segments and forming stronger hard domains. The polyurethane-urea chain extended with IPDA exhibits over 95% shape recovery as well as excellent resistance to stress relaxation. The recovery stress values of IPCL10000-D-40 and IPCL10000-I-40 are 2.5 and 4.2 MPa, respectively, which are comparable to those of the aromatic T_m-SMPUs and T_m-SMPUUs under the same conditions.

Finally, a series of segmented polyurethanes with urethane chains as soft segments were prepared. In order to strengthen the physical cross-links, the rigid chain extenders HQEE and MOCA were incorporated into the hard segments of the copolymers. DSC, DMA, SAXS and thermo-mechanical tests were conducted to investigate the structure and shape memory properties of the segmented polyurethanes. As compared with a typical two-step copolymerization, employing urethane chains as soft segments can greatly enhance phase separation. Increasing the urethane chain length or using rigid chain extenders can improve phase separation and thus increase deformation stress, but bring little change to T_g. The segmented polyurethanes exhibit more than 90% shape recovery under T_g + 15°C, which is reduced with increasing fraction of the hard phase and raised with the increasing strength of the hard domains.

The shape fixity of the segmented polyurethanes can descend more or less after removal of the external stress and be more than 85% at 120 minutes after removal. All the segmented polyurethanes recover most deformation strain (>80%) in narrow temperature regions. The segmented polyurethanes have 61 to 72% of deformation stress converted into recovery stress, which varies from 2 to 6 MPa with the changing lengths of soft and hard segments and types of chain extenders. The maximum recovery stress can be elevated by increasing the length of soft and hard segments or by employing rigid chain extenders. In short, employing the urethane chains as soft segments, T_g and shape memory properties can be adjusted and thus expand the molecular design for developing high-performance T_g-SMPUs.

4.6 References

Allara, D. L., Swalen, J. D. (1982), An infrared reflection spectroscopy study of oriented cadmium arachidate monolayer films on evaporated silver. *J. Phys. Chem.*, **86**, 2700–4.

Asif, A., Shi, W., Shen, X., Nie, K. (2005), Physical and thermal properties of UV curable waterborne polyurethane dispersions incorporating hyperbranched aliphatic polyester of varying generation number, *Polymer*, **46**, 11066–78.

Bogart, V., John, W. C., Gibson, P. E., Cooper, S. L. (1983), Structure–property relationships in polycaprolactone-polyurethanes, *J. Polym. Sci., Polym. Phys. Ed.*, **21**, 65–95.

Bogdanov, B., Toncheva, V., Schacht, E., Finelli, L., Sartib, B., Scandola, M. (1999), Physical properties of poly(ester-urethanes) prepared from different molar mass polycaprolactone-diols, *Polymer*, **40**, 3171–82.

Coleman, M. M., Lee, K. H., Skrovanek, D. J., Painter, P. C. (1986), Hydrogen bonding in polymers, Part IV: Infrared temperature studies of a simple polyurethane, *Macromolecules*, **19**, 2149–57.

Curgul, S., Yilgör, I., Yilgör, E., Erman, B., Cakmak, M. (2004), Effect of chemical composition on large deformation mechanooptical properties of high strength thermoplastic poly(urethane urea)s, *Macromolecules*, **16**, 8676–85.

Fu, B. X., Hsiao, B. S., Pagola, S., Stephens, P., White, H., *et al.* (2001), Structural development during deformation of polyurethane containing polyhedral oligomeric silsesquioxanes (POSS) molecules, *Polymer*, **42**, 599–611.

Garrett, J. T., Runt, J., Lin, J. S. (2000), Microphase separation of segmented poly(urethane urea) block copolymers, *Macromolecules*, **33**, 6353–9.

Garrett, J. T., Lin, J. S., Runt, J. (2002), Influence of preparation conditions on microdomain formation in poly(urethane urea) block copolymers, *Macromolecules*, **35**, 161–8.

Garrett, J. T., Xu, R., Cho, J., Runt, J. (2003), Phase separation of diamine chain-extended poly(urethane) copolymers: FT-IR spectroscopy and phase transitions, *Polymer*, **44**, 2711–19.

Hu, J. L., Yang, Z. H., Yeung, L., Ji, F. L., Liu, Y. Q. (2005a), Cross-linked polyurethanes with shape memory properties, *Polym. Int.*, **54**, 854.

Hu, J. L., Ji, F. L., Wong, Y. W. (2005b), Dependency of the shape memory properties of a polyurethane upon thermo-mechanical cyclic conditions, *Polym. Int.*, **54**, 600.

Huh, J. H., Kim, H. I., Rahman, M. M., Kim, H. D. (2008), Preparation and properties of MDI/h12MDI-based water-borne poly(urethane-urea)s-effects of MDI content and radiant exposure, *J. Appl. Polym. Sci.*, **110**, 3655–63.

Jena, K. K., Chattopadhyay, D. K., Raju, K. V. S. N. (2007), Synthesis and characterization of hyperbranched polyurethane-urea coatings, *Eur. Polym. J.*, **43**, 1825–37.

Jeong, H. M., Lee, J. B., Lee, S. Y., Kim, B. K. (2000a), Shape memory polyurethane containing mesogenic moiety, *J. Mater. Sci.*, **35**, 279.

Jeong, H. M., Lee, S. Y., Kim, B. K. (2000b), Shape memory polyurethane containing amorphous reversible phase, *J. Mater. Sci.*, **35**, 1579.

Kim, B. K., Lee, S. Y., Xu, M. (1996), Polyurethanes having shape memory effects, *Polymer*, **37**, 5781.

Kim, B. K., Lee, S. Y., Lee, J. S., Baek, S. H., Choi, Y. J., *et al.* (1998), Polyurethane ionomers having shape memory effects, *Polymer*, **39**, 2803.

Kim, B. K., Lee, J. S., Lee, Y. M., Shin, J. H., Park, S. H. (2001), Shape memory behavior of amorphous polyurethanes, *J. Macromol. Sci., Phys.*, **40**, 1179.

Lee, B. S., Chun, B. C., Chung, Y. C., Sul, K. I., Cho, J. W. (2001), Structure and thermo-mechanical properties of polyurethane block copolymers with shape memory effect, *Macromolecules*, **34**, 6431.

Li, F., Zhang, X., Hou, J., Xu, M., Luo, X., *et al.* (1997), Studies on thermally stimulated shape memory effect of segmented polyurethanes, *J. Appl. Polym. Sci.*, **64**, 1511.

Lin, J. R., Chen, L. W. (1998), Study on shape-memory behaviours of polyether-based polyurethanes, Part I: Influence of the hard-segment content, *J. Appl. Polym. Sci.*, **69**, 1563–74.

Luo, N., Wang, D. N., Ying, S. K. (1996), Crystallinity and hydrogen bonding of hard segments in segmented poly(urethane urea) copolymers, *Polymer*, **37**, 3577–83.

Luo, N., Wang, D. N., Ying, S. K. (1997), Hydrogen-bonding properties of segmented polyether poly(urethane urea) copolymer, *Macromolecules*, **30**, 4405–9.

Meng, Q., Hu, J., Mondal, S. (2008), Thermal sensitive shape recovery and mass transfer properties of polyurethane/modified MWNT composite membranes synthesized via *in situ* solution pre-polymerization, *J. Membrane Sci.*, **319**, 102–10.

Petrovic, Z. S., Ferguson, J. (1991), Polyurethane elastomers, *Prog. Polym. Sci.*, **16**, 695–836.

Ping, P., Wang, W., Chen, X., Jing, X. (2005), Poly(ε-caprolactone) polyurethane and its shape-memory property, *Biomacromolecules*, **6**, 587.

Ping, P., Wang, W., Chen, X., Jing, X. (2007), The influence of hard-segments on two-phase structure and shape memory properties of PCL-based segmented polyurethanes, *J. Polym. Sci., Polym. Phys.*, **45**, 557–70.

Rabani, G., Luftmann, H., Kraft, A. (2006), Synthesis and characterization of two shape-memory polymers containing short aramid hard segments and poly(ε-caprolactone) soft segments, *Polymer*, **47**, 4251.

Rogulska, M., Podkoscielny, W., Kultys, A., Pikus, S., Pozdzik, E. (2006), Studies on thermoplastic polyurethanes based on new diphenylethane-derivative diols, Part I: Synthesis and characterization of nonsegmented polyurethanes from HDI and MDI, *Eur. Polym. J.*, **42**, 1786–97.

Sheth, J. P., Klinedinst, D. B., Wilkes, G. L., Iskender, Y., Yilgör, I. (2005), Role of chain symmetry and hydrogen bonding in segmented copolymers with monodisperse hard segments, *Polymer*, **45**, 7317–22.

Szycher, M. (1999), *Szycher's Handbook of Polyurethanes*, CRC, Washington DC.

Takahashi, T., Hayashi, N., Hayashi, S. (1996), Structures and properties of shape-memory polyurethane block copolymers, *J. Appl. Polym. Sci.*, **60**, 1061.

Tobushi, H., Hara, H., Yamada, E., Hayashi, S. (1996), Thermo-mechanical properties in a thin film of shape memory polymer of polyurethane series, *Smart Mater. Struct.*, **5**, 483.

Wang, Z., Xie, X., Yang, R., Wang, X. F. (2006), Crystalline morphologies and molecular orientation of poly(ε-caprolactone) thin films, *Chem. J. Chinese U.*, **21**, 161–5.

Wei, Z. G., Sandstrom, R., Miyazaki, S. (1998), Review of shape-memory materials and hybrid composites for smart systems, *J. Mater. Sci.*, **33**, 3743.

Yang, J. H., Chun, B. C., Chung, Y. C., Cho, J. H. (2003), Comparison of thermal-mechanical properties and shape memory effect of polyurethane block-copolymers with planar or bent shape of hard segment, *Polymer*, **44**, 3251.

Yeh, F., Hsiao, B. S., Sauer, B. B., Michel, S., Siesler, H. W. (2003), *In-situ* studies of structure development during deformation of a segmented poly(urethane-urea) elastomer, *Macromolecules*, **36**, 1940–54.

Yen, F. S., Lin, L. L., Hong, J. L. (1999), Hydrogen-bond interactions between urethane–urethane and urethane–ester linkages in a liquid crystalline poly(ester-urethane), *Macromolecules*, **32**, 3068–79.

Yilgör, E., Yilgör, I., Yurtsever, E. (2002a), Hydrogen bonding and polyurethane morphology, Part I: Quantum mechanical calculations of hydrogen bond energies and vibrational spectroscopy of model compounds, *Polymer*, **43**, 6551–9.

Yilgör, E., Yurtsever, E., Yilgör, I. (2002b), Hydrogen bonding and polyurethane morphology, Part II: Spectroscopic, thermal and crystallization behavior of polyether blends with 1,3-dimethylurea and a model urethane compound, *Polymer*, **43**, 6561–8.

Yilgör, I., Mather, B. D., Unal, S., Yilgör, E., Long, T. E. (2004), Preparation of segmented, high molecular weight, aliphatic poly(ether-urea) copolymers in isopropanol. *In-situ* FT-IR studies and polymer synthesis, *Polymer*, **45**, 5829–36.

Yilgör, I., Yilgör, E., Guler, I. G., Ward, T. C., Wilkes, G. L. (2006), FT-IR investigation of the influence of diisocyanate symmetry on the morphology development in model segmented polyurethanes, *Polymer*, **47**, 4105–14.

5
Supramolecular shape memory polymers

DOI: 10.1533/9780857098542.111

Abstract: There is an important relationship between shape memory functionality and supramolecular chemistry, both in terms of shape memory effects and molecular interaction. The use of a supramolecular switch to fabricate smart materials is currently an area of significant interest. Thanks to investigations into the non-covalent interactions in supramolecules, particularly hydrogen bonding, researchers have been able to invent novel shape memory polymers (SMPs). This chapter reviews the relationship between SMPs and supramolecules and examines the promising potential offered by the use of supramolecular switches in the creation of SMPs.

Key words: supramolecular shape memory polymer, pyridine, N,N-bis(2-hydroxyl ethyl) isonicotinamide (BINA)-based shape memory polymer, phase structure, ureidopyrimidinone.

5.1 Introduction

5.1.1 Supramolecular chemistry and hydrogen bonding

The field of supramolecular chemistry (Lehn, 1985, 1993, 2005, 2007) focuses on the construction of highly complex chemical systems and advanced materials by designing arrays of components held together through inter-molecular forces. The dominant type of inter-molecular force in supramolecular chemistry is non-covalent bonding (Ujiie and Iimura, 1992; Lehn, 2002), including ionic bonds (Braga *et al.*, 2004), hydrophobic interactions, hydrogen bonds, Van der Waals forces (Hyla-Kryspin *et al.*, 2009) and dipole–dipole bonds (Cicoira *et al.*, 2008). These non-covalent interactions in smart materials or in stimuli-responsive polymers permit a rapid response to a change in environment (Wilson, 2007).

A perfect example of this is the assembly and disassembly of structural motifs such as DNA (Pontius and Berg, 1992; Shih, 2008), as shown in Fig. 5.1. Polynucleotide chains form a DNA double helix structure as result of the hydrogen bonding between the four different types of nucleotides. For example, A (adenine) forms two hydrogen bonds with T (thymine) on the opposite strand, and G (guanine) forms three hydrogen bonds with C (cytosine) on the opposite strand. The linking of monomers through non-covalent interactions therefore constitutes an attractive method for constructing stimuli-responsive soft materials. An alternative, as shown in Fig. 5.2, is the use of non-covalent interactions of functionalized side chains (Brunsveld *et al.*, 2001; Armstrong and Buggy, 2005).

The individual non-covalent interactions in question have low bond energies compared to covalent bonds. The latter typically have energies of 250 to 800 kJ mol^{-1}, whereas the strongest hydrogen bonds have energies of 210 kJ mol^{-1}, with

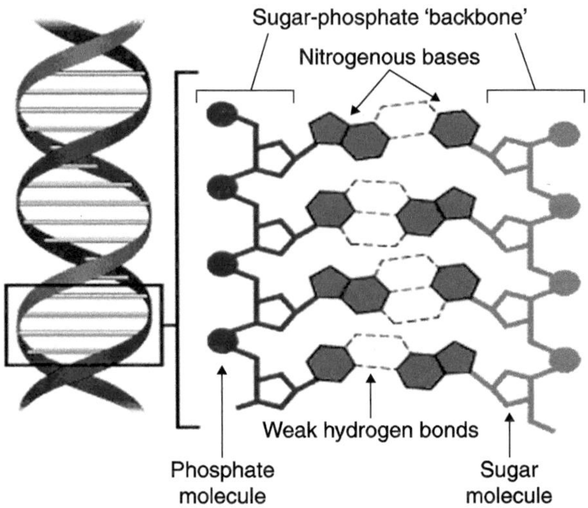

5.1 Supramolecular chemistry in the DNA structure (download from the web site: http://evolution.berkeley.edu/evosite/history/dna2.shtml).

12 to 25 kJ mol^{-1} being more usual for OH–O and NH–N hydrogen bonds (Armstrong and Buggy, 2005). Hydrogen bonds occur between hydrogen atoms and other electronegative atoms such as oxygen, nitrogen and fluorine. Unlike covalent and ionic bonds, they do not involve the exchange or sharing of electrons. Hydrogen bonds occur over short distances and can be easily formed and broken.

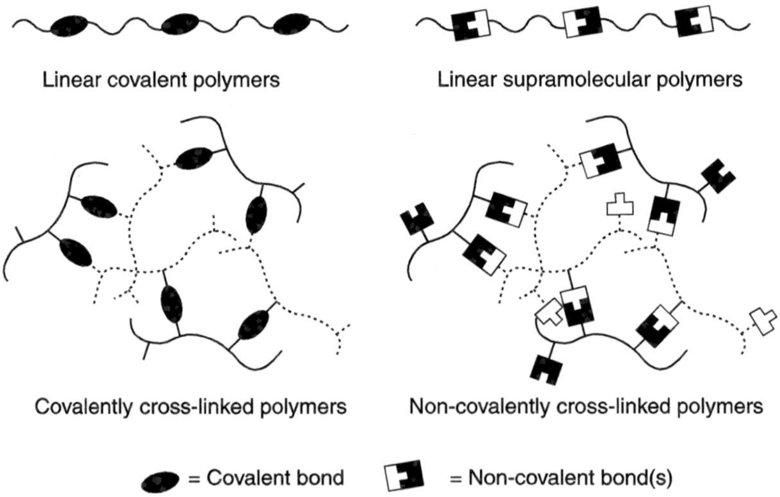

5.2 Diagram of polymer architectures made using covalent and non-covalent links between building blocks (Wilson, 2007).

They can also stabilize a molecule and vary in strength from very weak (1–2 kJ mol^{-1}) to extremely strong (>155 kJ mol^{-1}). The bond strength itself is dependent on temperature, pressure, bond angle and environment, usually characterized by a local dielectric constant. For example, the typical length of a hydrogen bond in water is 1.97 Å (Neves *et al.*, 2008).

The construction of stimuli-responsive soft materials requires non-covalent interactions that are both specific and directional to ensure that the required molecular assembly forms as intended, and that the assembly is reversible. The finished assembly may be annealed, or alternatively, defects can be self-healed. The use of hydrogen bonds in the construction of stimuli-responsive soft materials offers several advantages: they are highly directional and allow strong binding when several bonds are used together. When built into an array, they allow the creation of selectivity through the placement of donor and acceptor functionalities (Armstrong and Buggy, 2005). Hydrogen-bonded polymers have consequently been the subject of a great deal of recent research, since strength and selectivity of interaction are crucial in determining the properties of resulting polymers.

5.1.2 Overview of polymers containing functional pyridine

Some typical examples of pyridine-containing polymers include thermally stable and easy-to-process resins containing pyridine units apyrrolonebenzimidazole moieties (Mehdipour-Ataei and Heidaris, 2003); polyamides based on pyridine-containing monomers (Horsburgh *et al.*, 1999); conjugated polymers with electro-conductive and photoconductive properties such as poly (2,6- or 2,5-)pyridine; alternating copolymers such as poly-pyridine-N-alkyl carbazole (Zheng *et al.*, 2002); photo-conductive poly-thienyl-pyridine (Aly and Khalaf, 2000); and conjugated semiconductor polyazomethine containing pyridine units. In addition, polyureas incorporating 2,6-pyridine rings were synthesized and used for the production of light emitting diodes based on the terbium immobilized polyurea chelates (Mwaura *et al.*, 2003).

One important method for preparing pyridine-containing polymers with various functions is self-assembly; for example, different polymers functionalized with 2,6-diaminopyridine were synthesized by self-assembly with complementary thymine-functionalized gold (Au) nanoparticles (Carroll *et al.*, 2004) for preparing nanocomposites or for selective sorption of Au(III) from solutions containing the related metal ions (Drechsler *et al.*, 2002). Polymers containing structures able to non-covalently cross-link were prepared through ring opening metathesis polymerization of the corresponding norbornenes. Side-chain groups from diaminopyridine, uracil or diaminotriazine acted as recognition elements and the polymers self-assembled into spherical polymersomes. ABC tri-block copolymers with self-complementary abilities were prepared via ring opening metathesis polymerization, using three monomers derived from exo-7-oxabicyclo hept-5-ene-2,3-dicarboximide, one of which contained an acetylated diaminopyridine moiety (Bazzi *et al.*, 2003).

The polyamide obtained by polycondensation of 2,6-diaminopyridine and 2,6-pyridine dicarboxylic acid was the first polymer to assemble itself into a double helix (DNA-type) in solution. The synthesis and physicochemical characterization of some polymer-supported rhodium catalysts based on polyamides containing 2,6- and 2,5-pyridine units were reported by Michalska and Strzelec (2000); these catalysts were used for the hydrosilylation of vinyl compounds such as phenylacetylene. Chevallier *et al.* (2002) prepared polyamide-esters from 2,6-pyridine dicarboxylic acid and thanolamine derivatives and investigated their polymer sorption behavior towards heavy metal ions. Finally, Scortanu *et al.* (2006) also prepared a polymer with improved performance based on polyureas containing 2,6-pyridine moiety and polyparabanic acids, and polyurethane-ureas containing 2,6-pyridine rings.

5.2 Synthesis of polymers containing pyridine moieties

A number of polymers containing pyridine moieties have been synthesized in recent years (Scortanu *et al.*, 2006), usually on the basis of monomers containing pyridine moieties. These monomers include vinyl pyridines, such as 2-vinyl-pyridine and 4-vinyl-pyridine; pyridine containing glycols such as 3,5-pyridine dimethanol pyridine and 2,6-dimethanol pyridine; acrylate or methacrylate monomers with pendant pyridine groups such as 4-(3-methacryloyl-propyl) pyridine; bromo-pyridines such as 2,5-bis(5-bromo-thiophen-2-yl) pyridine; aminopyridines such as 2,6-diamino-pyridine; and some pyridine monomers without reactive groups such as 4,4-bipyridyl, among others. A variety of polymerization methods are therefore employed to synthesize these pyridine-containing polymers from the monomers and the most widely used of these are discussed in the following sections.

5.2.1 Radical polymerization

Radical polymerization, as shown in Fig. 5.3, is the most widely employed method for the synthesis of polymers containing pyridine, such as amphiphilic 4-vinyl pyridine and n-vinyl pyrrolidone copolymer (Sambasivudu *et al.*, 2006). The 4VP/NVP copolymers with various ratios are prepared with azobisisobutyronitrile (AIBN) as the free radical initiator and cyclohexane as the diluent system. A homopolymer P4VP is also prepared following the same procedure. However, the synthesis of the copolymer is difficult due to the significant difference in reactivity ratios.

Well-defined pyridine-containing polymers and copolymers with controlled molecular weight and narrow molecular weight distribution have a potential number of promising applications, for example as coordination reagents for

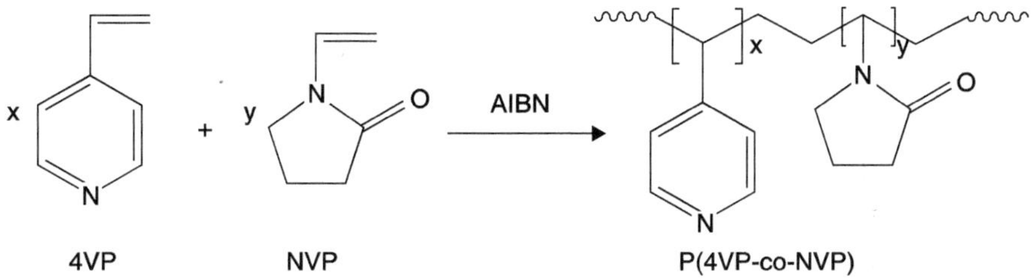

5.3 Synthesis of pyridine-containing polymers by radical polymerization.

transition metals. However, the typical radical polymerization method shown in Fig. 5.3 cannot produce polymers and copolymers with the necessary properties: the synthesis of well-defined functional polymers instead requires living polymerization methods including living-anionic polymerization, group-transfer polymerization (GTP) and atom radical polymerization (ATRP). For example, poly(4-vinyl pyridine) (P4VP) is synthesized by ATRP in propanol using 1-PECl as the initiator, with CuCl/Me$_6$ and N$_4$ as the catalyst and ligand. Figure 5.4 shows the synthesis of polyimide (PI) films with antibacterial surfaces. First, the PI chains are chloromethylated in order to introduce ATRP initiators onto the PI film surfaces; functional polymer brushes of 4-vinylpyridine (4VP)

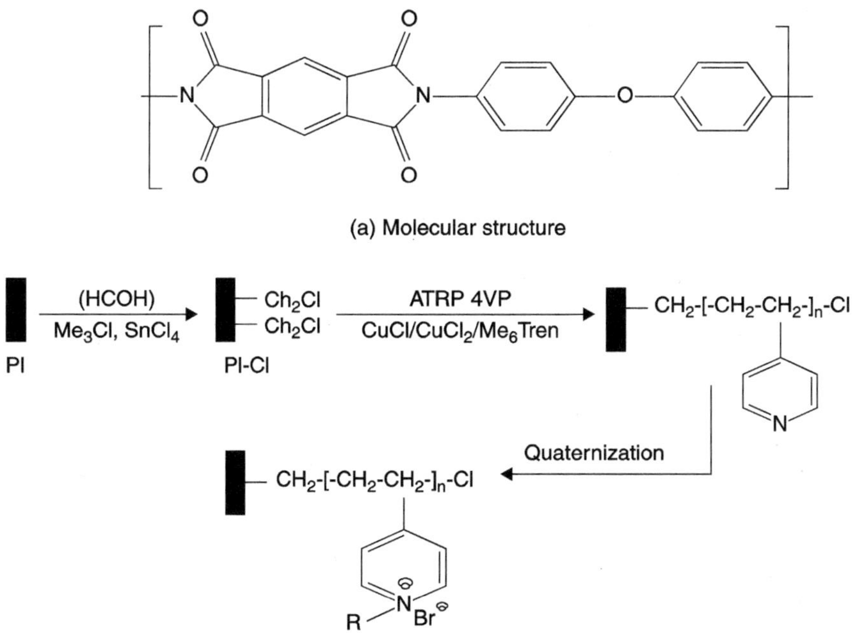

5.4 Synthesis of pyridine-containing polymers using ATRP.

are then prepared via surface initiated ATRP, followed by the alkylation of the grafted P4VP with hexyl bromide (Li *et al.*, 2008).

5.2.2 Condensation polymerization

Pyridine-containing polymers can also be synthesized using the condensation polymerization method. For example, a novel series of benzimidazole-containing sulfonated polyarylene sulfones with a controllable amount of basic 2,6-bis(benzimidazol-2-yl)pyridine (BIP) and sulfonic acid groups have been synthesized by the copolycondensation of a new BIP-containing arylene difluoride monomer (DFSBIP) with a sulfonated arylene difluoride (DSDFS) and 4,4′-biphenol (BP) (Qi *et al.*, 2009). Another example of the condensation polymerization method is the preparation of a series of new polyimides by reacting 2,6-diaminopyridine with various aromatic dianhydrides in dimethyl formamide (DMF) in a 1:1 mole ratio.

5.2.3 Self-assembly

In addition to the pyridine-containing polymers synthesized on the basis of covalent bonds, there are also a number of non-covalent linked supramolecular polymer assemblies based on arrays of hydrogen bonds and prepared through self-assembly. For example, hydrogen-bonded liquid crystal polymers (LCPs) with well defined structures including main-chain, side-chain, combined, and network structures have been prepared through the self-assembly of polymers and small molecules (Kato and Frechet, 1995; Kato *et al.*, 1998). Even cross-linked polymers or polymer networks can be obtained through the self-assembly of pyridine-based hydrogen bonding.

Segmented polyurethanes (PUs) have been the subject of research in many fields, not only because of their excellent properties, but also because of their high reactivity with diisocyanate, further enhanced by the addition of glycols. Several PUs containing pyridine moieties have hence been synthesized. For example, using chain extenders containing a pyridine ring, PU can be synthesized to contain a tertiary nitrogen atom. During the synthesis of PU, as shown in Fig. 5.5, 0.02 M poly (tetramethylene oxide) glycol (PTMG 1000) is added dropwise to toluene diisocyanate (TDI) (0.04 M) in a 250 ml 3-neck round bottomed flask at 40°C in nitrogen atmosphere for 1 hour. The temperature is then raised to 70°C and the reaction is allowed to proceed until the isocyanate content reaches half of the initial value as determined by dibutylamine titration. The temperature is reduced to 65°C and the chain extender (0.02 m) in 20 ml DMF is added dropwise. Then 0.01 g of dibutylin dilaurate (DBTDL) catalyst is added, the temperature is raised to 70°C and the reaction is carried out for another four hours (Sriram *et al.*, 2005; Scortanu *et al.*, 2006). Similarly, polyurethane-ureas containing pyridine units are also synthesized by the urethane polyaddition reaction between a macro-diol of

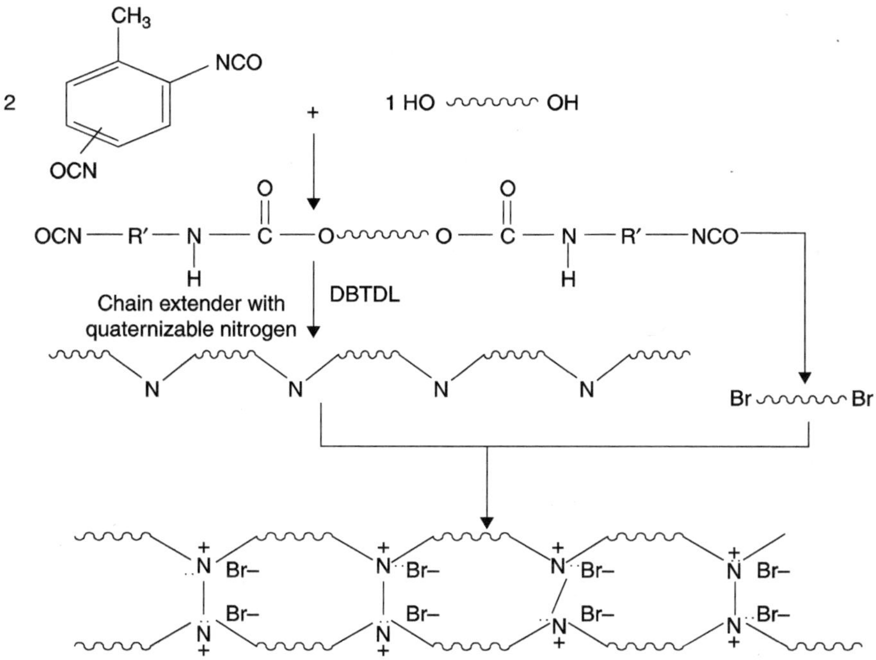

5.5 Synthesis of pyridine-containing polyurethane (PU) using the urethane polyaddition reaction.

polyether (PTMG) or polyester (PEA) type, and an aromatic diisocyantate such as methylene diphenyl diisocyanate (MDI), with 2,6-diamino pyridine (DAPy) as the chain extender (Scortanu *et al.*, 2006).

5.3 Supramolecular polymers containing functional pyridine

5.3.1 Carboxyl-pyridyl hydrogen bonding

The first use of arrays of hydrogen bonds in chain polymers and cross-linked polymers was reported by Lehn and co-workers, as described in Wilson (2007). A number of methods for forming functional polymers based on hydrogen bonds have since been developed. For example, several hydrogen bond systems are employed in the production of liquid crystalline polymers, such as acid-pyridine hydrogen bonding, acid-aminopyridine hydrogen bonding, acid-imidazole hydrogen bonding, pyridyl-phenol hydrogen bonding, and carboxyl-pyridyl hydrogen bonding. The latter has been of particular interest, since carboxyl can form stable hydrogen bonds with a pyridyl group with 1:1 stoichiometry. Carboxyl-pyridyl bonds have higher hydrogen bond energy (>45 kJ/mol^{-1}) and a better association constant compared to carboxyl dimerization.

The pyridyl ring, which plays a key role in the formation of strong hydrogen bonds (Scortanu *et al.*, 2006) is a rigid hetero-aromatic molecule. Polymers containing a pyridyl ring, particularly in the backbone of the polymer, tend to show good heat tolerance and chemical stability. The dissolvability of a polymer in a polar solvent can be significantly improved through the protonization of the nitrogen in the pyridine ring. The complexity of nitrogen makes polymers containing a pyridine ring suitable for use as the catalyst carrier. The improved mechanical properties, heat tolerance, chemical stability and dissolvability of these pyridine-containing polymers make them suitable for a wide variety of applications in the mechanical, electrical and aerospace fields, and even some special functions such as liquid crystalline (as discussed in Section 5.4) and optical-electrical properties can be obtained.

5.3.2 Studies of molecular recognition in supramolecular polymer architecture

Many studies have been carried out on molecular recognition in the construction of supramolecular polymer architecture through non-covalent bonds. For example, telechelic macromonomers end-capped with either nucleobase (adenine, thymine) derivatives or a tridentate ligand, for example BIP, show polymer-like properties in solid state or in the presence of metal ion solutions, respectively. Scortanu *et al.* (2006) reported the synthesis of thermally-reversible polymers built by non-covalent cross-linking, showing that a soluble polymer containing diacyl-diamidopyridine moieties can be precipitated by non-covalent cross-linking at room temperature using bivalent molecules such as bisthymine. The micron-scale spherical aggregates obtained can be completely dissolved by heating at 50°C. A cyclobarbital-imprinted polymer was prepared from a fluorescent functional monomer such as 2,6-bis(acrylamino) pyridine. This showed not only selective binding based on multiple hydrogen bonds, but also enhancement of fluorescence intensity, suggesting that the polymer could be used as a selective fluorescence probe (Ihata *et al.*, 2000). Finally, Kato *et al.* (2006) also reported the presence of macrocycles (crown-ether type) in hydrogen-bonded polymer complexes, containing a pyridyl moiety and carboxyl-functionalized polystyrenes.

5.4 Supramolecular liquid crystalline polymers containing pyridine moieties

Specific inter-molecular interactions play a major role in the supramolecular assembly of liquid crystals. Hydrogen-bonded materials such as benzoic acids have been used in the production of synthetic liquid crystals since the early twentieth century (Capitani *et al.*, 2001; Kato *et al.*, 2006); however, the potential offered by hydrogen-bonded materials was not recognized until supramolecular approaches involving hydrogen bonding were introduced into the design of liquid crystals in 1989 by Kato and Frechet. Vera *et al.* (2008) and Gulikkrzywicki *et al.* (1993) reported that complementary inter-molecular hydrogen bonding between different molecules leads to the formation of well-defined structures of mesogenic complexes. The first examples of supramolecular liquid crystals were hydrogen-bonded complexes, which contain either single-hydrogen-bonded one, or triple-hydrogen-bonded one. Thereof, single hydrogen bond complex exhibits smectic and nematic phase, while triple-hydrogen-bonded complex forms a columnar phase. Mesogens with rigid rod structures were obtained through the self-assembly of complementary molecular components based on pyridyl moieties.

Another study examined a liquid-crystalline phase induced by the 2:1 complexation of non-mesomorphic components, 4-methoxybenzoic acid and 4,4′-bipyridine (Kihara *et al.*, 1996). In this case, nematic mesogen is obtained through the simple assembly of two independent components. A nematic phase was observed from 153 to 163°C in the monomer. When the methoxybenzoic acid was replaced by a trialkyl-substituted molecule, the resulting complex showed a columnar phase up to 98°C (Sautter *et al.*, 2001). The materials could also be functionalized by the incorporation of a dye as a hydrogen-bond-proton acceptor. Koh *et al.* (1995) also observed fluorescence in the supramolecular LCP complex and found that LCP properties may be fine-tuned by changing the hydrogen-donor components. A variety of mesogenic complexes, such as cones and bananas, have been obtained by connecting two different components through carboxyl-pyridyl hydrogen bonding (Kraft *et al.*, 2000).

Pyridyl-based hydrogen bonding was also developed for the side-chain LCPs, where polymers with mesogenic side chains were prepared by combining functionalized polymers and small molecules through acid-pyridyl interactions. For example, the benzoic acid/pyridine mesogenic complexes were formed through the formation of single hydrogen bonds in the side chains (Kato *et al.*, 1992, 1996). This type of structure exhibited thermally stable smectic phases, while another structure with doubly hydrogen-bonded side-chain mesogens showed columnar liquid-crystalline phases (Kato *et al.*, 1995; Kawakami and Kato, 1998). Polymeric complexes were obtained by the complexation of end-functionalized mesogenic molecules and backbone-functionalized polymers such as poly(vinylpyridine) and poly(acrylic acid) (Alder *et al.*, 1995; Stewart and Imrie, 1995). Finally, simple alkylphenols, which are not mesogenic, have also been shown to form supramolecular LCP complexes (Kato *et al.*, 2006).

Recently, alternative supramolecular LCP complexes have been prepared using pyridyl-acid hydrogen bonds, in which the pyridyl unit was introduced into the backbone of polymer chain (Paleos and Tsiourvas, 1995). Kato *et al.* (1998) described the first examples of this type of supramolecular LCP with the complexation of 4-alkyloxybenzoic acid derivatives and polyamides containing 2,6-diaminopyridine moiety in the main chain (Ambrozic *et al.*, 2002). Hydrogen-bonded liquid crystalline polyurethane complexes with 4-aklyoxybenzoic acid were also prepared with 2,6-bis(hydroxymethyl)pyridine isonicotinamide (BINA) (Ambrozic and Zigon, 2005a,b). In these systems, the formation of complexes was also primarily a result of the hydrogen bonding between pyridine as the hydrogen-acceptor and the carboxylic acid group as the hydrogen-donor. The urethane groups on the polymer backbone contributed to the binding of 4-dodecyloxybenzoic acid (DOBA) molecules to a minor extent.

5.5 Supramolecular polymers and shape memory polymers

Generally, polymers exhibit shape memory functionality if the material can be stabilized in the deformed state within the temperature range of the intended application. This can be achieved through the use of polymer network chains equipped with a specific type of molecular switch. The flexibility of chains should be a function of the temperature in the thermal-induced shape memory polymers (SMPs). One possible method of implementing a switch function is a thermal transition of the network chains. At a temperature above T_{trans}, the chains are flexible; while the flexibility of the chains is partly limited at below T_{trans}. This can be achieved either through:

1. the glass transition between the rubber-elastic or viscous state and the glassy state in the T_g-type-SMPs; or
2. the strain-induced crystallization caused by material cooling in the T_m-type-SMPs.

However, the permanent shape of SMP networks is stabilized by covalent or non-covalent net points with the higher transition temperature (T_{perm}).

Both the glass transition and the strain-induced crystallization are related to the inter-molecular force, which plays a key role in the phase transition of reversible phases and in the formation of hard domains. In particular, inter-molecular hydrogen bonding has a significant influence on the thermo-mechanical properties of SMPs. In an SMPU composed of hard and soft segments, the phase separation and formation of hard domains is affected by the hydrogen bonds between the C=O group and N=H group. Furthermore, the formation of hydrogen bonds is highly sensitive to changes in temperature, ion concentration and other parameters. A useful example of this sensitivity can be found in the inter-molecular complexes that poly(carboxylic acid)s form with poly(ethylene glycol)(PEG). These complexes form due to the hydrogen bonds between the carboxyl groups of poly(carboxylic acid)s and the ether oxygen atoms of the PEG, and are extremely sensitive to any change in the concentration or molecular weight of PEG, as well as to temperature and other changes. Shape memory properties with a shape recovery of 99% can therefore be observed in poly(acrylic acid-co-methyl methacrylate)/PEG complexes, due to the significant difference in their storage modulus below and above T_{trans}.

It has also been proposed that the thermo-reversible, non-covalent interactions could be utilized in SMPs, as many kinds of supramolecular polymers were developed from small molecules or oligomers. This proposal was recently confirmed for the cross-linked polymer networks that contain only a small fraction of reversibly associating side-groups like ureidopyrimidinone (UPy). In this system, hydrogen-bonding interactions can mechanically stabilize the strained states in polymer elastomers. Li *et al.* (2007) introduced a UPy unit into a polymer network by radical polymerization. As shown in Fig. 5.6, a novel type of SMP was fabricated, containing site-specific, thermo-reversible interactions. The shape-memory effects can be observed in a lightly cross-linked polymer network containing only a small fraction (~2 mol.%) of UPy pendent side-groups. According to the findings of the study, the mechanically strained states in polymer elastomers can be stabilized by H-bonding interactions alone, resulting in strain fixity of about 90% and strain recovery of about 100%. The lack of complete strain fixity contributes to the elasticity of the material itself. The study also found that the unique feature of this new shape-memory polymer is the dynamics of its shape-memory response.

By introducing self-complementary quadruple hydrogen bonding units (UPy) into soft segments, Zhu *et al.* (2009) fabricated a type of thermoplastic shape memory polyurethane that displays a significant shape memory effect (SME) under the usual thermodynamic programming conditions (Fig. 5.7). The introduction of UPy into soft segments was found to increase the glass transition temperature from 28.3°C in the control sample to 73.3°C. The temporary deformation can be thus fixed well after cooling at room temperature. At a later stage, thermally-induced shape memory recovery can be triggered by raising the temperature to 86°C. The immediate shape recovery ratio and shape fixity ratio can be up to 95.8 and 95.9%, respectively. Even after 24 hours relaxation for the stretched films, these values can remain as high as 94 and 60%. In contrast, in the sample without quadruple hydrogen bonding, elasticity and the deformation cannot be fixed after 24 hours' relaxation.

Ware *et al.* (2012) introduced a UPy unit into T_g-SMPU, thereby creating a triple SMP (Fig. 5.8). Triple shape properties arise from the combination of the

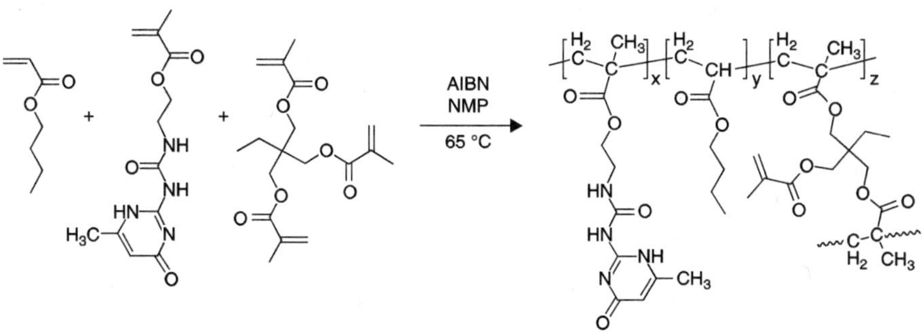

5.6 Synthesis of a SMP containing UPy side-groups (Li *et al.*, 2007).

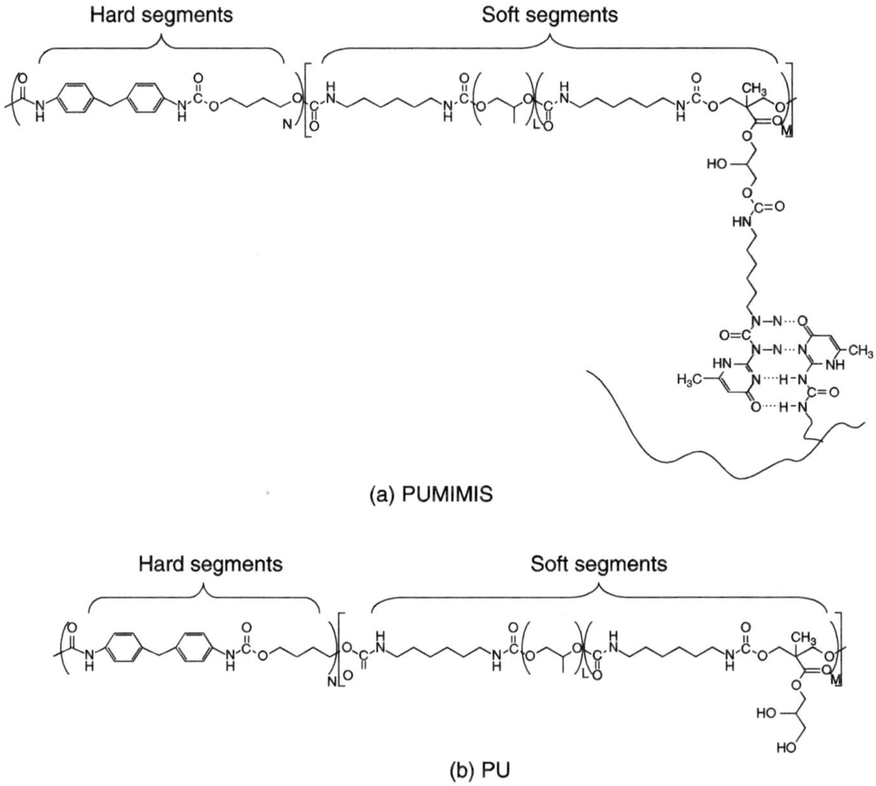

5.7 Molecular model of SMPU with UPy switch (Zhu *et al.*, 2009).

glass transition of (meth)acrylate copolymers and the dissociation of self-complementary hydrogen bonding moieties, enabling broad and independent control of both glass transition temperature (T_g) and cross-link density. Specifically, ureidopyrimidone methacrylate and a novel monomer, ureidopyrimidone acrylate, were copolymerized with various alkyl acrylates and bisphenol A ethoxylate diacrylate. Control of T_g from 0 to 60°C is demonstrated: concentration of hydrogen bonding moieties is varied from 0 to 40 wt.%; concentration of the diacrylate is varied from 0 to 30 wt.%. Toughness ranges from 0.06 to 0.14 MPa and is found to peak near 20 wt.% of the supramolecular cross-linker. A widely tunable class of amorphous triple-SMPs has been developed and characterized through dynamic and quasi-static thermo-mechanical testing to gain insights into the dynamics of supramolecular networks.

In addition, Zhang *et al.* (2008) also achieved supramolecular physical net points through the inclusion of α-CD or γ-CD with PCL or PEG (Fan *et al.*, 2009; Luo *et al.*, 2008, 2009). In their system, the inclusion of α-CD or γ-CD crystallites with PEG or PCL served as a fixing phase, while the naked PEG crystallites and PCL crystallites served as a reversible phase, as shown in Fig. 5.9.

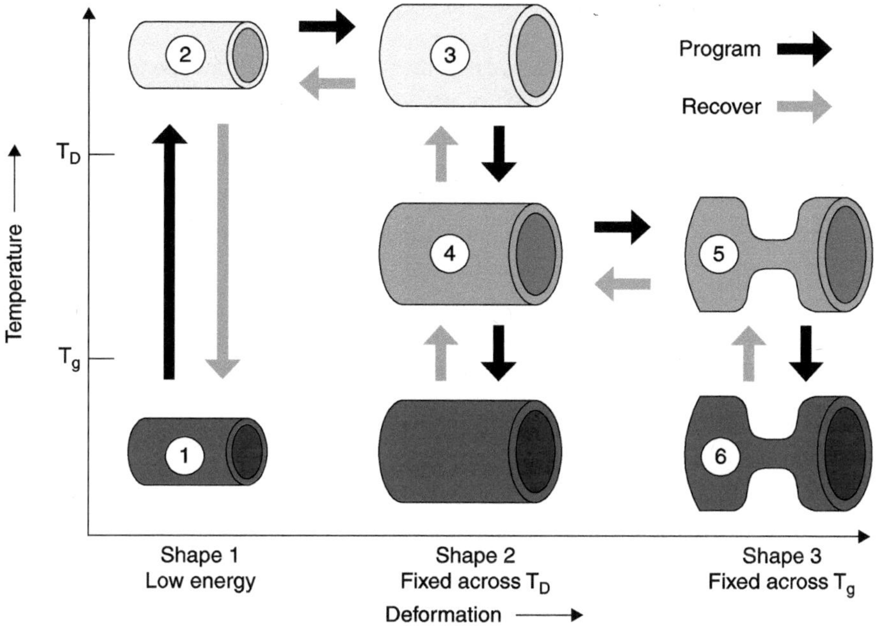

5.8 A triple-shape memory polymer containing a UPy unit (Ware *et al.*, 2012).

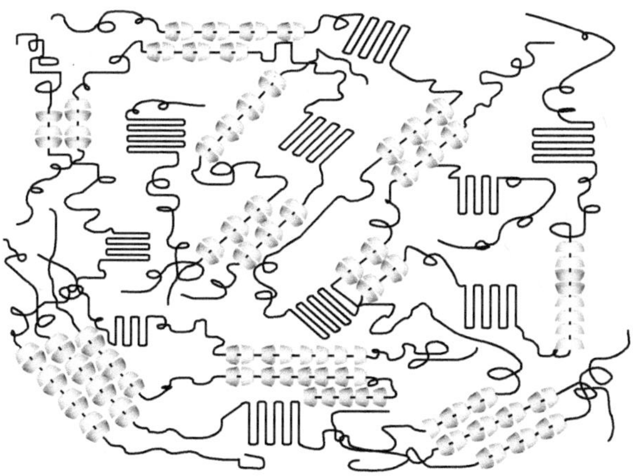

5.9 Schematic representation of the proposed chain structures (adapted from Li *et al.*, 2007; Luo *et al.*, 2009; Ware *et al.*, 2012; Wilson, 2007; Zhu *et al.*, 2009).

5.6 Conclusions

SMPs are an emerging class of active polymers with applications spanning various areas of everyday life. Many kinds of non-covalent bonds are used to construct highly complex chemical systems and advanced materials. Pyridine moieties in particular have been widely used as hydrogen acceptors in the fabrication of functional, supramolecular and liquid crystalline polymers. However, as yet these have not been used in the production of SMPs, and their SMEs have not been reported. Going forward, the molecular design of SMPs should take into account non-covalent interactions, particularly hydrogen bonding. The introduction of a large fraction of hydrogen-acceptors and hydrogen-donors in the polymer chain would have a significant effect on the shape memory properties of SMP networks.

Shape memory functionality has been shown to have a close relationship with supramolecular chemistry, and the use of a supramolecular switch for fabricating smart materials shows great potential. Shape recovery mostly results from a change in the surrounding conditions such as temperature, light, and magnetic or electric fields. However, to date, there have been fewer studies on SMEs that can be achieved without the heat stimulus. So far, studies on water-driven SMPs have mainly concentrated on the change in morphology from a physics perspective, while investigations into the molecular design and synthesis of water-driven and moisture-sensitive SMPs remain relatively few.

5.7 References

Alder, K. I., Stewart, D., Imrie, C. T. (1995), Towards supramolecular side-chain liquid-crystal polymers, Part III: Miscibility of mesogenic acids and amorphous polymers, *Journal of Materials Chemistry*, **5**, 2225–8.

Aly, K. I., Khalaf, A. A. (2000), New polymer syntheses, Part IX: Synthesis and properties of new conducting polyazomethine polymers containing main chain cycloalkanone and pyridine moieties, *Journal of Applied Polymer Science*, **77**, 1218–29.

Ambrozic, G., Mavri, J., Zigon, M. (2002), Liquid-crystalline complexes of polyurethane containing an isonicotinamide moiety with 4-dodecyloxybenzoic acid, *Macromolecular Chemistry and Physics*, **203**, 439–47.

Ambrozic, G., Zigon, M. (2005a), Hydrogen-bonded liquid-crystalline polyurethane complexes with 4-dodecyloxybenzoic acid, *Acta Chimica Slovenica*, **52**, 207–14.

Ambrozic, G., Zigon, M. (2005b), Hydrogen-bonded polyurethane complexes based on 4-alkoxybenzoic acids as the low molar mass components, *Polymer International*, **54**, 606–13.

Armstrong, G., Buggy, M. (2005), Hydrogen-bonded supramolecular polymers: A literature review, *Journal of Materials Science*, **40**, 547–59.

Bazzi, H. S., Bouffard, J., Sleiman, H. F. (2003), Self-complementary ABC triblock copolymers via ring-opening metathesis polymerization, *Macromolecules*, **36**, 7899–902.

Braga, D., Maini, L., Polito, M., Grepioni, F. (2004), Hydrogen bonding interactions between ions: A powerful tool in molecular crystal engineering, *Supramolecular Assembly Via Hydrogen Bonds II*, **111**, 1–32.

Brunsveld, L., Folmer, B. J., Sijbesma, R. P. (2001), Supramolecular polymers, *Chemical Reviews*, **101**, 4071–97.

Capitani, D., De Prisco, N., Laurienzo, P., Malinconico, M., Proietti, N., Roviello, A. (2001), A new class of poly(ester hydrazide) copolymers with liquid crystalline properties: Synthesis and characterization, *Polymer Journal*, **33**, 575–83.

Carroll, J. B., Frankamp, B. L., Srivastava, S., Rotello, V. M. (2004), Electrostatic self-assembly of structured gold nanoparticle/polyhedral oligomeric silsesquioxane (POSS) nanocomposites, *Journal of Materials Chemistry*, **14**, 690–4.

Chevallier, P., Soutif, J. C., Brosse, J. C., Brunelle, A. (2002), Poly(amide ester)s from 2,6-pyridinedicarboxylic acid and ethanolamine derivatives: Identification of macrocycles by matrix-assisted laser desorption/ionization mass spectrometry, *Rapid Communications in Mass Spectrometry*, **16**, 1476–84.

Cicoira, F., Santato, C., Rosei, F. (2008), Two-dimensional nanotemplates as surface cues for the controlled assembly of organic molecules, *STM and AFM Studies on (Bio) Molecular Systems: Unravelling the Nanoworld*, **285**, 203–67.

Drechsler, U., Thibault, R. J., Rotello, V. M. (2002), Formation of recognition-induced polymersomes using complementary rigid random copolymers, *Macromolecules*, **35**, 9621–3.

Fan, M. M., Yu, Z. J., Luo, H. Y., Zhang, S., Li, B. J. (2009), Supramolecular network based on the self-assembly of gama-cyclodextrin with poly(ethylene glycol) and its shape memory effect, *Macromolecular Rapid Communications*, **30**, 89–903.

Gulikkrzywicki, T., Fouquey, C., Lehn, J. M. (1993), Electron-microscopic study of supramolecular liquid-crystalline polymers formed by molecular-recognition-directed self-assembly from complementary chiral components, *Proceedings of the National Academy of Sciences of the USA*, **90**, 163–7.

Horsburgh, L. E., Monkman, A. P., Samuel, I. D. W. (1999), Synthesis and characterisation of polypyridines, *Synthetic Metals*, **101**, 113–14.

Hyla-Kryspin, I., Grimme, S., Djukic, J. P. (2009), The CR-MN interaction in syn-facial [tricarbonyl(benzyl)chromium]manganesetricarbonyl complexes: A non-covalent metal–metal bond, *Organometallics*, **28**, 1001–13.

Ihata, O., Yokota, H., Kanie, K., Ujiie, S., Kato, T. (2000), Induction of mesophases through the complexation between benzoic acids with lateral groups and polyamides containing a 2,6-diaminopyridine moiety, *Liquid Crystals*, **21**, 69–74.

Kato, T., Frechet, J. M. J. (1989), New approach to mesophase stabilization through hydrogen-bonding molecular-interactions in binary-mixtures, *Journal of the American Chemical Society*, **111**, 8533–4.

Kato, T., Kihara, H., Uryu, T., Fujishima, A., Frechet, J. M. J. (1992), Molecular self-assembly of liquid-crystalline side-chain polymers through intermolecular hydrogen-bonding – polymeric complexes built from a polyacrylate and stilbazoles, *Macromolecules*, **25**, 6836–41.

Kato, T., Frechet, J. M. J. (1995), Hydrogen-bonding and the self-assembly of supramolecular liquid-crystalline materials, *Macromolecular Symposia*, **98**, 311–26.

Kato, T., Nakano, M., Moteki, T., Uryu, T., Ujiie, S. (1995), Supramolecular liquid-crystalline side-chain polymers built through a molecular recognition process by double hydrogen bonds, *Macromolecules*, **28**, 8875–6.

Kato, T., Kihara, H., Ujiie, S., Uryu, T., Frechet, J. M. J. (1996), Structures and properties of supramolecular liquid-crystalline side-chain polymers built through intermolecular hydrogen bonds, *Macromolecules*, **29**, 8734–9.

Kato, T., Ihata, O., Ujiie, S., Tokita, M., Watanabe, J. (1998), Self-assembly of liquid-crystalline polyamide complexes through the formation of double hydrogen bonds between a 2,6-bis(amino)pyridine moiety and benzoic acids, *Macromolecules*, **31**, 3551–5.

Kato, T., Mizoshita, N., Kishimoto, K. (2006), Functional liquid-crystalline assemblies: Self-organized soft materials, *Angewandte Chemie-International Edition*, **45**, 38–68.

Kawakami, T., Kato, T. (1998), Use of intermolecular hydrogen bonding between imidazolyl moieties and carboxylic acids for the supramolecular self-association of liquid-crystalline side-chain polymers and networks, *Macromolecules*, **31**, 4475–9.

Kihara, H., Kato, T., Uryu, T., Frechet, J. M. J. (1996), Supramolecular liquid-crystalline networks built by self-assembly of multifunctional hydrogen-bonding molecules, *Chemistry of Materials*, **8**, 961–8.

Koh, K. N., Araki, K., Komori, T., Shinkai, S. (1995), Thermotropic liquid-crystal construction through the hydrogen-bonds between a stilbazole calix[4]arene and carboxylic-acids, *Tetrahedron Letters*, **36**, 5191–4.

Kraft, A., Reichert, A., Kleppinger, R. (2000), Supramolecular liquid crystals with columnar mesophases through self-assembly of carboxylic acids around a tribasic core, *Chemical Communications*, **12**, 1015–16.

Lehn, J. M. (1985), Supramolecular chemistry – receptors, catalysts, and carriers, *Science*, **227**, 849–56.

Lehn, J. M. (1993), Supramolecular chemistry, *Science*, **260**, 1762–3.

Lehn, J. M. (2002), Toward complex matter: Supramolecular chemistry and self-organization, *Proceedings of the National Academy of Sciences of the USA*, **99**, 4763–8.

Lehn, J. M. (2005), Dynamers: Dynamic molecular and supramolecular polymers, *Progress in Polymer Science*, **30**, 814–31.

Lehn, J. M. (2007), From supramolecular chemistry towards constitutional dynamic chemistry and adaptive chemistry, *Chemical Society Reviews*, **36**, 151–60.

Li, J. H., Viveros, J. A., Wrue, M. H., Anthamatten, M. (2007), Shape-memory effects in polymer networks containing reversibly associating side-groups, *Advanced Materials*, **19**(19), 2851–5.

Li, L., Ke, Z. J., Yan, G. P., Wu, J. Y. (2008), Polyimide films with antibacterial surfaces from surface-initiated atom-transfer radical polymerization, *Polymer International*, **57**, 1275–80.

Luo, H. Y., Liu, Y., Yu, Z. J., Zhang, S., Li, B. J. (2008), Novel biodegradable shape memory material based on partial inclusion complex formation between alpha-cyclodextrin and poly(epsilon-caprolactone), *Biomacromolecules*, **9**, 2573–7.

Luo, H. Y., Fan, M. M., Yu, Z. J., Meng, X. W., Li, B. J., Zhang, S. (2009), Preparation and properties of degradable shape memory material based on partial alpha-cyclodextrin-poly(epsilon-caprolactone) inclusion complex, *Macromolecular Chemistry and Physics*, **210**, 669–76.

Mehdipour-Ataei, S., Heidaris, H. (2003), Synthesis and characterization of novel soluble and thermally stable polyamides based on pyridine monomer, *Macromolecular Symposia*, **193**, 159–67.

Michalska, Z. M., Strzelec, K. (2000), Synthesis and physicochemical characterization of polymeric supports and polymer-supported rhodium catalysts based on polyamides having a pyridine moiety, *Reactive & Functional Polymers*, **44**, 189–99.

Mwaura, J. K., Mathai, M. K., Chen, C. Q., Papadimitrakopoulos, F. (2003), Light emitting diodes prepared from terbium-immobilized polyurea chelates, *Journal of Macromolecular Science – Pure and Applied Chemistry*, **A40**, 1253–62.

Neves, R. S., Motheo, A. J., Fartaria, R. P. S., Fernandes, F. M. S. (2008), Modelling water adsorption on Au(210) surfaces, Part II: Monte Carlo simulations. *Journal of Electroanalytical Chemistry*, **612**, 179–85.

Paleos, C. M., Tsiourvas, D. (1995), Thermotropic liquid-crystals formed by intermolecular hydrogen interactions, *Angewandte Chemie – International Edition in English*, **34**, 1696–711.

Pontius, B. W., Berg, P. (1992), Rapid assembly and disassembly of complementary-DNA strands through an equilibrium intermediate state mediated by a1 hnrnp protein, *Journal of Biological Chemistry*, **267**, 13815–18.

Qi, Y. H., Gao, Y., Tian, S. H., Hill, A. R., Gaudet, J., *et al.* (2009), Synthesis and properties of novel benzimidazole-containing sulfonated polyethersulfones for fuel cell applications, *Journal of Polymer Science Part A – Polymer Chemistry*, **47**, 1920–29.

Sambasivudu, K., Maheedhar, G., Rao, V. S., Kumar, M. V. S., Shailaja, D. (2006), Synthesis and amphiphilic 4-vinyl pyridine and N-vinyl pyrrolidone copolymer beads, *Journal of Applied Polymer Science*, **102**, 192–7.

Sautter, A., Thalacker, C., Wurthner, F. (2001), Control of liquid crystallinity of diazadibenzoperylene dyes by covalent and hydrogen-bonded attachment of mesogens, *Angewandte Chemie – International Edition*, **40**, 4425.

Scortanu, E., Prisacariu, C., Hitruc, E. G., Caraculacu, A. A. (2006), New heterocyclic polyurethane-ureas based on 4,4′-dibenzyl diisocyanate, Part II: Influence of pyridine units on the mechanical properties, *High Performance Polymers*, **18**, 877–91.

Shih, W. (2008), Biomolecular assembly – dynamic DNA, *Nature Materials*, **7**, 98–100.

Sriram, V., Sundar, S., Dattathereyan, A., Radhakrishnan, G. (2005), Synthesis and characterization of cationomeric AB cross-linked polyurethane polymers based on different chain extenders, *Reactive & Functional Polymers*, **64**, 25–34.

Stewart, D., Imrie, C. T. (1995), Supramolecular side-chain liquid-crystal polymers, Part I: Thermal-behavior of blends of a low molar-mass mesogenic acid and amorphous polymers, *Journal of Materials Chemistry*, **5**, 223–8.

Ujiie, S., Iimura, K. (1992), Thermal-properties and orientational behavior of a liquid-crystalline ion complex polymer, *Macromolecules*, **25**, 3174–8.

Vera, F., Almuzara, C., Orera, I., Barbera, J., Orio, L. L., *et al.* (2008), Side-chain supramolecular polymers with induced supramolecular chirality through H-bonding interactions, *Journal of Polymer Science Part A – Polymer Chemistry*, **46**, 5528–41.

Ware, T., Hearon, K., Lonnecker, A., Wooley, K. L., Maitland, D. J., Voit, W. (2012), Triple-shape memory polymers based on self-complementary hydrogen bonding, *Macromolecules*, **45**(2), 1062–9.

Wilson, A. J. (2007), Non-covalent polymer assembly using arrays of hydrogen-bonds, *Soft Matter*, **3**, 409–25.

Zhang, S., Yu, Z., Govender, T., Luo, H., Li, B. (2008), A novel supramolecular shape memory material based on partial A-CD-PEG inclusion complex, *Polymer*, **49**, 3205–10.

Zheng J., Zhan, C., Qin, J., Zhan, R. (2002), Synthesis and fluorescence of pyridine-n-alkyl carbazole copolymer prepared by oxidative-coupling reaction, *Chemistry Letters*, **12**, 1222–3.

Zhu, Y., Hu, J. L., Liu, Y. J. (2009), Shape memory effect of thermoplastic segmented polyurethanes with self-complementary quadruple hydrogen bonding in soft segments, *European Physical Journal E*, **28**, 3–10.

6
Supramolecular shape memory polymers containing pyridine

DOI: 10.1533/9780857098542.128

Abstract: This chapter introduces the molecular design, structure and morphology of supramolecular shape memory polymers (SMPs) containing pyridine moieties, particularly shape memory polyurethane (SMPU). The structure of SMPU based on N,N-bis(2-hydroxyl ethyl) isonicotinamide (BINA) is discussed and examined by a number of methods: molecular analysis; theoretical analysis using the G03W program; Fourier transform infrared (FT-IR) analysis; and nuclear magnetic resonance (NMR) analysis. From the morphological perspective, the thermal and dynamic mechanical properties are presented, along with a detailed account of the current understanding of the molecular design, structure and morphology of SMPU.

Key words: supramolecular shape memory polyurethane, pyridine, thermal-responsive shape memory effect, shape memory mechanism, N,N-bis(2-hydroxyl ethyl) isonicotinamide.

6.1 Introduction

Over the past decade, research into the molecular design of shape memory polymers (SMPs) has mainly focused on the crystallization and vitrification of the soft phase. Recently, thermal-reversible non-covalent interactions such as strong hydrogen bonding have been proposed for use in SMPs. One use of reversible hydrogen-bonding is in the stabilization of mechanical strain in polymer elastomers: shape recovery is achieved upon heating due to the dissociation of non-covalent bonds. One method relies on hydrogen-bonded semi-interpenetration networks (semi-IPNs) such as polyethylene glycol (PEG), polymethyl methacrylate (PMMA), PEG/P(MMA-co-AA(acrylate acid)) and PEG/poly (MMA-co-VP (N-vinyl-2-pyrrolidone)) (Liu *et al.*, 2004, 2005, 2006).

Li *et al.* (2007) also described a type of supramolecular shape memory elastomer that contains reversibly associating side groups, namely ureidopyridinone (UPy). In addition to its excellent shape memory effects (SMEs) – about 90% strain fixity and 100% strain recovery – a unique feature of this new SMP is the dynamics of its shape-memory response. More recently, Zhu *et al.* (2009) synthesized a novel supramolecular polyurethane for utilization as a SMP, by grafting the UPy unit to the elastic polyurethane. Supramolecular SMPs based on α-cyclodextrin (α-CD) network through the inclusion of either PEG or polycaprolactone (PCL) with α-CD. In this system, the α-CD-PEG inclusion crystallites and α-CD-PCL crystallites serve as a fixing phase, while the naked PEG crystallites and PCL crystallites act as a reversible phase (Luo *et al.*, 2008; Zhang *et al.*, 2008). Fan *et al.* (2009) also prepared another novel supramolecular SMP based on the self-assembly of γ-CD with PEG according to the same principle.

Supramolecular complexes formed with pyridine moieties as hydrogen acceptors have been widely studied (Millaruelo *et al.*, 2006; Meudtner and Hecht, 2008). Many kinds of supramolecular polymer networks, including functional polymers and supramolecular liquid crystal polymers (LCPs), can be achieved through the strong hydrogen bonding between carboxyl or phenolyl and pyridine moieties. For example, Micozzi *et al.* (2005) synthesized conjugated polymers with electro-conductive and photo-conductive properties such as poly(2,6- or 2,5-), while Chang *et al.* (2007) also reported the synthesis of conjugated semi-conductor polyazomethine containing pyridine units. Polyureas incorporating 2,6-pyridine rings have been synthesized and used for the production of light emitting diodes based on terbium immobilized polyurea chelates (Tamami *et al.*, 2005). Furthermore, a large number of polyurethanes and polyurethane-ureas containing pyridine moieties have been synthesized through the use of pyridine-containing chain extenders such as N,N-bis(2-hydroxyl ethyl) isonicotinamide (BINA) (O'Connell *et al.*, 1996; Sriram *et al.*, 2005) and 2,6-diamino pyridine (Scortanu, 2006). In the preparation of hydrogen-bonded LCP complexes, polyurethane containing complementary pyridyl units in the main chain and side chain has also been synthesized from 2,6-bis(hydroxymethyl) pyridine and BINA, respectively (Ambrozic and Zigon, 2005a,b).

It is suggested that the use of non-covalent interactions such as hydrogen bonds for the production of SMPs can offer substantial improvements. Moreover, studies have revealed that the fabrication of SMPs with supramolecular entities is feasible and advantageous, and that supramolecular complexes, particularly those containing pyridine moieties such as pyridine-containing polyurethanes (PUPys) have great potential for use in SMPs. However, the structure and morphology of PUPys have not been widely investigated. This study aims to fill that gap by examining the synthesis of shape memory polyurethane (SMPU) using the pyridine derivative BINA in combination with hexamethylene diisocyanate (HDI). In order to gain a greater understanding of the supramolecular structure and morphology of BINA-based SMPUs (BIN SMPUs), butanediol (BDO) and methylene diphenyl diisocyanate (MDI) are also used to control the pyridine content and the fraction of hard domains. A series of experiments are then described, focusing on:

1. the investigation of supramolecular structure;
2. the effect of pyridine content; and
3. the effect of hard domains.

A number of techniques were employed in these studies, including Fourier transform infrared (FT-IR) spectroscopy, differential scanning calorimetry (DSC), dynamic mechanical analysis (DMA) and wide-angle X-ray diffraction (WAXD).

6.2 Synthesis of shape memory polyurethanes (SMPUs) containing pyridine moieties

In order to investigate the supramolecular structure of BINA-based SMPU, as well as the influence of the pyridine content or BINA content and the hard segment on its morphology and properties, two series of BINA-based SMPUs were investigated. The PUPy-BDO series of SMPU with various BINA contents and the PUPy-MB series of SMPU with various MDI-BDO contents were synthesized in this study. Table 6.1 presents the principal raw materials used in the study, while the chemical compositions of the BIN SMPUs are given in Table 6.2.

For comparison, one SMPU based on PCL and one based on poly(hexylene adipate) glycol (PHAG) were synthesized by bulk polymerization. For the sake of convenience, the PUPy-BDO series of samples are coded as PUPy-BDO??, where '??' is the BINA content, and the PUPy-MB series are coded as PUPy-MB##,

Table 6.1 Materials for synthesis of SMPU containing pyridine moieties

Code	Full name
HDI	1,6-hexandiisocyanate
BINA	N,N-bis(2-hydroxyethyl)isonicotinamide
BDO	1,4-butandiol
MDI	4,4′-diphenylmethane diisocyanate
DMF	N,N-dimethyl formamide

Table 6.2 Chemical composition of samples used in this project

Series	Samples	BINA g	HDI g	BDO g	MDI g	BINA wt%	HDI wt%	MB wt%
PUPy-BDO series	PUPyBDO53	5.24	4.52	0	0	53.7	46.3	0
	PUPyBDO45	4.5	5.22	0.66	0	43.3	50.3	0
	PUPyBDO40	4.0	5.4	1.0	0	38.5	51.9	0
	PUPyBDO30	3.0	5.8	1.6	0	28.8	55.8	0
	PUPyBDO20	2.0	6.22	2.23	0	19.1	59.5	0
	PUPyBDO10	1.0	6.63	2.86	0	9.5	63.2	0
PUPy-MB series	PUPyMDII5	5.24	4.52	0.46	1.29	45.5	39.3	15.2
	PUPyMDI25	5.24	4.52	0.86	2.38	40.3	34.8	24.9
	PUPyMDI35	5.24	4.52	1.39	3.86	34.9	30.1	35.0
	PUPyMDI45	5.24	4.52	2.12	5.87	24.5	25.5	45.0
	PUPyMDI55	5.24	4.52	3.17	8.74	24.2	20.9	55.0

BINA wt.% is the weight fraction of BINA, which is calculated by W_{BINA}/W_{total} *100%
HDI wt.% is the weight fraction of HDI, which is calculated by $W_{HD}I/W_{total}$ *100%
MB wt.% is the weight fraction of MDI + BDO, which is calculated by (W_{MDI} + W_{BDO}/W_{total} *100%

where the '##' is the MDI-BDO content. For example, PUPyBDO45 indicates that the sample contains 45 wt.% BINA; and PUPyMB25 indicates that the sample contains 25 wt.% MDI-BDO. Extra pure grade of 1,6-HDI, 1,4-BDO, 4,4′-MDI and BINA were all purchased from the Aldrich Chemical Company, USA. During the synthesis of PU, they were used without any treatment. High pressure liquid chromatography (HPLC) grade dimethyl formamide (DMF) was also purchased from the Aldrich Chemical Company. DMF is usually dried with 4 Å molecular sieves before use to eliminate any water.

6.2.1 BINA-HDI copolymers (PUPyA)

The preparation of BINA-HDI copolymer was carried out in a 500 ml flask filled with nitrogen and equipped with a mechanical stirrer, a thermometer and a condenser. The process used to synthesize this copolymer is presented in detail in Fig. 6.1. BINA powder (5.24 g) was added to the flask according to the composition, as given in Table 6.2. To dissolve the BINA, 10 mL DMF was added using mechanical stirring. The equal molar HDI (4.19 g) was then added to the flask. The oil temperature was increased to about 70°C in the first two hours. After 30 minutes, 0.02 wt.% of dibutyltin dilaurate (DBTDL) was added to the solution as a catalyst. The reaction started immediately and a significant increase in viscosity

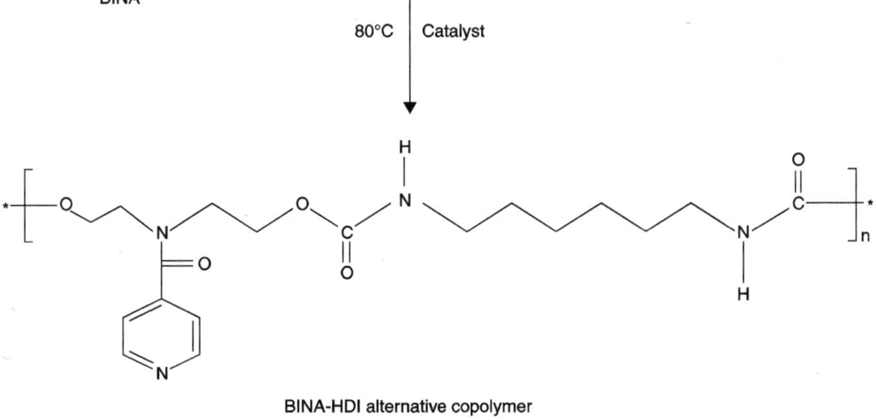

6.1 Process for synthesising a BINA-HDI copolymer.

was observed. At this stage, another 10 mL DMF was added to the solution to control the viscosity during the reaction. After two hours, the reaction temperature was raised to 80°C and the reaction continued for another four hours. Another 8% excess HDI (i.e., 0.334 g) was then added to the solution, and the reaction continued for another 12 hours, until a polymer solution diluted to 10 wt.% was obtained.

6.2.2 BINA-HDI-BDO series (PUPy-BDO series) SMPU

The preparation of the PUPy-BDO series SMPU was carried out in a 500 ml flask filled with nitrogen and equipped with a mechanical stirrer, a thermometer and a condenser. The process used to synthesize this SMPU is presented in detail in Fig. 6.2. BINA powder (4 g) was added to the flask, according to the composition given in Table 6.2. DMF (10 mL) was added in order to dissolve the BINA using mechanical stirring. Followed by the addition of an excess molar of HDI (5.0 g) to the flask, the oil temperature was raised to about 70°C in the first two hours. After 30 minutes, 0.02 wt.% DBTDL was added to the solution as a catalyst. The

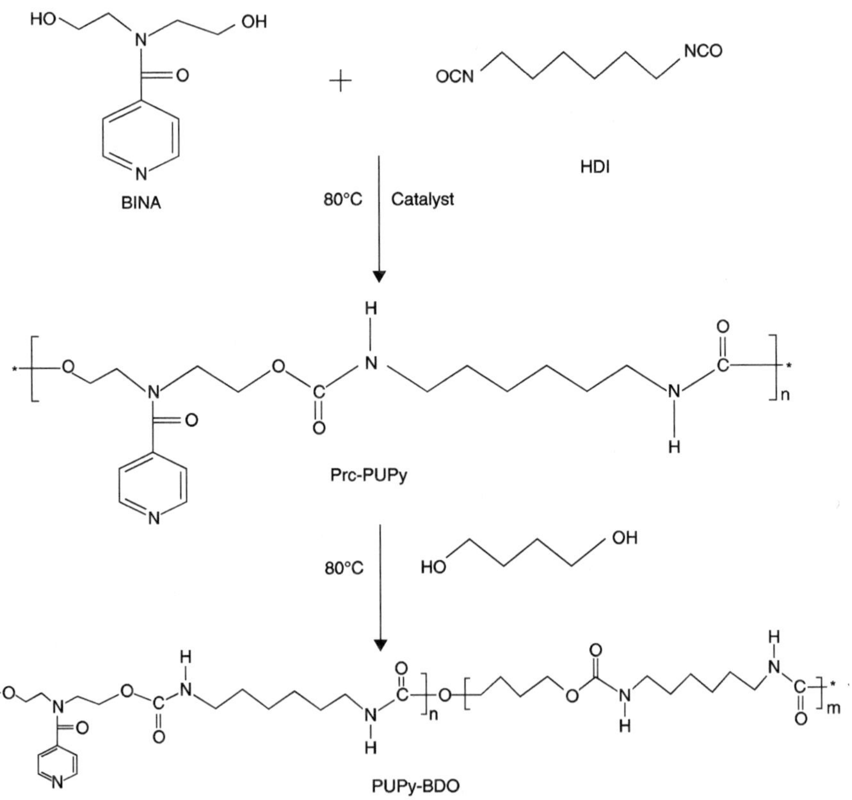

6.2 Process for synthesizing a PUPy-BDO series SMPU.

reaction started immediately and a significant increase in viscosity was observed. At this stage, 10 mL DMF was added to the solution to control the viscosity during the reaction. After two hours, 1.0 g of BDO was added to the solution and the reaction temperature was raised to 80°C for another four hours. Another 8% excess HDI (0.4 g) was then added to the reaction, which then continued for another 12 hours until a polymer solution diluted to 10 wt.% was obtained.

6.2.3 BINA-HDI-BDO-MDI series (PUPy-MB series) SMPU

The preparation of PUPy-MB series SMPU was carried out in a 500 mL flask filled with nitrogen and equipped with a mechanical stirrer, a thermometer and a condenser. The process used to synthesize the PUPy-MB series SMPU is presented in detail in Fig. 6.3. BINA powder (5.24 g) was added to the flask according to the composition, as shown in Table 6.2. DMF (10 mL) was added in order to dissolve the BINA using mechanical stirring. The equal molar of HDI (4.18 g) was then

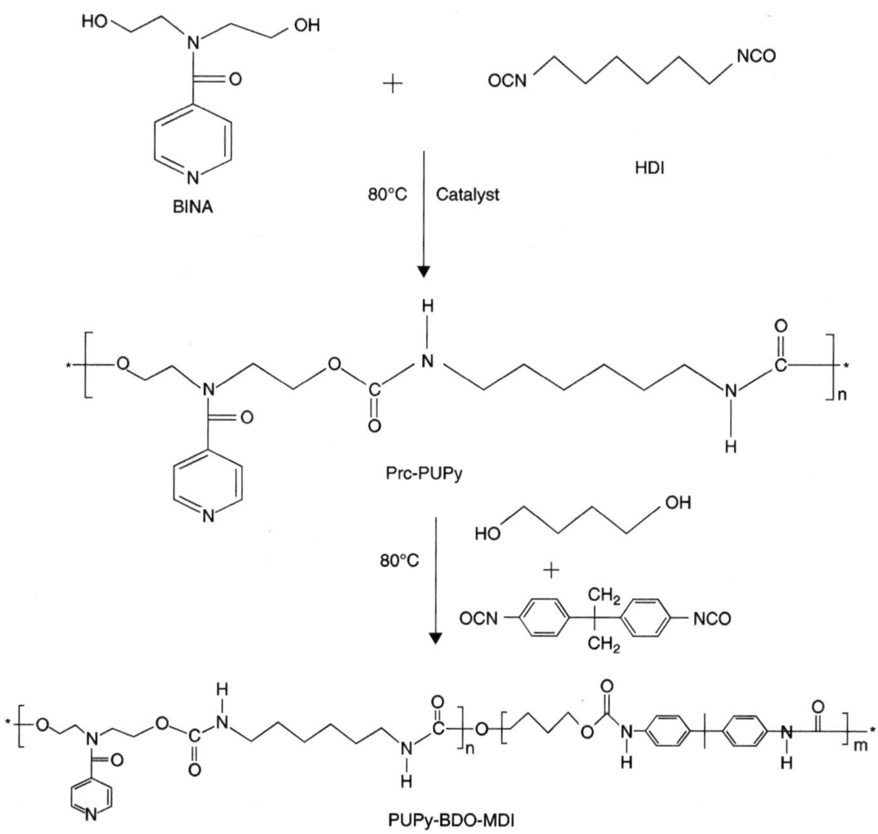

6.3 Process for synthesizing a PUPy-MB series SMPU.

added to the flask. The oil temperature was increased to about 70°C in the first two hours. After 30 minutes, 0.02 wt.% of dibutyltin dilaurate was added to the solution as a catalyst. The reaction started immediately and a significant increase in viscosity was observed. At this stage, 10 mL DMF was added to the solution to control the viscosity during the reaction. After two hours, 1.39 g of BDO and 3.86 g of MDI were added to the reaction. The reaction temperature was raised to 80°C, and the reaction was maintained at this temperature for another four hours. Another 8% excess HDI (0.334 g) was then added to the solution, and the reaction continued for another 12 hours until a polymer solution diluted to 10 wt.% was obtained.

6.3 The molecular structure of BINA-based SMPUs (BIN-SMPUs)

Figure 6.4 presents the possible inter-molecular hydrogen bonding present in BIN-SMPUs. BINA is known to be a hydroxyl end-capped pyridine derivative, and is widely used as the chain extender in polyurethane synthesis (O'Connell *et al.*, 1996; Sriram *et al.*, 2005). By observing the molecular structure of BIN-SMPU, it is clear that the pyridine ring acting as a pendant is attached to the backbone of the polyurethane, as shown in Fig. 6.4. It is well-known that the polar urethane groups have a significant influence on the stabilization or destabilization of polyurethane (Brunette *et al.*, 1982). The urethane groups (–NH–COO–), which have a partially negative charge on the oxygen atom, can interact with either the hydrogen donor or the hydrogen acceptor. Hence, hydrogen bond associations are predicted between the C=O (A1) and N–H (D1) of the urethane groups. These hydrogen bonds are beneficial to the formation of hard segment domains, as shown in Fig. 6.4(a). Similarly, the C=O group (A2) next to the pyridine ring can also form hydrogen bonds with the N–H of the urethane groups, as shown in Fig. 6.4(b).

However, the most favorable hydrogen acceptor in this system is the nitrogen in the pyridine ring (Ambrozic and Zigon, 2005a,b). Figure 6.4(c) shows the formation of strong hydrogen bonds between the nitrogen in the pyridine ring as hydrogen acceptor (A3), and the N–H of the urethane group as hydrogen donor. In addition, the N (A4) next to the C=O also acts as a hydrogen acceptor, forming hydrogen bonds with the N–H in the urethane. Hydrogen bonds are therefore an important factor in the formation of urethane groups.

6.4 Theoretical calculations of BIN-SMPU properties and performance

To estimate the interaction energies of the hydrogen bonds involved, *ab initio* calculations were performed using a truncated molecule containing one BINA unit and one urethane unit. The compounds considered in the theoretical calculations contain the same hydrogen-bonded functional groups as the real

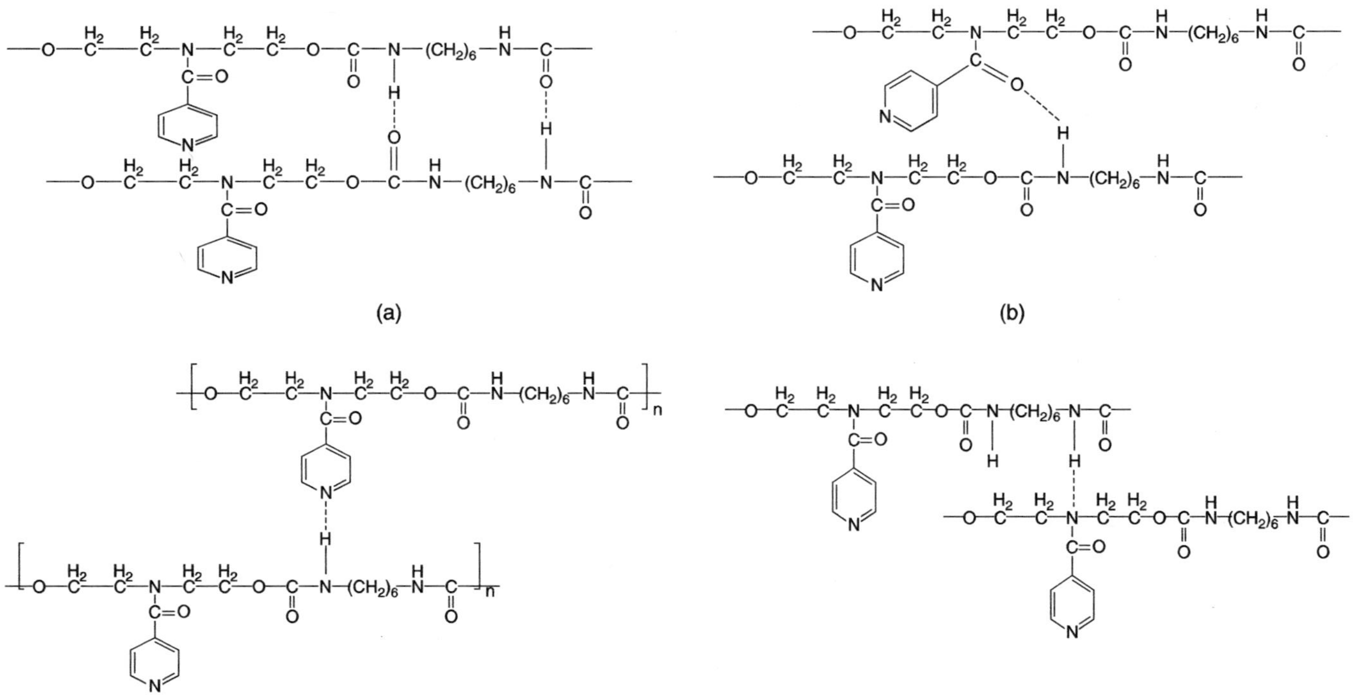

6.4 Inter-molecular hydrogen bonding in the BIN-SMPU.

species used in this experiment. The calculations took into account the hydrogen bonds formed:

1. between D1 and A1 (N–H. . . .O=C–O) marked as a-H;
2. between D1 and A2 (N–H. . . .O=C–Py) marked as b-H;
3. between D1 and A3 (N–H. . . .N–Py) marked as c-H; and
4. between D1 and A4 (N–H. . . .N–C=O) marked as d-H.

Figure 6.5 shows the optimized binding geometry of hydrogen bonds, while the calculated interaction energies of hydrogen bonds are summarized in Table 6.3.

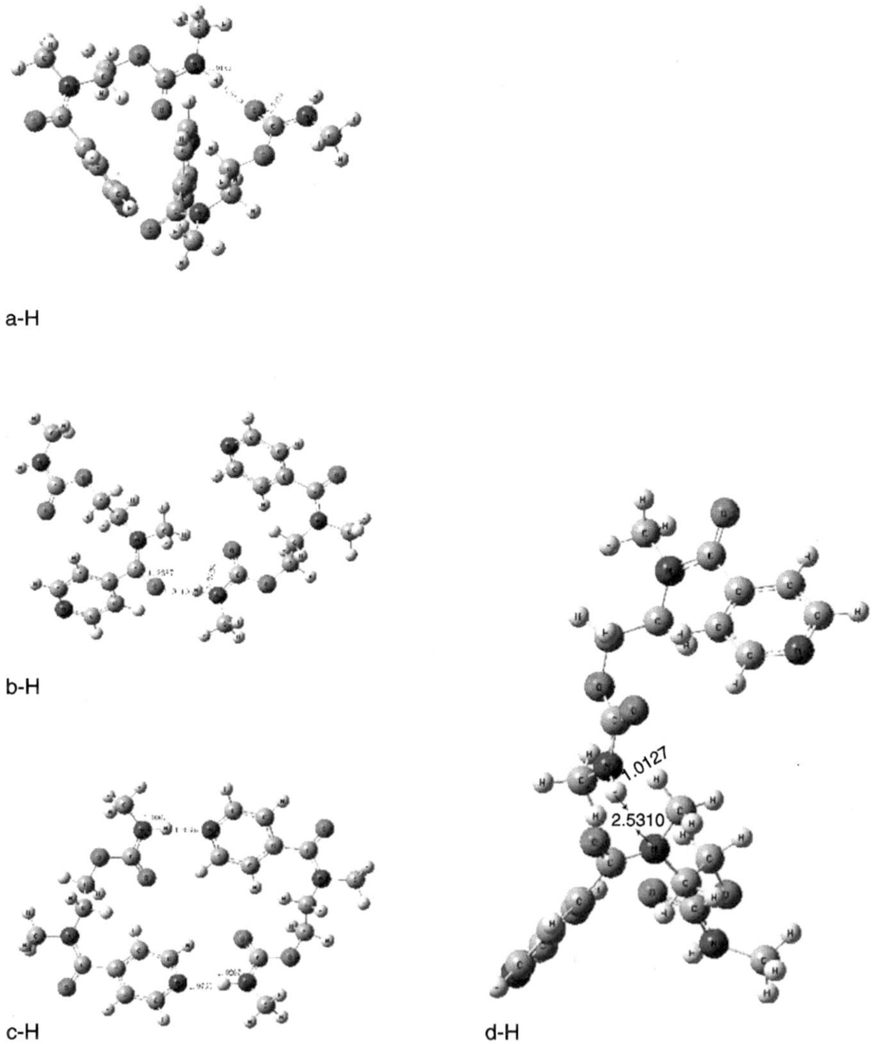

6.5 Optimized binding geometries at the b3lyp/6-31G level using the G03W program.

Table 6.3 Energy of hydrogen bond at the b3lyp/6-31G level using the G03W program

Type	Hydrogen bonding	E_H/kJ,mol^{-1}
a-H	between the urethane group (D1-A1)	−48.47
b-H	between NH of urethane with C=0 of pyridine (D1-A2)	−43.49
c-H	between N–H of urethane with N of pyridine (DI-A?)	−82.49
d-H	between NH of urethane with N beside CO (DI-A4)	−15.45

EH is energy of hydrogen bond; D1 is hydrogen-donor (N–H); and A1, A2, A3, A4 are four kinds of different hydrogen-acceptors

Figure 6.5 shows that the length of the hydrogen bond between hydrogen donor and hydrogen acceptor is 1.91 Å in a-H; 1.90 Å in b-H; 1.98 Å in c-H; and 2.53 Å in d-H. In addition, Table 6.3 shows that the hydrogen bond energy is 48.47 kJ mol^{-1} in a-H; 43.49 kJ mol^{-1} in b-H; and 82.49 kJ mol^{-1} in c-H. These can be classified as medium-strong hydrogen bonds. However, the hydrogen bond (d-H) between the N–H of the urethane and the N next to the carboxyl of the BINA unit is weak, as its bond energy is only 15.45 kJ mol^{-1}. These calculated interaction energies indicate that the hydrogen bonds present in the urethane group and pyridine ring are the most favorable hydrogen bonds in IN-SMPUs, which can therefore be expected to have a hydrogen-bonded supramolecular structure.

6.5 Fourier transform infrared (FT-IR) analysis of BIN-SMPUs

6.5.1 FT-IR spectroscopy study of BIN-SMPUs and polycaprolactone (PCL)-SMPUs

Infrared spectroscopy is a powerful tool used for the study of specific interactions in the hydrogen-bonded system, and the frequency shift has been generally accepted as a measure of the strength of the hydrogen bond (Brunette *et al.*, 1982). Figure 6.6 presents the FT-IR spectra of a BINA-HDI copolymer compared with its monomer, BINA and a typical PCL-based SMPU (Zhu *et al.*, 2007). When BINA is copolymerized with HDI to form the BINA-HDI copolymer, two obvious frequencies appear in the spectrum of the HDI-BINA copolymer at 3317 and 1696 cm^{-1}: these are widely accepted as the stretching vibration of N–H and the stretching vibration of C=O, respectively and represent the formation of the urethane group (–NH–COO–). At the same time, some closed frequencies between

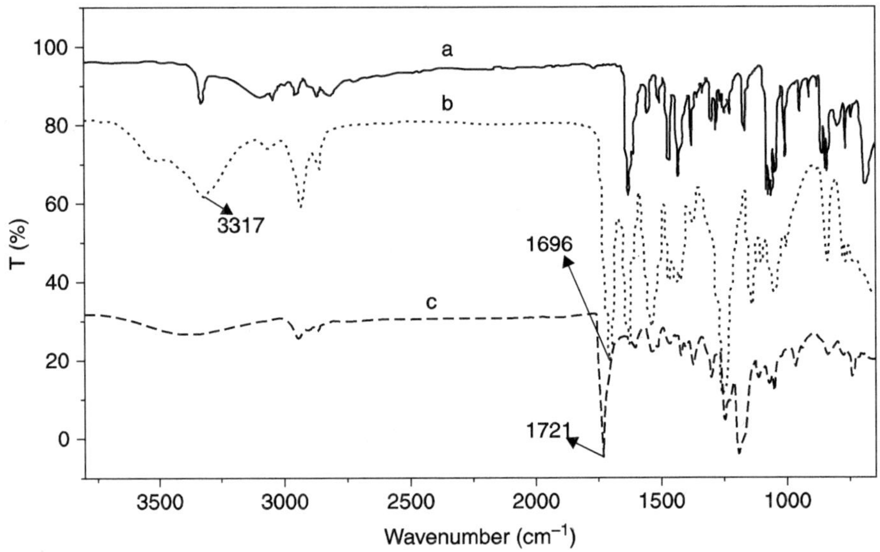

6.6 FT-IR spectra of (b) BINA-HDI copolymer compared with (a) BINA and (c) PCL-based SMPU.

the BINA-HDI copolymer spectrum and the BINA spectrum were observed to shift, as follows: 1623 to 1626 cm^{-1}, 1603 to 1606 cm^{-1}, 1503 to 1500 cm^{-1}; 1464 to 1463 cm^{-1}; 1374 to 1369 cm^{-1}, 1001 to 998 cm^{-1}, and 757 to 758 cm^{-1}. This confirms the formation of urethane groups, and shows that the pyridine ring is attached to the polyurethane chain as a pendant in the BIN-SMPU. In addition, the frequency at about 3059 cm^{-1} should be assigned to the C–H vibration in the pyridine ring, since it appears in the spectrum of both the monomer BINA and the BINA-HDI copolymer, while no absorption is observed at the same frequency in the PCL-based SMPU.

In previous studies (Kim *et al.*, 1996; Chen *et al.*, 2006, 2007a,b), SMPUs are usually designed as segmented block copolymers composed of a hard segment and a soft segment with a long chain. The urethane group fraction is low, and the movement of the polymer chain is principally determined by the ratio of flexible soft segment to rigid hard segment, or by the flexibility of the soft segment. However, in the BIN-SMPUs, the urethane group fraction and pyridine ring fractions are much higher since both HDI and BINA have a lower molecular weight. The hydrogen bonds are also significantly stronger in this system, such that they play an important role in determining the movement of the polymer chain. The density or peak area of frequency at 3317 cm^{-1} for N–H stretching vibration is much higher in the spectrum of the BINA-HDI copolymer in Fig. 6.6. This implies that the urethane group can provide a large number of hydrogen donors for the formation of hydrogen bonds.

6.5.2 FT-IR spectroscopy study of PUPys with various BINA contents

In order to investigate the effect of pyridine content, a series of PUPys with different BINA contents were synthesized by adding BDO to the BINA-HDI prepolymer. Figure 6.7 presents the FT-IR spectra showing the results of this analysis.

Table 6.4 presents the frequencies of PUPys with different BINA contents for some typical vibrations, and shows that as the BDO content increases, the BINA content – which reflects the pyridine ring fraction – decreases. However, the HDI content, reflecting the urethane group fraction, increases as BINA content decreases. Hence, Fig. 6.7(a) shows that the band density at 3313 to 3320 cm^{-1}, representing the N–H stretching vibration of the urethane group, also increases as the BINA content drops. With regard to the band density, decreases are observed at about 3059 cm^{-1} for C–H stretching vibration of the pyridine ring and about 2935 and 2868 cm^{-1} for CH_2 vibration as the BINA content decreases. However, it should be observed that a relatively strong density is still achieved at this frequency in the sample of PUPyBDO10 that contains a small pyridine ring fraction. This implies that the frequency in the range of 3000 to 2800 cm^{-1} results from the stretching vibration of CH_2 from both the BINA unit and the HDI unit.

Figure 6.7(b) also shows that the C=O stretching vibration of the urethane group shifts to a low frequency as the BINA content is decreasing, and a relatively high density is obtained when the BINA content is lower than 30 wt.%, due to the much higher urethane groups fraction. However, the density of the C=O stretching vibration at about 1635 cm^{-1}, as well as the density of the C–N–C stretching vibration at about 1600 cm^{-1}, drops significantly as the BINA content decreases. Finally, density of the C–N–C stretching vibration at about 1600 cm^{-1} tends to be overlapped by the stretching vibration of C=O of urethane group in the sample of PUPyBDO10 due to the low BINA fraction. This again confirms that the frequency at 1600 cm^{-1} should be assigned to the C–N–C stretching vibration of the BINA unit. The density increases at about 1530 cm^{-1} as HDI content increases, suggesting that this frequency should be assigned specifically to the stretching vibration of

Table 6.4 Typical frequencies of PUPys with various BINA contents

Sample	BINA wt%	HDI wt%	NH (cm^{-1})	C=N (cm^{-1})	C=O(Py) (cm^{-1})	C–N amide (cm^{-1})	CH(py) (cm^{-1})
PUPyBDO53	53.7	46.3	3319.5	1695.9	1628.2	1530.4	832.3
PUPyBDO40	43.3	50.3	3316.8	1693.3	1630.8	1530.7	832.7
PUPyBDO30	38.5	51.9	3315.2	1682.0	1632.2	1531.8	833.1
PUPyBDO20	28.8	55.8	3316.3	1681.3	1637.9	1532.6	833.7
PUPyBDO10	19.1	59.5	3312.8	1680.6	–	1532.8	834.5

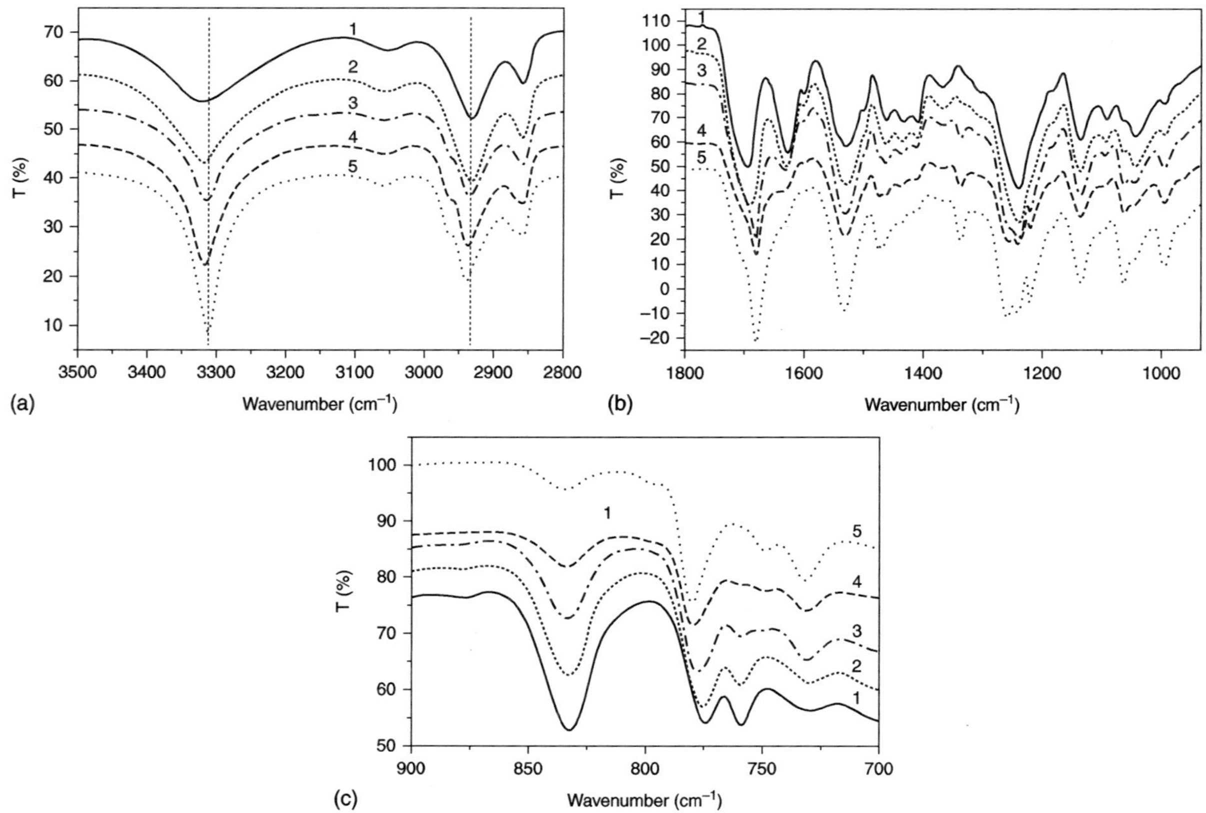

6.7 FT-IR spectra of PUPy with various BINA contents, 1: PUPyBDO53; 2: PUPyBDO40; 3: PUPyBDO30; 4: PUPyBDO20; and 5: PUPyBDO10 at the frequency range of: (a) 3500–2800 cm^{-1}; (b) 1800–900 cm^{-1}; and (c) 900–700 cm^{-1}.

the C–N in amide II in the urethane group. At the same time, the frequencies at 1434 and 1411 cm^{-1} can be assigned to the stretching vibrations of the pyridine ring, because their density is found to decrease as BINA content decreases.

The frequencies at 1474, 1337, 1258, 1241 and 1220 cm^{-1} may result from the urethane groups, as these become clearer as the HDI content increases, particularly in the spectra of samples PUPyBDO30, PUPyBDO20 and PUPyBDO10. Moreover, a reduction in the density of the frequencies at 1094 and 1054 cm^{-1} is observed as BINA content decreases. Figure 6.7(c) shows that the frequencies at 835 and 759 cm^{-1} also drop as the BINA content decreases, and can therefore be assigned to the vibration of the BINA unit. In addition, the pyridine monomer, 4-aminopyridine (DMAP), also shows a strong frequency at 759 cm^{-1}, which is therefore also assigned to the C–H stretching vibration of the pyridine ring.

Table 6.4 shows that as the BINA content decreases, the stretching vibration frequencies of the N–H of the urethane group and of the C=O of the urethane group both shift to lower frequencies. However, the frequencies of the stretching vibration of the C=O next to the pyridine ring and the out-of-plane C–H stretching vibration both shift to high frequencies. For example, the 832.3 cm^{-1} vibration frequency of the C–H in the pyridine ring in PUPyBDO53 shifts to 832.7 cm^{-1} in PUPyBDO40; to 833.1 cm^{-1} in PUPyBDO30; to 833.7 cm^{-1} in PUPyBDO20; and to 834.5 cm^{-1} in PUPyBDO10. Similarly, another vibration frequency of the C–H in the pyridine ring at 774 cm^{-1} in PUPyBDO53 shifts to 775 cm^{-1} in PUPyBDO40; to 777 cm^{-1} in PUPyBDO30; to 779 cm^{-1} in PUPyBDO20; and to 780 cm^{-1} in PUPyBDO53. It has previously been reported that the hydrogen-bonded N–H vibration and C=O vibrations appeared at a lower frequency (Luo *et al.*, 1997; Yen *et al.*, 1999). These results therefore indicate that as the BINA content is dropping, the strength of the hydrogen bonding present in the urethane group is increasing, while the strength of hydrogen bonding present in the pyridine ring is decreasing.

6.5.3 FT-IR analysis for pure-PUPyA, protonized-PUPyA and cross-linked-PUPyA

The nitrogen atom of the pyridine ring has a strong tendency to accept free H^+ or to form a hydrogen bond with –COOH. The pyridine ring is therefore easily protonized by the H^+ of an acid such as H_2SO_4 or linked physically by the agent with two or more carboxyl groups such as oxalic acid. FT-IR spectra of the BINA-HDI copolymer (temporarily referred to here as PUPyA), protonized-PUPyA and cross-linked-PUPyA are presented in Fig. 6.8. The data show that the frequencies at 1600, 1502 and 992 cm^{-1} in the spectrum of PUPyA tend to disappear in the protonized-PUPyA. A theoretical and experimental FT-IR spectroscopy study of isonicotinamide suggests that frequencies at 1600 and 1502 cm^{-1} can be assigned to the stretching vibration of C–N–C in the pyridine ring, and the frequency at 992 cm^{-1} is the result of the ring breathing vibration (Bakiler *et al.*, 2007). This

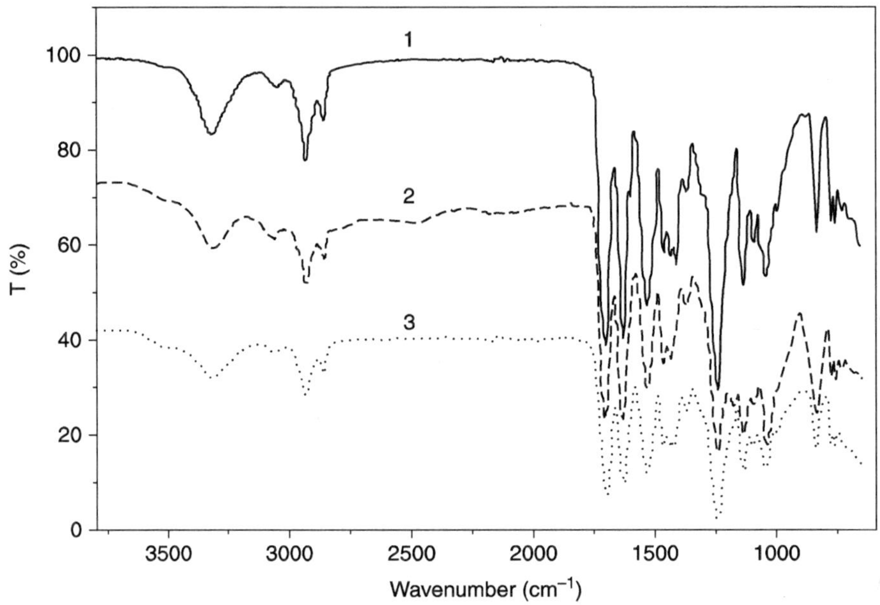

6.8 Spectra of (1) pure-PUPyA compared with (2) protonized PUPyA and (3) cross-linked-PUPyA.

means that the electric structure of the pyridine ring is influenced by the protonization reaction with H_2SO_4. Frequency shifts are also observed between the pure-PUPyA, protonized-PUPyA and cross-linked-PUPyA, as follows: 3314 cm^{-1} in pure-PUPyA shifts to 3320 cm^{-1} in protonized-PUPyA and 3317 cm^{-1} in cross-linked-PUPyA; 3050 to 3061 and 3059 cm^{-1}; and 1697 to 1702 and 1694 cm^{-1}.

The frequency at 3314 cm^{-1} is assigned to the N–H stretching vibration of urethane groups; the frequency at 3050 cm^{-1} to the C–H stretching vibration of the pyridine ring; and the frequency at 1697 cm^{-1} to the C=O stretching vibration of urethane groups. This means that the stretching vibration frequency of the N–H of the urethane group and the stretching vibration frequency of the C–H of the pyridine ring both shift to higher frequencies in the protonized-PUPyA and in the cross-linked-PUPyA. The stretching vibration frequency of the C=O of the urethane group shifts to a higher frequency in the protonized-PUPyA but to a lower frequency in the physically cross-linked-PUPyA. These results indicate that the strength of the hydrogen bonds present in the N–H and pyridine ring decreases after protonization by H_2SO_4 and physical linking by oxalic acid. The strength of the hydrogen bonds in the C=O of the urethane group also decreases after protonization, but actually increases after cross-linking. This again confirms that the pyridine ring is the key hydrogen acceptor for the formation of hydrogen bonds in PUPys.

6.6 Thermal properties of BIN-SMPUs

Traditional polyurethanes are composed of alternating soft and hard segments. The soft segments are usually long-chain polyols such as polyether glycols and polyester glycols, while the hard segments are usually the reaction products of diisocyanates and short-chain extenders. Due to the thermodynamic incompatibility between the soft and hard segments, micro-phase separation occurs, resulting in the formation of hard segment-rich and soft segment-rich matrices. The phase separation is also influenced by a number of factors such as chemical composition, sequence length of the hard segment and hydrogen bonding. It is generally accepted that the hydrogen bonding present in polyurethane is closely related to the phase separation. The thermal properties of a polymer are crucial, since they provide information about the structure and morphology of the polymer. In BIN-SMPUs, there are strong hydrogen bonds between the urethane groups and between the urethane group and the pyridine ring; these strong hydrogen bonds can lead to differences in the thermal properties of BIN-SMPUs when compared to traditional polyurethanes. The following sections will provide a detailed account of the various thermal properties of BIN-SMPUs.

6.6.1 Effect of pyridine ring protonization

Figure 6.9 presents the DSC curves of pure PUPyA compared with protonized-PUPyA and cross-linked-PUPyA, and Table 6.5 summarizes their DSC results.

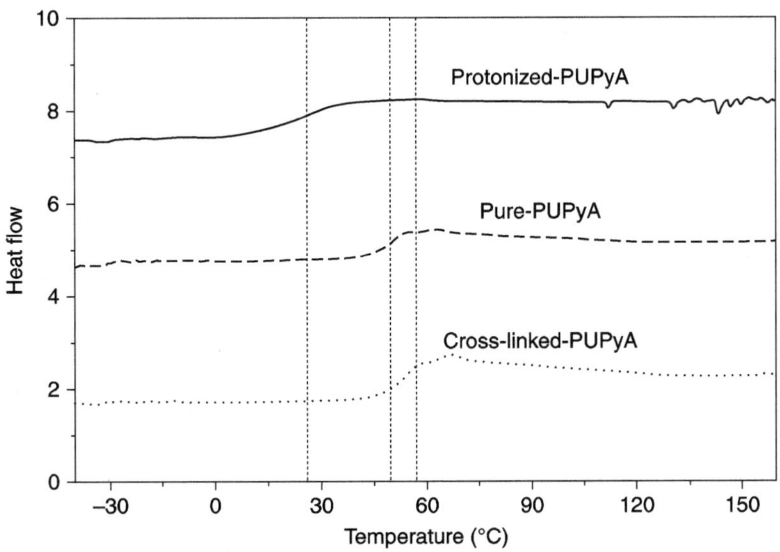

6.9 DSC curves of pure-PUPyA compared with protonized-PUPyA and cross-linked-PUPyA.

Table 6.5 DSC results of pure-PUPyA compared with protonized-PUPyA and cross-linked-PUPyA

Samples	T_g (°C)	ΔC_p (J/g)
Pure-PUPyA	48.50	0.451
Protonized-PUPyA	25.51	0.397
Cross-linked-PUPyA	54.09	0.424

The conditions were maintained for all three samples: drying and annealing were carried out in an oven at 100°C. It is clear that the glass transition temperature (T_g) of the soft phase in the pure-PUPyA is 48.50°C; but it decreases to only 25.51°C in the protonized-PUPyA, and increases to 54.09°C in the cross-linked-PUPyA. This suggests that the inter-molecular interaction of PUPyA is weakened when the pyridine ring is protonized by H^+, but becomes stronger when the pyridine ring is physically cross-linked by the oxalic acid. As previously stated, the movement of the polymer chain in SMPU is usually determined by the ratio of flexible soft segment to rigid hard segment, but the stronger hydrogen bonds and increased urethane group and pyridine ring fractions in this system mean that the hydrogen bonding of PUPy plays an important role in the movement of the polymer chain: namely, the T_g of the soft phase is controlled by the hydrogen bonding of polyurethane, particularly in the pyridine ring.

6.6.2 Effect of BINA content

Figures 6.10 to 6.12 present the first DSC cooling curves, the first DSC heating curves and the second DSC heating curves of PUPys with different BINA contents, respectively. In Fig. 6.10, an obvious re-crystalline peak appears at 103.3°C on the cooling curve of PUPyBDO20 and at 131.8°C on the cooling curve of PUPyBDO10. However, PUPyBDO53, PUPyBDO40 and PUPyBDO30 only show a slight glass transition at a lower temperature range with a scan rate of 10°C/min. In the first heating curves, as shown in Fig. 6.11, the crystal melting peak can still be measured clearly in the PUPyBDO10, PUPyBDO20 and PUPyBDO30 samples, and even in PUPyBDO40, since all four recrystallize after a long period at room temperature.

Interestingly, in addition to the lower temperature glass transition upon cooling, the PUPyBDO53 also shows a clear glass transition with a high specific heat capacity (Cp) at a higher temperature range in the first heating curves. This implies that the PUPyBDO53 has two amorphous phases. However, this exothermic process, which occurred in PUPyBDO53, PUPyBDO40 and PUPyBDO30, disappears in the DSC trace in the second scan, as shown in Fig. 6.12. Robert and Stuart (1973) suggested that the three characteristic endothermic transitions

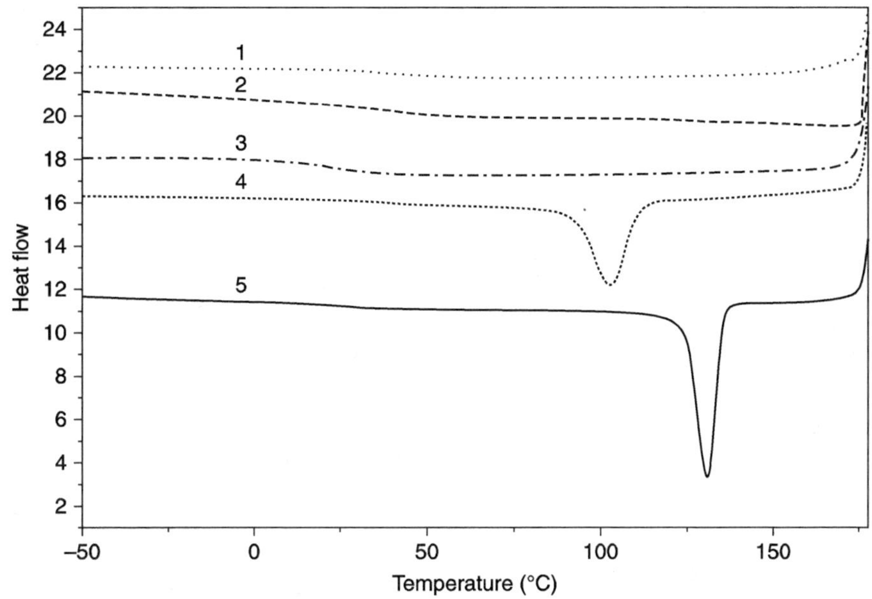

6.10 The first DSC cooling curves of PUPys with various BINA contents: (1) PUPyBDO53; (2) PUPyBDO40; (3) PUPyBDO30; (4) PUPyBDO20; and (5) PUPyBDO10.

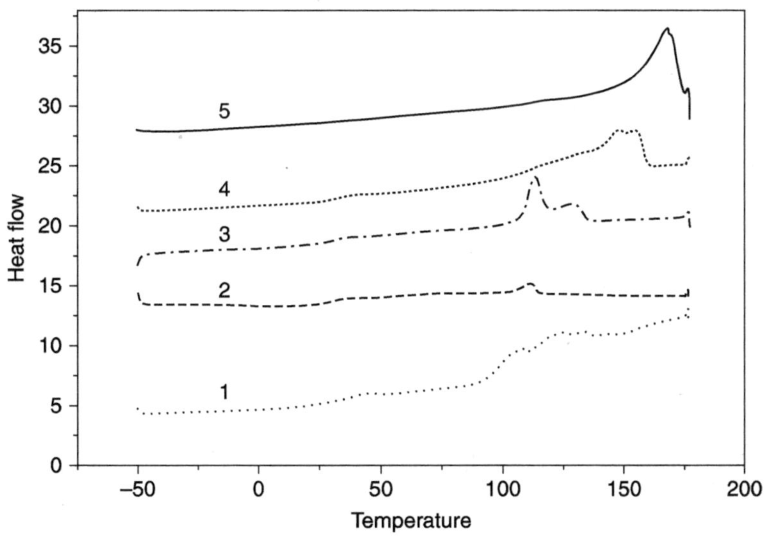

6.11 The first DSC heating curves of PUPys with various BINA contents: (1) PUPyBDO53; (2) PUPyBDO40; (3) PUPyBDO30; (4) PUPyBDO20; and (5) PUPyBDO10.

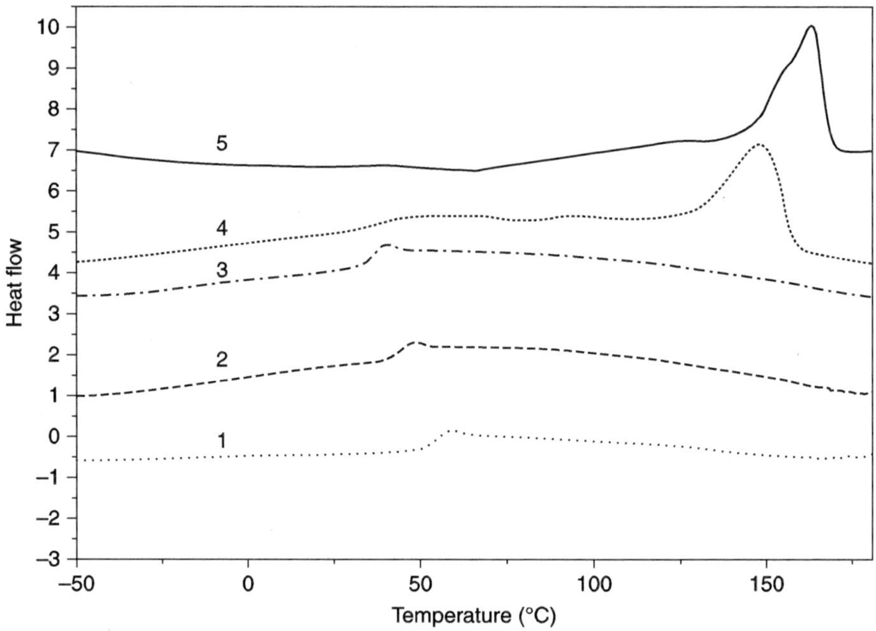

6.12 The second DSC heating curves of PUPys with various BINA contents: (1) PUPyBDO53; (2) PUPyBDO40; (3) PUPyBDO30; (4) PUPyBDO20; and (5) PUPyBDO10.

observed in MDI-BDO based polyurethane at temperatures of 60 to 80°C, 120 to 180°C and above 200°C were correlated with the short-range, long-range and microcrystalline ordering of hard segment units, respectively. The high temperature transition observed here should therefore result from the long-range disordered hard segment units formed by the strong hydrogen bonding between the urethane groups (Korley *et al.*, 2006).

Figure 6.12 also shows that the T_g of PUPyBDO53 is 55.4°C, while it is 43.8°C for PUPyBDO40 and 33.7°C for PUPyBDO30, i.e. that the T_g of the soft phase decreases as the BINA content decreases. In addition, it can be observed that both PUPyBDO20 and PUPyBDO10 show a significant crystal melting peak at 148.2 and 163.7°C, respectively. However, no clear glass transition is determined at lower temperatures. Moreover, the melting temperature (T_m) and critical temperature (T_c) tend to be higher in the samples with higher HDI-BDO content or lower BINA content, the melting temperature is influenced by the BINA unit and that part of the hard phase results from the HDI-BDO unit in BIN-SMPUs. This can be explained by the fact that the pyridine ring in the BINA unit, which attaches to the polyurethane chain as a pendant, prevents the aggregation of the polymer chain and hence the formation of crystals. However, more HDI-BDO units, which are reported to form crystals easily, are aggregated as the BINA fraction is small in both PUPyBDO20 and PUPyBDO10. In addition to the amorphous soft phase based on the BINA unit, the urethane unit of HDI-BINA and the HDI-BDO unit also tend to form an amorphous hard phase or even a semi-crystalline hard phase when the HDI-BDO unit fraction is high.

6.6.3 Effect of the rigid hard segment

As mentioned previously, the MDI-formed hard segment of polyurethanes contains not only inter-molecular hydrogen bonds, but also dipole–dipole interaction and induced dipole–dipole interaction (Lee *et al.*, 2001). If the HDI of the BINA-HDI-BDO system is partially replaced by MDI to form a BINA-HDI-BDO-MDI copolymer, a more rigid MDI-BDO segment will be formed (Chen *et al.*, 2007a). Moreover, the aggregation of relatively regular HDI-BDO units will be interrupted by the irregular MDI-BDO unit. This type of colpolymer consequently displays different thermal properties to those observed in the PUPy-BDO series of PUPys.

Figures 6.13 to 6.15 respectively present the first DSC heating curves, the second DSC heating curves and the first DSC cooling curves of PUPys with different BINA contents. As shown in Figs 6.13 and 6.15, no crystal melting peaks or recrystalline peaks can be observed on the first heating curves or first cooling curves. However, a clear glass transition can be determined at the lower temperature range on the second heating curves in all the PUPy-MB series of

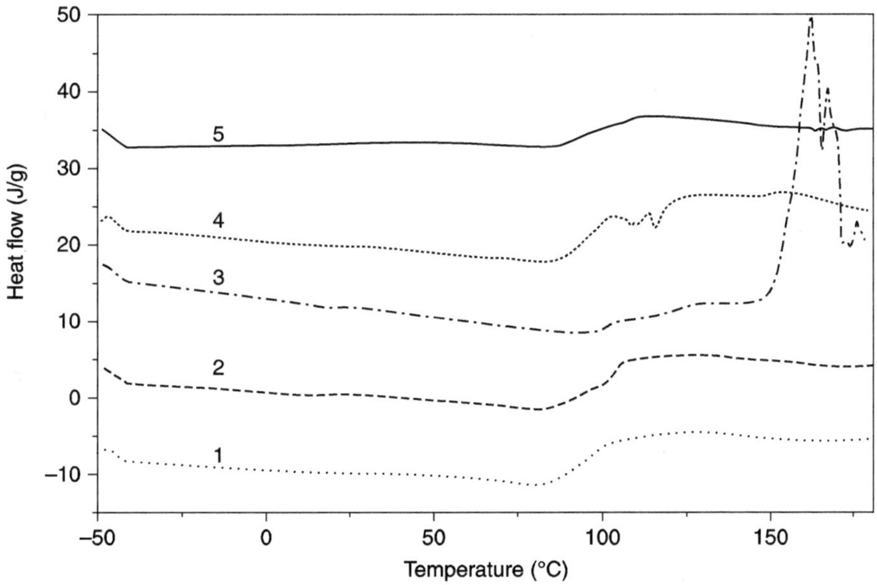

6.13 The first DSC heating curves of PUPys with various MDI-BDO contents: (1) PUPyMB15; (2) PUPyMB25; (3) PUPyMB35; (4) PUPyMB45; and (5) PUPyMB55.

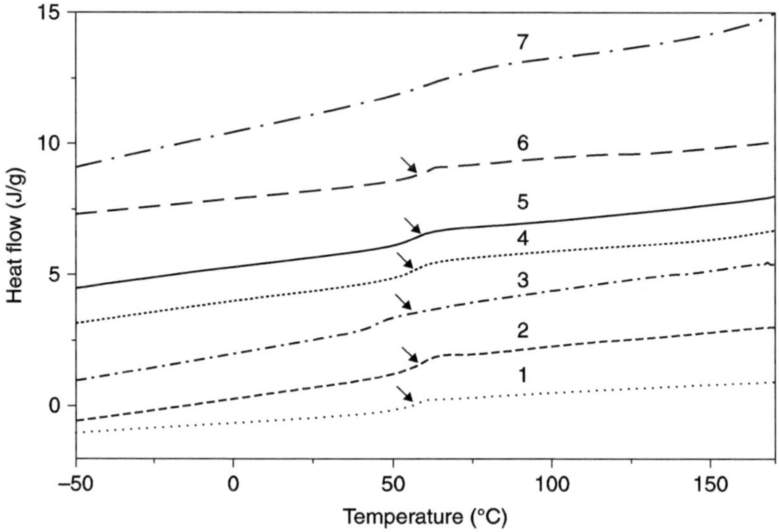

6.14 The second DSC heating curves of PUPys with various MDI-BDO contents: (1) PUPyMB0; (2) PUPyMB15; (3) PUPyMB20; (4) PUPyMB25; (5) PUPyMB30; (6) PUPyMB35; and (7) PUPyMB45.

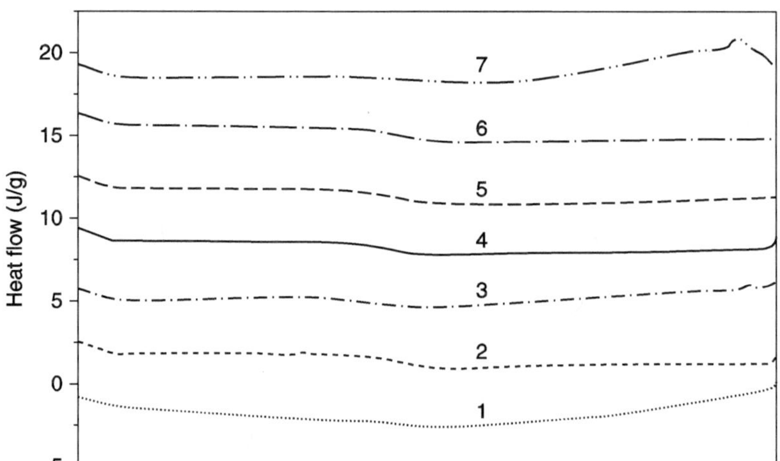

6.15 The first DSC cooling curves of PUPy-MB series with various MDI-BDO contents: (1) PUPyMB0; (2) PUPyMB15; (3) PUPyMB20; (4) PUPyMB25; (5) PUPyMB30; (6) PUPyMB35; and (7) PUPyMB45.

PUPys. Moreover, as the MDI-BDO content increases, the T_g of the soft phase increases slightly. The increased MDI-BDO hard segment not only improves the interaction of the polymer, but also improves the rigidity of the polyurethane chain (Chen *et al.*, 2007a). In the PUPy-MB series, there is a drop in the hydrogen bonding present in the pyridine ring as the BINA content decreases, confirming that the increased MDI-BDO segment plays the main role in the movement of the polymer chain.

Similar to what was observed with the PUPy-BDO series, the phase transition with high Cp appears on the first heating curves in the PUPy-MB series, as shown in Fig. 6.13. All phase transitions disappear in the DSC trace when the specimen undergoes the second scan. This again confirms that the hard phase results from the disordered urethane units by means of hydrogen bonds, and that the two amorphous phases exist in the PUPy-MB series. As is the case in the PUPy-BDO series, the addition of MDI-BDO units promotes the formation of hard phase.

6.7 Wide angle X-ray diffraction (WAXD) studies of BIN-SMPUs

Figure 6.16 presents WAXD traces of PUPys with various BINA contents. Similar to the results obtained using the DSC method, no crystalline peaks can be observed in the WAXD traces of PUPys with more than 35 wt.% BINA content in the PUPy-BDO series. When the BINA content drops to below 35 wt.%, some clear peaks appear at about $2\theta = 13.48°$ and $24.13°$ in the PUPyBDO25, and at

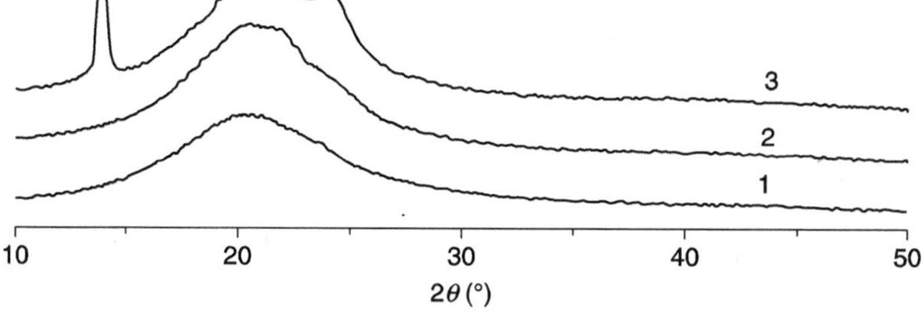

6.16 WAXD traces of PUPys with various BINA contents: (1) PUPyBDO53; (2) PUPyBDO45; (3) PUPyBDO35; (4) PUPyBDO25; and (5) PUPyBDO15.

$2\theta = 13.55°$ and $24.32°$ in the PUPyBDO15. In PUPyBDO35, diffraction peaks were observed at $2\theta = 13.86°$ and $24.00°$, but also at about $2\theta = 20.30°$ and $21.97°$. In PUPyBDO45, peaks at about $2\theta = 20.30°$ and $21.85°$ are difficult to observe, while in the case of the PUPyBDO53, there is only a broad non-crystalline peak at about $2\theta = 20.30°$.

It has been established that semi-crystalline hard phases are formed in the PUPy-BDO series with more than 30 wt.% BINA, and the soft phase is actually amorphous in all the PUPys samples. This suggests that the diffraction peaks at about $2\theta = 13.48°$ and $24.13°$ are to be assigned to the crystals of the HDI-BDO hard segment. As the BINA content decreases, first from 35 to 25 wt.% and then to 15 wt.%, the diffraction peak (2θ) shifts from 24.00° to 24.13° and then to 24.32°. This indicates that the distance d ($d = n\lambda/2\sin\theta$) between the parallel planes in the HDI-BDO crystallites decreases with the increase in HDI-BDO content, implying that the hard phase increases in stability in the PUPys with higher HDI-BDO content. As the BINA content drops from 53 to 15 wt.%, the soft phase tends to change gradually from a continuous amorphous phase above 30 wt.% to a droplet-like dispersion phase below 15 wt.%. However, the hard phase moves from a droplet-like dispersion amorphous phase above 40 wt.% to a continuous crystalline phase below 30 wt.%. This pattern of change is similar to that observed in the micro-structures in blends of PCL with styrene butadiene styrene (SBS) (Wang *et al.*, 2006).

It has also been reported that the HDI-BDO hard segment does not form a crystalline phase until the content of HDI-BDO is above a critical value: above 28.6 wt.% in the PLA3000-based polyurethane; above 19.4 wt.% in the PLA5000-based polyurethane; and above 10.7 wt.% in the PLA10000-based polyurethane (Wang *et al.*, 2006). This means that the lower limit of HDI-BDO content required for the formation of the crystalline hard phase is higher if the soft segment is shorter in the polyurethane containing the HDI-BDO hard segments. Since the BINA is a small molecular monomer in the present BINA-HDI-BDO system, the HDI-BDO segment cannot form the crystalline hard phase until the HDI-BDO content is above 65 wt.%, i.e. the BINA content is less than 35%. This provides further confirmation that the soft phase is determined by the BINA unit, while the hard phase is influenced by the HDI-BDO unit.

Figure 6.17 shows the WAXD traces of PUPys with different MDI-BDO contents. In the PUPy-MB series, no diffraction peaks are observed in any specimen. This means that both the soft phase and the hard phase present in all the PUPy-MB series are amorphous. This result is consistent with the DSC testing results, and confirms that the aggregation of the relatively regular HDI-BDO unit is interrupted by the irregular MDI-BDO unit present in the PUPy-MB series. Furthermore, the non-crystalline peak becomes broader in PUPys with higher MDI-BDO content, with the peak width at half height related to the domain size (Zhou and Wang, 2003). The enlargement of the peak width therefore reflects the growth of the hard domain with the increase in MDI-BDO content, showing that the addition of the MDI-BDO segment reinforces the hard domains in the PUPy-MB.

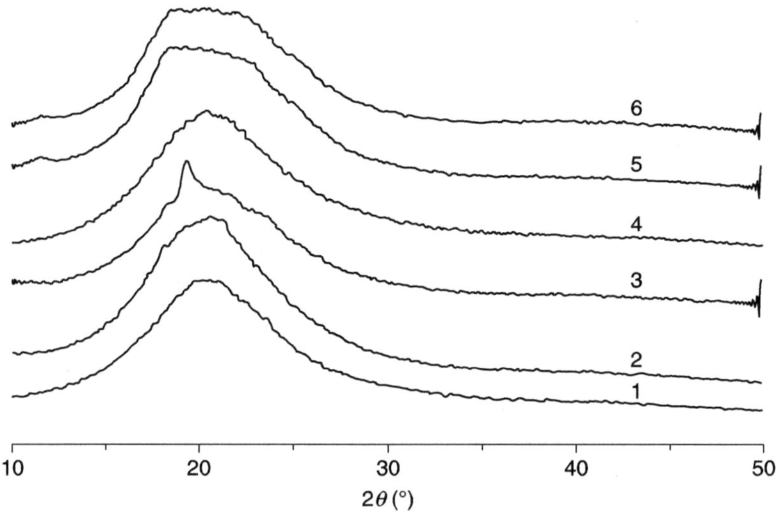

6.17 WAXD traces of PUPys with various MDI-BDO contents: (1) PUPyMB0; (2) PUPyMB15; (3) PUPyMB25; (4) PUPyMB35; (5) PUPyMB45; and (6) PUPyMB55.

6.8 Dynamic mechanical properties of BIN-SMPUs

Deformation of polymers at high temperature is usually an easy process due to the lower rubbery modulus (E_r), which makes the orientation of the polymer more feasible. However, this orientation will be partly relaxed before the structure is frozen during the subsequent cooling cycle. However, deformation at low temperature is usually more difficult due to the polymer's higher glassy modulus (E_g). However, chain orientation will remain at a higher degree as the relaxation process is slowed down. It is therefore expected that a high E_g will provide the material with good shape fixity during simultaneous cooling and unloading, whereas a high E_r will provide the material with a good rate of elastic recovery at higher temperature (Ratna and Karger-Kocsis, 2008).

6.8.1 Comparison of BIN-SMPUs with other SMPUs

Figure 6.18 presents the dynamic mechanical analysis (DMA) curves of T_m-type SMPU, synthesized from a poly(hexylene adipate) glycol (PHAG) soft segment and MDI-BDO hard segment (Chen *et al.*, 2007b). In addition to the crystal melting transition at about 55.6°C, a glass transition can also be observed on the tanδ curves at about −30°C. Moreover, the storage modulus decreases from 4.0 GPa at −50°C to 0.57 GPa at 0°C, i.e. a seven-fold decrease in modulus resulting from the glass transition. This decrease continues to only 0.041 GPa at 80°C (i.e. to 14 times below the modulus at −50°C) after the crystal melting transition.

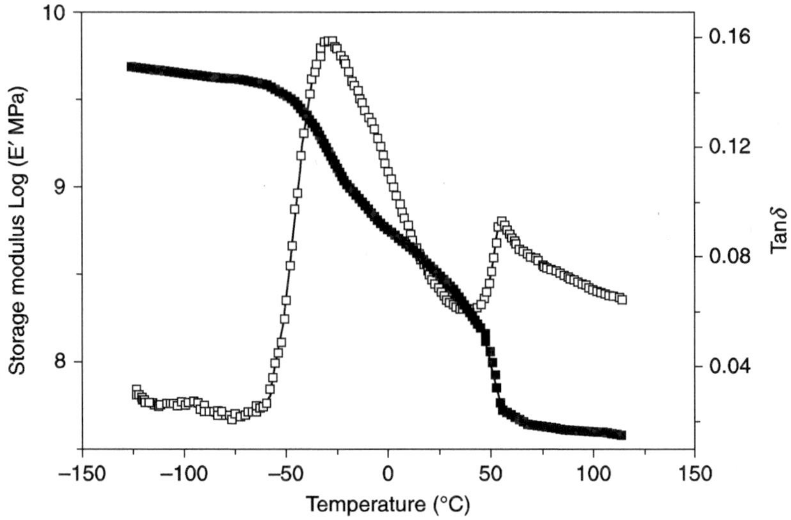

6.18 Storage modulus (E′) and tanδ of the T_m-type SMPU as a function of temperature.

The T_g-type SMPU was synthesized from a PHAG soft segment and MDI-BDO hard segment (Chen *et al.*, 2007b). In Fig. 6.19, no crystals are formed in the soft segment domains of T_g-type SMPU, and the dynamic mechanical properties are mainly influenced by the glass transition of the soft segment. During the glass transition process, the modulus ratio (E_g/E_r) is above 25. However, it should be noted that the T_g-type SMPU usually has a wide transition range. In order to clearly demonstrate the SME, its T_g was designed to be above room temperature.

For the BIN-SMPU, it is impossible to measure the crystals at a lower temperature range on the tanδ curves. However, strong hydrogen bonds are formed in both the soft phase and hard phase. The movement of the polymer chain is significantly influenced by the inter-molecular force and particularly by the hydrogen bonding present in the pyridine ring, as described previously. The dynamic mechanical properties of BIN-SMPUs are therefore clearly different from the dynamic mechanical properties of traditional T_g-type and T_m-type SMPU. Figure 6.20 clearly shows that there is a large decrease in modulus at 40 to 60°C, while below 20°C a plateau can be observed, with an E_g of 4.0 GPa. This means that the BIN-SMPU is rigid at room temperature compared to the T_m-type and T_g-type SMPUs. However, the E_r is only 6.5 MPa at above 80°C, meaning that BIN-SMPU is very soft after the glass transition. This softness is a result of hydrogen bonding, which causes a substantial inter-molecular force with a high glassy storage modulus macroscopically at lower temperatures. However, after the dissociation of the hydrogen bonds, the weak inter-molecular force results in substantial movement of the polymer chain. In addition, the pyridine ring is

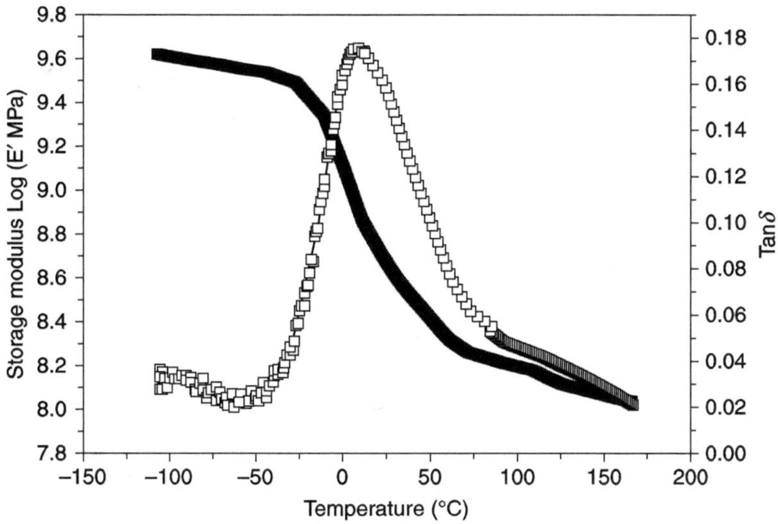

6.19 Storage modulus (E′) and tanδ of the T_g-type SMPU as a function of temperature.

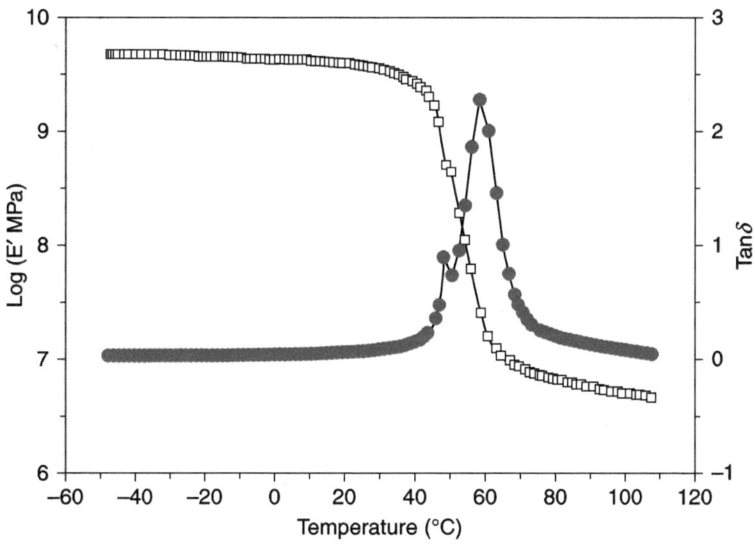

6.20 Storage modulus (E′) and tanδ of the BIN-SMPU as a function of temperature.

attached to the polymer chain as a pendant, suggesting that the pendant pyridine ring also affects the stiffness of polyurethane. A large modulus ratio ($E_g/E_r > 615$) is therefore observed in the BIN-SMPU. It has also been observed that the transition range of BIN-SMPU is much narrower than that of T_g-type SMPU, meaning that fast strain recovery is expected in the former.

Another unique dynamic mechanical property of BIN-SMPU is its high maximum tanδ, as shown in Fig. 6.20. For example, the tanδ at 58°C is about 2.3 in the HDI-BINA copolymer, while the tanδ of T_m-type SMPU is only 0.09 during the crystal melting phase transition, and only 0.16 during the glass transition. In addition, the maximum tanδ of the traditional T_g-type SMPU is also only 0.18. This means that BIN-SMPU loses a large amount of energy during the glass transition process; it should therefore be suitable for use as a damping material.

6.8.2 Effect of pyridine ring protonization

To further investigate the role of the pyridine ring in the dynamic mechanical properties of BIN-SMPUs, a study was carried out on the protonization of the pyridine ring using the HDI-BINA copolymer (PUPyA). Figures 6.21 and 6.22 present the DMA curves and tanδ curves of pure PUPyA compared with protonized-PUPyA and cross-linked-PUPyA. The FT-IR analysis presented above has already shown that the high fraction of strong inter-molecular hydrogen bonds is associated at a lower temperature range in the PUPyA. In Fig. 6.21 it can be observed that the pure-PUPyA shows a high E_g, which remains constant at above 4.3 GPa at temperatures below 20°C. However, when the pyridine ring is protonized by the H^+, the E_g decreases significantly from 4.2 GPa at −80°C to only 2.0 GPa at 20°C in the protonized-PUPyA. This decrease in modulus is observed particularly between −80 to 50°C. In contrast, when the pyridine ring is linked by oxalic acid, no obvious decrease in modulus can be observed during the glassy state.

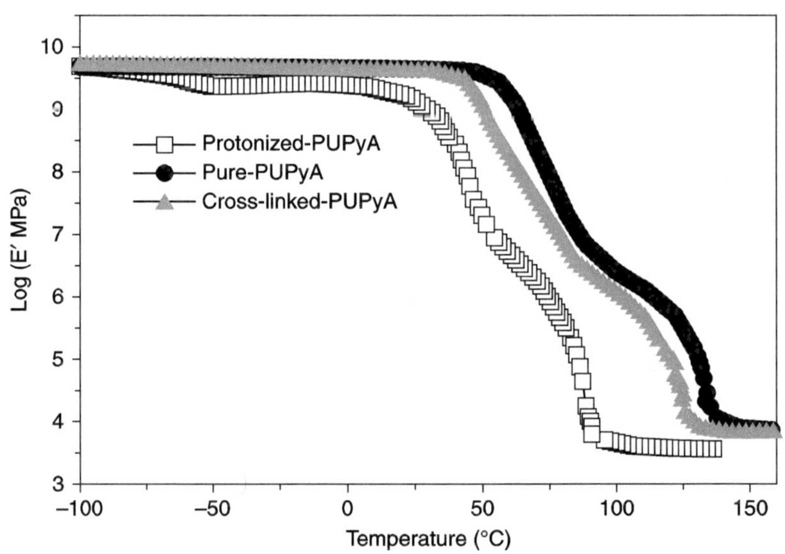

6.21 Storage modulus (E′) of pure-PUPy compared with protonized-PUPy and cross-linked-PUPy as a function of temperature.

These results confirm that the hydrogen bonding present in the pyridine ring plays an important role in inter-molecular interaction in the PUPy. This can be explained by the fact that the inter-molecular forces are strong in the pure-PUPyA and in the cross-linked-PUPyA, but the hydrogen bonding present in the pyridine ring is easily destroyed by the H^+ in the protonized-PUPyA. As the temperature is raised to above −60°C, the protonized pyridine ring starts to shake, and the E_g also decreases due to the drop in inter-molecular force. As a result, Fig. 6.22 also shows that a β transition related to the relaxation of the protonized pyridine ring in the side-chain appears at about −52°C on the tanδ curve. This provides further evidence that the pyridine ring influences the movement of the polymer chain through the formation of strong inter-molecular hydrogen bonds.

Similar to other SMPs, the storage modulus of PUPyA also undergoes two stages of decrease upon heating. For the pure-PUPyA, the first decrease in modulus occurs at 44 to 95°C, while the second decrease occurs at 115 to 129°C, suggesting that phase separation occurs in the PUPyA. The modulus ratio (E_g/E_r) reaches 400 within the first stage with a high E_r (>2.5 MPa) of 100. However, in the protonized-PUPyA, the decrease in modulus occurs at 23 to 54°C (first stage) and 78 to 94°C (second stage). The protonized-PUPyA is consequently soft at room temperature but becomes very viscous at above 100°C. Although protonized-PUPyA is easy to

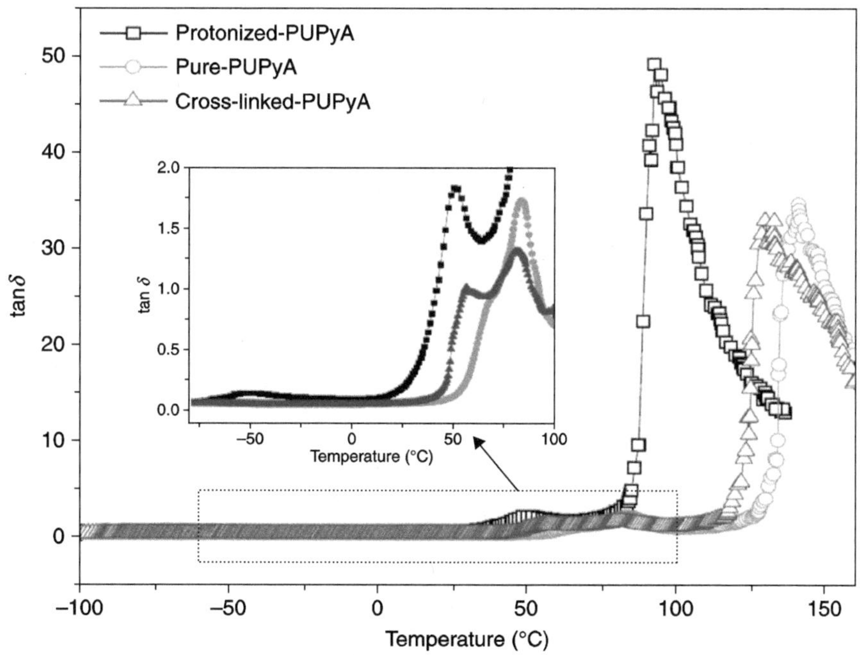

6.22 Tanδ curves of pure-PUPyA compared with protonized-PUPyA and cross-linked-PUPyA as a function of temperature.

manufacture into multiple different shapes at high temperature, its recovery stress is still weak, so it demonstrates little to no SMEs (while pure-PUPyA can be expected to show good SMEs).

In the cross-linked-PUPyA, it can unexpectedly be observed that the decrease in modulus starts at a lower temperature, but the E_r is still much higher than that of the protonized-PUPyA. This is because the strong hydrogen bond that existed between the N–H of the urethane group and the :N< of the pyridine ring is partly replaced by the hydrogen bond between the carboxyl group and the :N< of the pyridine ring. Moreover, the inter-molecular distance of the polymer chain increases as oxalic acid is added. The inter-molecular force in the cross-linked-PUPyA should therefore be weaker than that in the pure-PUPyA, but still stronger than that in the protonized-PUPyA. The peak temperatures of tanδ, as shown in Fig. 6.22, which represent the T_g of the soft phase and the hard phase, are both observed to shift to a lower temperature when PUPyA is protonized or linked. For example, in the protonized-PUPyA, the second peak temperature of tanδ appears below 90°C. This means that the hydrogen bonds in both the soft phase and hard domains will be dissociated when the temperature is raised to above 90°C. Thereafter, the polymer chain can move freely without net points. The pre-deformed strain may also be damaged completely without strain recovery at that temperature.

6.8.3 Effect of BINA content

Figure 6.23 presents the dynamic mechanical properties of PUPys with different BINA contents. Similarly to what is observed for other kinds of SMPs, Fig. 6.23(a) shows that in the glassy state, i.e. at below 47°C, all the BIN-SMPU samples have a high E_g followed by a significant decrease in modulus during the glass transition process. The polymer then enters into a rubber state with a lower E_r, particularly when the BINA content is high: for instance, if the BINA content is above 40 wt.%

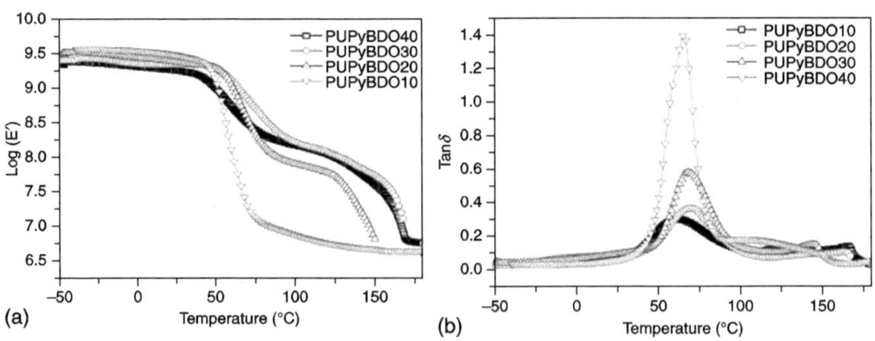

6.23 DMA curves of (a) modulus (b) tanδ of PUPy samples with various BINA contents.

by weight, the modulus ratio is also high, at $E_g/E_r > 300$ in the PUPyBDO40. However, as the BINA content drops to below 30 wt.%, the E_r increases significantly. As a result, the modulus ratio is reduced: it is only 100 in the PUPyBDO10. These results show that when the BINA content drops to below 30 wt.%, the BINA-HDI-BDO polyurethane tends to form a clearer micro-phase separation structure consisting of both a soft phase and a hard phase. Moreover, the soft phase mainly results from the congregation of the HDI-BINA unit while the HDI-BDO unit mainly forms the hard phase. This observation is consistent with the results of the DSC testing presented previously.

In addition, Fig. 6.23(b) also demonstrates that two transitions appear on the $\tan\delta$ curve of the PUPyBDO30, PUPyBDO20 and PUPyBDO10. These transitions may be attributed to the T_g transition of the soft phase and the T_m or T_g transition of the hard phase. Interestingly, as the BINA content drops, the maximum $\tan\delta$ decreases from 1.40 in the PUPyBDO40 sample to only 0.31 in the PUPyBDO10 sample, suggesting that the BINA-HDI formed from the amorphous phase significantly affects energy loss in BIN-SMPUs, i.e. the higher the BINA content, the greater the $\tan\delta$. This is due to the fact that higher fractions of hydrogen bonds formed in the pyridine ring cause increased energy losses. The influence of hydrogen bonding on the T_g of the soft phase in the BINA-HDI-BDO system is therefore confirmed.

6.8.4 Effect of rigid hard domains

In order to further confirm the prediction of the morphology of BIN-SMPUs, according to the analysis of the BINA-HDI-BDO system described above, this study also investigated the DMA properties of the BINA-HDI-BDO-MDI copolymer. Figure 6.24 shows the DMA curves of PUPys with different MDI-BDO contents, and Table 6.6 summarizes the results. It is well-known that MDI-BDO is the main component of the hard segment, which constitutes the

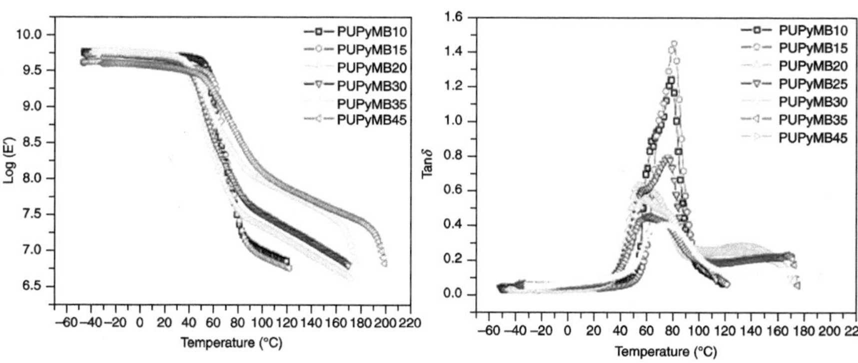

6.24 DMA properties of PUPys with various MDI-BDO contents.

Table 6.6 DMA results of PUPys with various MDI-BDO contents

Samples	BINA wt.%	E_g at 20°C	E_r at 90°C	E_g/E_r	tanδ
PUPyMB0	53.7	3.77 + E09	6.88 + E06	548	2.28
PUPyMB10	48.3	5.14 + E09	1.21 + E07	425	1.23
PUPyMB15	45.5	3.66 + E09	1.06 + E07	345	0.64
PUPyMB25	40.3	4.56 + E09	2.54 + E07	180	0.56
PUPyMB30	37.8	4.67 + E09	4.01 + E07	117	0.44
PUPyMB35	34.9	5.51 + E09	1.14 + E08	46	0.42
PUPyMB45	29.5	3.45 + E09	1.79 + E08	19	0.40

physical net points of SMPU as a result of strong interactions (Lee *et al.*, 2001). These hard segments will probably lead to phase separation or hard domain formation, and will disperse among the soft segment phase as a filler to improve the mechanical properties of polyurethane. Figure 6.24 shows that as the MDI-BDO content increases, the E_r increases while the tanδ decreases. Table 6.6 also shows that the E_r increases with the MDI-BDO content. As a result, the E_g/E_r decreases significantly from 425 in the PUPyMB10 to only 19 in the PUPyMB45, indicating that the addition of the MDI-BDO segment reinforces the hard domains of PUPys. It can also be observed that the tanδ decreases significantly as the MDI-BDO content increases and the BINA content decreases.

This tendency is similar to what was observed in the PUPy-BDO series, again confirming that the hydrogen bonding present in the pyridine ring does affect energy loss in BIN-SMPUs. Based on the investigation of the DMA properties of the PUPy-BDO and the PUPy-MB series, it can be concluded that the dissociation of the hydrogen bonds present in the pyridine ring contributes to the chain movement at glass transition. However, the hydrogen bonds present in the urethane group, which have a higher dissociation temperature, act as physical net points. At the same time, if the dissociation of the hydrogen bonds in the pyridine ring is sharp, the modulus of the polyurethane will drop significantly between the glass state and the elastic state. A large modulus ratio can thus be observed in these supramolecular polyurethane systems. Moreover, the modulus ratio increases as the BINA content increases and as the urethane group content decreases in both the PUPy-BDO and PUPy-MB series of PUPys, i.e. it basically increases as the pyridine ring content decreases. It is expected that PUPys with higher pyridine content should display good SMEs.

6.9 Molecular model and morphology of BIN-SMPUs

The molecular structure of BIN-SMPUs can be illustrated by the molecular model shown in Fig. 6.25. At the molecular level, two kinds of strong hydrogen bonds are present in the polymer chain of BIN-SMPUs. One type is the 'lower dissociation temperature (T_l)' hydrogen bonds present in the pyridine ring, while

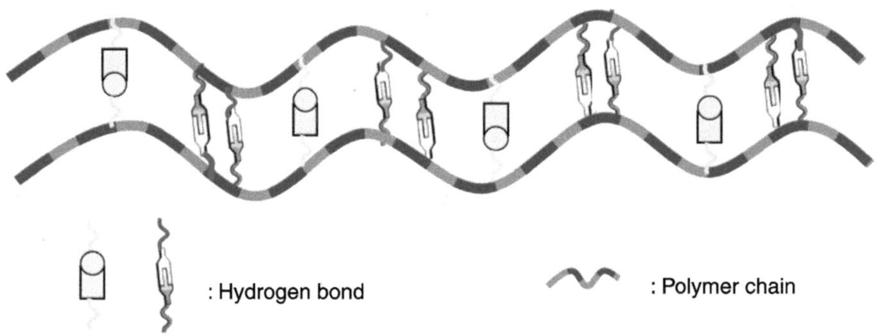

6.25 Molecular model of hydrogen-bonded BIN-SMPUs.

the other is the 'higher dissociation temperature (T_h)' hydrogen bonds present in the urethane groups. At temperatures above T_h, all hydrogen bonds are dissociated. During the cooling process, the higher dissociation temperature hydrogen bond associates, and results in a large fraction of physical net points in the polymer chain. As the temperature continues to cool to below T_l, the T_l hydrogen bonds will also associate, forming the strong inter-molecular force which will limit the movement of the polymer chain. However, the hydrogen bonds present in the pyridine ring are dissociated upon heating. When the temperature is raised to above T_r ($T_l < T_r < T_h$), the deformed chain is then released, whereas the T_h hydrogen bonds still provide the polymer with an elastic network.

With regard to the morphology of BIN-SMPUs, the T_h hydrogen bonds promote the aggregation of the urethane units, and also form the hard phase when the urethane unit fraction is high. At the same time, the T_l hydrogen bonds will promote the formation of the amorphous soft phase, causing phase separation to occur in the BIN-SMPUs. As the BINA content drops, the soft phase tends to change gradually from a continuous amorphous phase to a droplet-like dispersion phase, while the hard phase changes from a droplet-like dispersion amorphous phase to a continuous crystalline phase. This change in microstructure is illustrated in Fig. 6.26.

For example, in the PUPy-BDO series of BIN-SMPUs, the polymer is mainly composed of an amorphous soft phase, while the hard phase is in the form of a droplet-like dispersion amorphous phase in the PUPyBD053. In PUPyBDO45 and PUPyBDO40, as the BINA content drops, or the HDI-BDO content increases, the amorphous hard phase fraction increases. When the BINA content drops to below 35 wt.%, and the HDI-BDO content increases to above 65 wt.%, the semi-crystalline hard phase is partly formed in PUPyBDO35. The hard phase continues to transform into a large crystalline phase in PUPyBDO30. However, the amorphous soft phase fraction decreases as the BINA content decreases. Finally, PUPy with very low BINA content forms a semi-crystalline continuous hard phase dispersed with a low fraction of droplet-like amorphous soft phase.

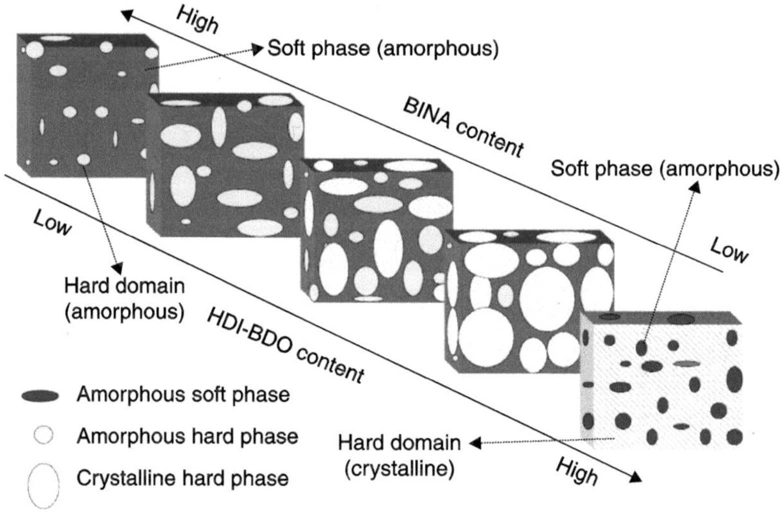

6.26 Illustration of morphology of BINA-based polyurethane.

6.10 Summary

In this chapter, a series of BIN-SMPU supramolecular networks were synthesized from the pyridine derivative and diisocyanate. The structure and morphology of BIN-SMPUs were investigated using a number of techniques, including FT-IR, DSC, WAXD and DMA, and the conclusions drawn can be summarized as follows:

1. Strong hydrogen bonds are formed in both the urethane group and the pyridine ring. As the BINA content drops, the hydrogen bond fraction in the urethane groups increases, while the hydrogen bond fraction in the pyridine ring decreases. Moreover, the strength of the hydrogen bonds increases in the urethane group but decreases in the pyridine ring.
2. DSC results show that an amorphous soft phase is present in PUPys, and that the T_g of the soft phase is controlled by the pyridine ring via hydrogen bonding. A semi-crystalline hard phase tends to form in the PUPy-BDO series PUPys when the BINA content drops to below 30 wt.%, while PUPys with more than 30 wt.% BINA content mainly form an amorphous hard phase. Furthermore, the addition of an MDI-BDO unit can promote the formation of the amorphous hard phase in the PUPy-MB series.
3. DMA testing confirms that phase separation structure does occur in the PUPys. The rubber modulus increases, and the modulus ratio (E_g/E_r) drops, as the BINA content decreases, and as the MDI-BDO content increases. In addition, tanδ drops as the pyridine content decreases in both the PUPy-BDO and PUPy-MB series.

4. An investigation into protonization using the DMA technique confirms that the pyridine ring influences the movement of the polymer chain through the formation of strong inter-molecular hydrogen bonds. The glassy modulus results from the association of a large hydrogen bond fraction, while the decrease in modulus is dependent on the dissociation of the hydrogen bonds present in the pyridine ring.
5. Finally, it is proposed that as the BINA content drops from 53 to 10 wt.%, the soft phase tends to change gradually from a continuous amorphous phase to a droplet-like dispersion phase. However, the hard phase changes from a droplet-like dispersion amorphous phase to a continuous semi-crystalline phase if no irregular unit such as MDI is present.

6.11 References

Ambrozic, G., Zigon, M. (2005a), Hydrogen-bonded liquid-crystalline polyurethane complexes with 4-dodecyloxybenzoic acid, *Acta Chimica Slovenica*, **52**, 207–14.

Ambrozic, G., Zigon, M. (2005b), Hydrogen-bonded polyurethane complexes based on 4-alkoxybenzoic acids as the low molar mass components, *Polymer International*, **54**, 606–13.

Bakiler, M., Bolukbasi, O., Yilmaz, A. (2007), An experimental and theoretical study of vibrational spectra of picolinamide, nicotinamide, and isonicotinamide, *Journal of Molecular Structure*, **826**, 6–16.

Brunette, C. M., Hsu, S. L., Macknight, W. J. (1982), Hydrogen-bonding properties of hard-segment model compounds in polyurethane block copolymers, *Macromolecules*, **15**, 71–7.

Chang, J., Han, J., An, J., Chan, I., Yu, Y.-J., Jin, J.-I. (2007), Photoluminescence properties of poly [2-(5′-cyano-5′-methyl-hexyloxy)-1,4-phenylene] and its copolymers with pyridine comonomer units, *Journal of the Korean Physical Society*, **51**, 1993–8.

Chen, S. J., Cao, Q., Liu, P. S. (2006), Studies on morphology and properties of phma liquified mdi/bdo shape memory polyurethanes (PU), *Acta Polymerica Sinica*, **1**, 1–5.

Chen, S. J., Hu, J. L., Liu, Y. Q., Liem, H. M., Zhu, Y., Liu, Y. J. (2007a), Effect of SSL and HSC on morphology and properties of PHA based-SMPU synthesized by bulk polymerization method, *Journal of Polymer Science Part B – Polymer Physics*, **45**, 444–54.

Chen, S. J., Hu, J. L., Liu, Y. Q., Liem, H. M., Zhu, Y., Meng, Q. H. (2007b), Effect of molecular weight on shape memory behavior in polyurethane films, *Polymer International*, **56**, 1128–34.

Fan, M. M., Yu, Z. J., Luo, H. Y., Zhang, S., Li, B. J. (2009), Supramolecular network based on the self-assembly of gamma-cyclodextrin with poly(ethylene glycol) and its shape memory effect, *Macromolecular Rapid Communications*, **30**, 89–903.

Kim, B. K., Lee, S. Y., Xu, M. (1996), Polyurethane having shape memory effect, *Polymer*, **37**, 5781–93.

Korley, L. T. J., Pate, B. D., Thomas, E. L., Hammond, P. T. (2006), Effect of the degree of soft and hard segment ordering on the morphology and mechanical behavior of semicrystalline segmented polyurethanes, *Polymer*, **47**, 3073–82.

Lee, B. S., Chun, B. C., Chung, Y.-C., Sul, K. I., Cho, J. W. (2001), Structure and thermo-mechanical properties of polyurethane block copolymers with shape memory effect, *Macromolecules*, **34**, 6431–7.

Li, J. H., Viveros, J. A., Wrue, M. H., Anthamatten, M. (2007), Shape-memory effects in polymer networks containing reversibly associating side-groups, *Advanced Materials*, **19**, 2851–4.

Liu, G. Q., Ding, X. B., Cao, Y. P., Zheng, Z. H., Peng, Y. X. (2004), Shape memory of hydrogen-bonded polymer network/poly(ethylene glycol) complexes, *Macromolecules*, **37**, 2228–32.

Liu, G. Q., Ding, X. B., Cao, Y. P., Zheng, Z. H., Peng, Y. X. (2005), Novel shape-memory polymer with two transition temperatures, *Macromolecular Rapid Communications*, **26**, 649–52.

Liu, G. Q., Guan, C. L., Xia, H. S., Guo, F. Q., Ding, X. B., Peng, Y. X. (2006), Novel shape-memory polymer based on hydrogen bonding, *Macromolecular Rapid Communications*, **27**, 1100–4.

Luo, H. Y., Liu, Y., Yu, Z. J., Zhang, S., Li, B. J. (2008), Novel biodegradable shape memory material based on partial inclusion complex formation between alpha-cyclodextrin and poly(epsilon-caprolactone), *Biomacromolecules*, **9**, 2573–7.

Luo, N., Wang, D. N., Ying, S. K. (1997), Hydrogen-bonding properties of segmented polyether poly(urethane urea) copolymer, *Macromolecules*, **30**, 4405–9.

Meudtner, R. M., Hecht, S. (2008), Responsive backbones based on alternating triazole-pyridine/benzene copolymers: From helically folding polymers to metallosupramolecularly cross-linked gels, *Macromolecular Rapid Communications*, **29**, 347–51.

Micozzi, A., Ottaviani, M., Giardina, G., Ricci, A., Pizzoferrato, R., *et al.* (2005), Use of the PD-promoted extended one-pot (EOP) synthetic protocol for the modular construction of poly-(arylene ethyylene) co-polymers $[-Ar-C{\equiv}C-Ar'-C{\equiv}C-]_n$, opto- and electro-responsive materials for advanced technology, *Advanced Synthesis & Catalysis*, **347**, 143–60.

Millaruelo, M., Chinelatto, L. S., Oriol, L., Pinol, M., Serrano, J. L., Tejedor, R. M. (2006), Synthesis and characterization of supramolecular polymeric materials containing azopyridine units, *Macromolecular Chemistry and Physics*, **207**, 2112–20.

O'Connell, E. M., Yang, C. Z., Root, T. W., Cooper, S. L. (1996), Spectroscopic studies of pyridine-containing polyurethanes blended with metal acetates, *Macromolecules*, **29**, 6002–10.

Ratna, D., Karger-Kocsis, J. (2008), Recent advances in shape memory polymers and composites: A review, *Journal of Materials Science*, **43**(1), 254–269.

Robert, W. S., Stuart, L. C. (1973), Thermal anlysis of polyurethane block polymers, *Macromolecules*, **6**, 48–53.

Scortanu, E. (2006), New heterocyclic polyurethane-ureas based on 4,4′-dibenzyl diisocyanate, Part II: Influence of pyridine units on the mechanical properties, *High Performance Polymers*, **18**, 877–91.

Sriram, V., Sundar, S., Aruna, D., Ganga, R. (2005), Synthesis and characterization of cationomeric AB cross-linked polyurethane polymers based on different chain extenders, *Reactive & Functional Polymers*, **64**, 25–34.

Tamami, B., Yeganeh, H., Koohmareh, G. A. (2005), Synthesis and characterization of polyureas derived from 4-aryl-2,6-bis(4-aminophenyl) pyridines and diisocyanates, *Iranian Polymer Journal*, **14**, 785–92.

Wang, W. S., Ping, P., Chen, X. S., Jing, X. B. (2006), Polylactide-based polyurethane and its shape-memory behavior, *European Polymer Journal*, **42**, 1240–9.

Yen, F. S., Lin, L. L., Hong, J. L. (1999), Hydrogen-bond interactions between urethane–urethane and urethane–ester linkages in a liquid crystalline poly(ester-urethane), *Macromolecules*, **32**, 3068–79.

Zhang, S., Yu, Z., Govender, T., Luo, H., Li, B. (2008), A novel supramolecular shape memory material based on partial A-CD-PEG inclusion complex, *Polymer*, **49**, 3205–10.

Zhou, J., Wang., H. J. (2003), The physical meanings of 5 basic parameters for an X-ray diffraction peak and their application, *Chinese Journal of Geochemistry*, **22**, 1–5.

Zhu, Y., Hu, J. L., Yeung, K. W., Choi, K. F., Liu, Y. Q., Liem, H. M. (2007), Effect of cationic group content on shape memory effect in segmented polyurethane cationomer, *Journal of Applied Polymer Science*, **103**, 545–56.

Zhu, Y., Hu, J. L., Liu, Y. J. (2009), Shape memory effect of thermoplastic segmented polyurethanes with self-complementary quadruple hydrogen bonding in soft segments, *European Physical Journal E*, **28**, 3–10.

7

Thermally-induced properties of supramolecular shape memory polymers containing pyridine

DOI: 10.1533/9780857098542.164

Abstract: This chapter presents the thermally-induced shape memory properties of supramolecular shape memory polyurethane (SMPU) containing pyridine moieties, and compares them with those of other types of SMPU. The influence of the BINA content and the MDI-BDO content on the thermally-induced shape memory effects (SMEs) of these SMPUs are discussed, including shape fixity, shape recovery and shape recovery stress, as well as the temperature-dependent strain recovery process. A description and analysis of the mechanism of thermally-induced SMEs is also provided, covering the effect of temperature and pyridine content on the hydrogen bonds present in SMPUs.

Key words: supramolecular shape memory polyurethane, pyridine, thermally-induced shape memory effect, shape memory mechanism.

7.1 Introduction

Shape memory effects (SMEs) are the fundamental properties of shape memory polymers (SMPs). SMEs can be divided into three major categories, according to the stimulus: thermally-induced SMEs (Yuan and Wu, 2009), light-induced SMEs (Lendlein *et al.*, 2005) and electro-active SMEs (Cho *et al.*, 2005). Some studies have also reported magnetically-induced SMEs (Jiang *et al.*, 2003) and water-influenced SMEs (Huang and Yang, 2005), among others. Thermally-induced SMEs are the most common type, as the shape recovery is then able to take place at a specific critical temperature. Since the first discovery of SMPs under the trade name of Polynorborene in 1984, thermally-induced SMPs have been the focus of increasing attention because of their scientific and technological significance. A number of different types of polymers with SMEs have been developed, such as polyacrylate copolymers, segmented polyurethanes and cross-linked polyethylenes. SMPU, composed of a hard segment and a soft segment, has been extensively researched, and will be the focus of this chapter.

SMPU is usually synthesized from long-chain polyols, diisocyanate and short-chain extenders. The polyols that have typically been used to date include polycaprolactone (PCL), polytetramethylene ether glycol (PTMG), polybutylene adipate (PBA), polyhexamethylene adipate (PHA) and polylactic acid (PLA)

(Ratna and Karger-Kocsis, 2008). Some examples used for the hard segments include: methylene diphenyl diisocyanate and butanediol (MDI-BDO); hexamethylene diisocyanate (HDI) with 4,4-dihydroxyl-biphenyl; MDI with 4,4′-bis-(2-hydroxyethoxy) biphenyl (BEBP) or 4,4′-bis-(2-hydroxyhexoxy) biphenyl (BHBP); and MDI with 2,2-dimethoxy-2-phenylacetophenone (DMPA) (Ratna and Karger-Kocsis, 2008). In the common segmented form of SMPU, the phase separation structure consisting of hard domains and soft domains can be microscopically observed. It is generally accepted that hard segments can bind themselves via hydrogen bonds or crystallization, with the result that the SMPU is solid at temperatures below the melting temperature (T_m). The reversible phase transformation of the soft segment, either the amorphous phase or the semi-crystalline phase, is reported to be responsible for the thermally-induced SMEs of SMPU. The SMEs of traditional SMPU can therefore be determined by a number of factors:

1. the molecular weight of the soft segment;
2. the mole ratio between hard and soft segment; and
3. the polymerization process (Lee *et al.*, 2001).

As well as T_g-type SMPs with an amorphous switching segment, and T_m-type SMPs with a crystalline switching segment, supramolecular SMPs based on thermally-reversible non-covalent bonding have recently been introduced. Previous studies have reported that a higher recovery rate can be achieved by the addition of poly(ethylene glycol) (PEG) with a higher molecular weight into the P(AA-co-MMA) networks, and that reversible phase transformation is responsible for the SMEs observed in the hydrogen-bonded PEG/P(AA-co-MMA) semi-interpenetrating polymer network (IPN) (Liu *et al.*, 2004). In polymer networks that contain a small fraction (~2 mol %) of 2-ureido-4[1H]-pyrimidinone (UPy) pendant side-groups, the shape recovery and viscosity of the sample are heavily dependent on the temperature, and the hydrogen bonding dissociation dynamics have a significant effect on mechanical relaxation. Li *et al.* (2007) have suggested that improved SMEs could be achieved through the use of a high fraction of hydrogen-bonded side groups.

Most recently, Zhu *et al.* (2009) and Chen *et al.* (2009) reported another new type of supramolecular polyurethane network that contains a relatively high fraction of UPy groups for shape memory materials (SMMs). However, to date there has been no systematic account of the SMEs of pyridine-containing polyurethanes (PUPys), and previous studies have not explained the relationship between SMEs and non-covalent bonds. This chapter aims to fill that gap with a detailed study of the thermally-induced SMEs of BIN-SMPUs based on an investigation of their morphology and properties. The influence of the BINA content and pyridine ring fractions on shape fixity, shape recovery, shape stability, thermal-dependent shape recovery and shape recovery force is also systematically examined.

7.2 Comparison of BINA-based shape memory polyurethane (BIN-SMPU) with other SMPUs

7.2.1 Thermally-induced shape memory behaviour of BIN-SMPUs

Shape memory polymers (SMPs) are attractive because of their lower cost, excellent process ability and good recovery ability: polymeric shape memory materials can be easily fabricated into a variety of products. Figure 7.1 shows the thermally-induced shape recovery behavior of coils made of BIN-SMPU upon heating, while the opening process of a shape memory flower at a high temperature (80°C) is given in Fig. 7.2. It is well established that SMPs can be fabricated into a temporary shape, which can then be eliminated and the original shape restored at a critical temperature (referred to as T_r). In the study discussed in this chapter, a rectangular film made of BIN-SMPU is fabricated into a temporary coil shape. This coil shape is fixed at a lower temperature (20°C) and is maintained for a long period of time. As the temperature is raised, the coil shape becomes relaxed before the temperature reaches 50°C, and quickly begins to return to the flat rectangular shape above 50°C. At 80°C, the original shape is recovered perfectly.

Another shape memory polymeric product is the shape memory flower shown in Fig. 7.2. The flower has eight petals and is cut from a flat square film made of BIN-SMPU. To begin with, the shape is that of an opening flower, but as the temperature is increased to over T_r, i.e. 80°C, the petals of the flower can be coiled by hand to make a closed flower, as shown in Fig. 7.2(a). Cooling to room temperature fixes the shape as a closed flower; however, when the closed flower is put into an oven at 80°C, the flower can be observed to open its petals one by one. Figure 7.2 shows that the petals are completely opened into a full flower shape within 5 seconds. These two examples therefore show that BIN-SMPUs have typical thermally-induced SMEs.

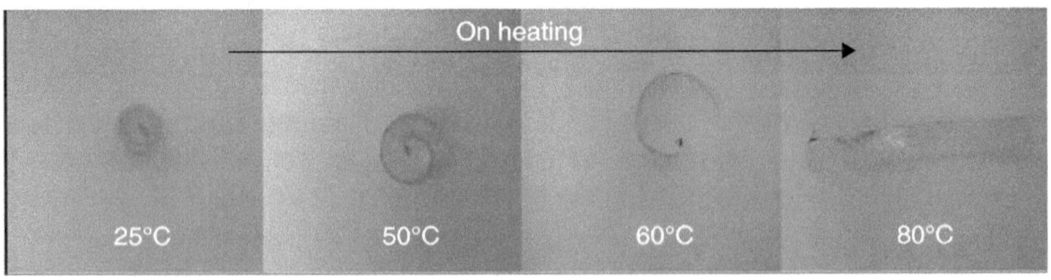

7.1 Opening process of shape memory coils upon heating.

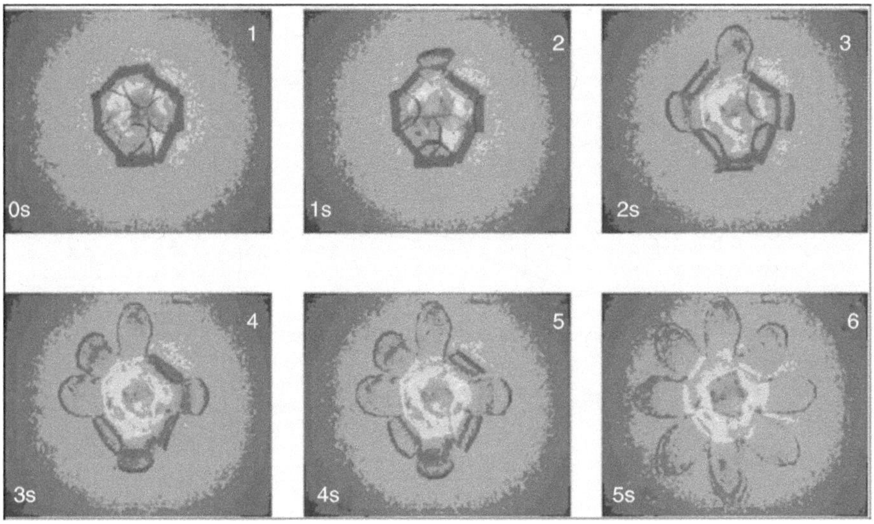

7.2 Opening process of shape memory flower at 80°C.

7.2.2 Comparison of the thermal-mechanical properties of different BIN-SMPUs

The key values to be determined in any description of the shape memory properties of materials are shape recovery (R_r) and shape fixity (R_f). Shape recovery qualifies the ability of SMMs to memorize their permanent shape, whereas shape fixity describes the ability of the switching segment to fix the mechanical deformation. In order to better understand the shape memory behavior of BIN-SMPU, thermal-mechanical tests were carried out in this study under the following conditions: 100% elongation at 45°C; fixing at 20°C; and recovering at 80°C. The resulting thermal-mechanical strain–stress curve of PUPyBDO53 is presented in Fig. 7.3, along with the equivalent strain–stress curves of T_m-type SMPU synthesized from poly(hexylene adipate) glycol (PHAG), MDI and BDO, and T_g-type SMPU synthesized from poly(1,4-butylene adipate) glycol (PBAG), MDI and BDO (Chen *et al.*, 2007b). T_m-type SMPU was found to display a small stress at 60°C due to its lower hard segment content (HSC), whereas the T_g-type SMPU displays higher stress because of its higher HSC. In PUPyBDO53, there are no soft segments: the BINA unit is the main component of soft phase and the HDI-BDO forms the hard phase via the hydrogen bonds present in the urethane group. PUPyBDO53 therefore also displays higher stress, as it has a higher modulus at 45°C.

Similar to the patterns observed for T_m-type SMPU, PUPyBDO53 also has a high shape fixity of above 98%, whereas the T_g-type SMPU has a shape fixity of only about 65% (Fig. 7.4). Shape fixation in the T_m-type SMPU results from the fast crystallization of the PHA soft segment. Most of the deformed strain in the

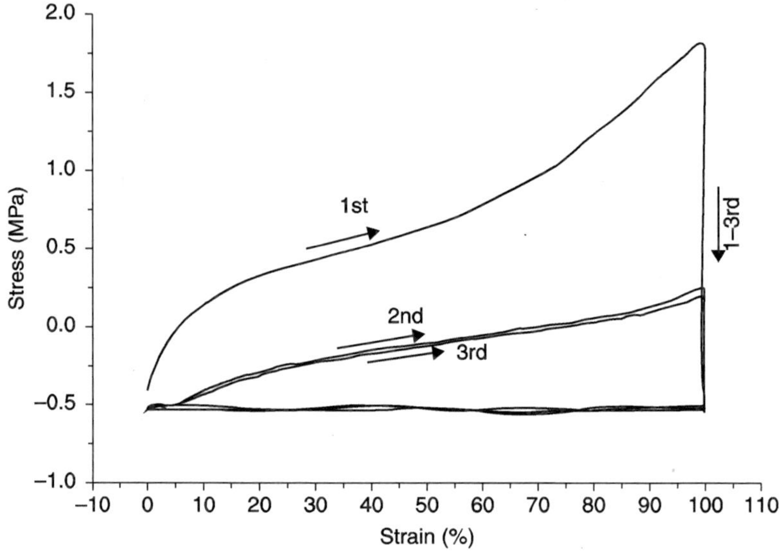

7.3 Cyclic tensile curves of typical T_m-type SMPU.

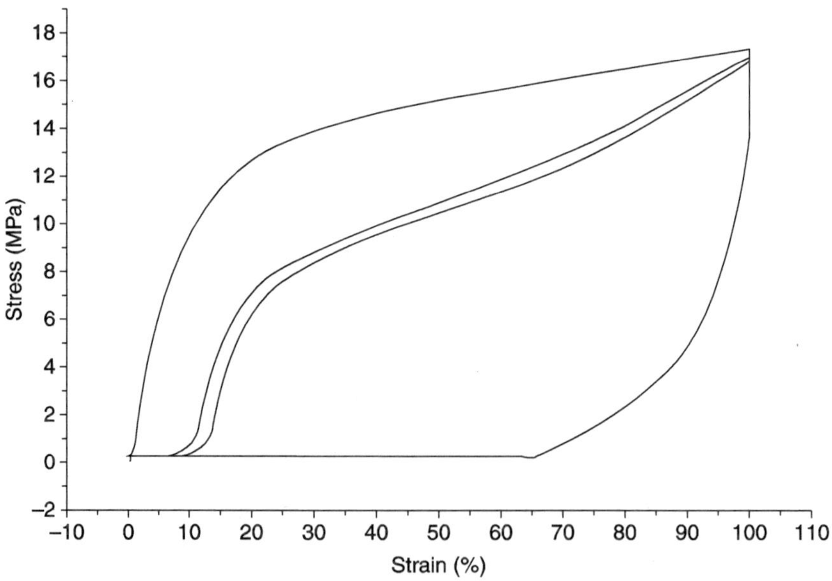

7.4 Cyclic tensile curves of typical T_g-type SMPU.

T_g-type SMPU is fixed when the polymer enters into the glass state, and the chain relaxation has a significant effect on strain fixation. However, shape fixation in PUPyBDO53 is mainly dependent on the thermally-reversible association of the hydrogen bonds present in the pyridine ring. Due to the association of the hydrogen bonds, the polymer chain is fixed by the strong interaction between molecules. The strain deformed at a higher temperature can be easily fixed when the temperature cools to below the association temperature of the hydrogen bonds present in the pyridine ring.

The T_m-type SMPU and the PUPyBDO53 both have a shape recovery of above 90%, while the T_g-type SMPU shows a lower shape recovery. The shape recovery of PUPyBDO53 is also observed to decrease as the test cycle time increases, but the T_m-type SMPU and T_g-type SMPU have a stable shape recovery after the second cycle, due to the presence of MDI and BDO in the hard phases of both types. The MDI-formed hard segment contains not only strong hydrogen bonds, but also dipole–dipole interaction and induced dipole–dipole interactions (Lee *et al.*, 2001). The physical net points present in the T_m-type SMPU and T_g-type SMPU are therefore relatively stable after the thermal training process in the first test cycle. In contrast, the hard phases acting as the physical net points in the BIN-SMPU are ordered urethane groups formed via the hydrogen bonds. A large pendant pyridine ring fraction in the PUPyBDO53 can prevent the aggregation of urethane units (Fig. 7.5). As the cycle time increases, the physical net points are destroyed easily, meaning that the shape recovery of BIN-SMPU decreases as the cycle time increases.

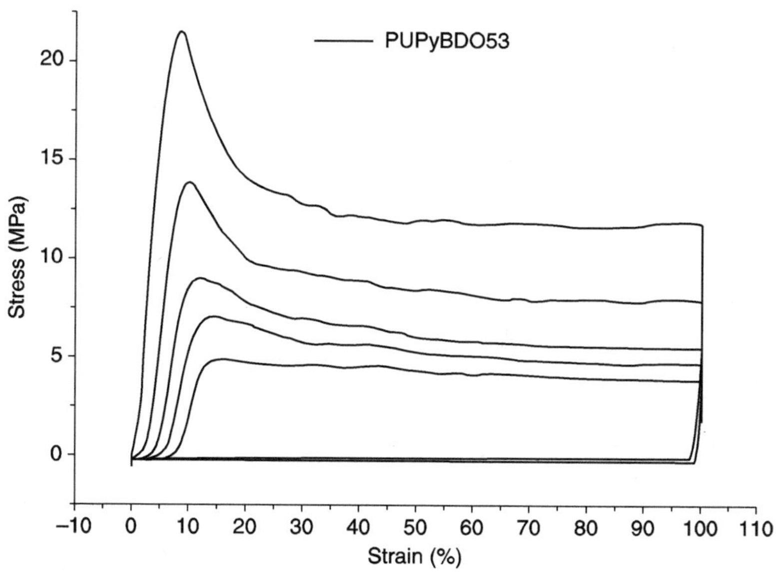

7.5 Cyclic tensile curves of typical BIN-SMPU (PUPyBDO53).

7.3 Influence of different factors on shape memory effects (SMEs) of BIN-SMPUs

7.3.1 Influence of pyridine content on SMEs

Shape memory effects (SMEs) are known to be significantly influenced by both the molecular weight of the soft segment and the soft segment content in traditional SMPU, such as T_g-type SMPU and T_m-type SMPU (Lee *et al.*, 2001). However, BIN-SMPUs contain a small unit of BINA, meaning that the SMEs are principally determined by the BINA content. Thermal-mechanical tests were carried out on BIN-SMPUs with different BINA contents, in order to investigate the effect of BINA content on SMEs. The resulting strain–stress curves are shown in Fig. 7.6, and their shape fixity and shape recovery are summarized in Table 7.1.

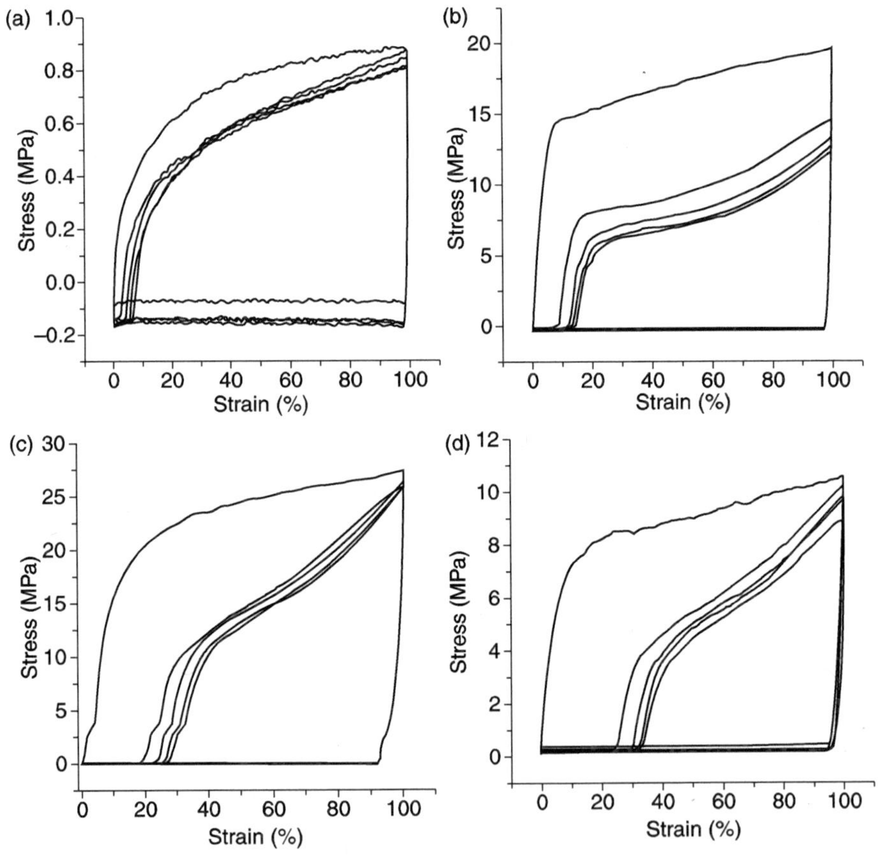

7.6 Cyclic tensile curves of BIN-SMPUs with various BINA contents: (a) PUPyBDO40; (b) PUPyBDO30; (c) PUPyBDO20; and (d) PUPyBDO10.

Table 7.1 Cycle tensile testing results of BIN-SMPU with various BINA contents

Samples	BINA wt.%	R_f (%)	1st R_r(%)	2nd R_r(%)	3rd R_r(%)	4th R_r(%)
PUPyBDO53	53.7	97.9–98.7	98.54	96.34	94.49	92.57
PUPyBDO40	38.5	99.1–99.2	97.96	96.03	94.68	93.80
PUPyBDO30	28.8	97.3–97.5	92.54	88.78	87.16	86.15
PUPyBDO20	19.1	91.7–92.5	81.01	76.94	74.36	72.73
PUPyBDO10	9.5	94.7–95.5	76.0	70.64	68.71	67.53

Figure 7.6 and Table 7.1 clearly show that all the BIN-SMPUs have good shape fixity. The samples with higher BINA content, namely PUPyBDO53 and PUPyBDO40, exhibit higher shape fixity due to the large fraction of hydrogen bonds present in the pyridine ring. However, significant differences are found with regard to shape recovery. As the BINA content drops, the shape recovery undergoes a gradual decrease: PUPyBDO53 containing 53.7 wt.% BINA has a shape recovery of 98.5%, and PUPyBDO40 containing 38.5 wt.% BINA has a shape recovery of 97.9%. When the BINA content drops to below 30 wt.%, the PUPyBDO20 containing 19.1 wt.% BINA has a shape recovery of 81%, and the PUPyBDO10 containing 9.5 wt.% BINA has a shape recovery of only 76%. Furthermore, the shape recovery of all BIN-SMPUs can be observed to decrease as the cyclic time increases, because the BINA content reflects the pyridine ring fraction, and the shape recovery is mainly because of the hydrogen bonds present in the pyridine ring. These results imply that the pyridine content has a significant influence on both shape recovery and shape fixity. The lowest BINA content that still allows good shape recovery is 30 wt.%.

7.3.2 Influence of rigid hard domains on SMEs

The hard segment, i.e., MDI-BDO, which acts as the physical net point, is responsible for the strain recovery of traditional SMPU. Previous studies have found that both the shape retention and shape recovery improved as the MB content increased within the range of 30 to 45 wt.% in the polytetramethylene glycol (PTMG)/MDI/BDO system of SMPUs (Lee *et al.*, 2001). An investigation into the influence of the soft segment length (SSL) and the HSC on the SME in the PHAG/MDI/BDO system of SMPU revealed that the shape fixity improved as the SSL increased, whereas the shape recovery decreased. However, samples with lower HSC SMPU display higher shape fixity and higher shape recovery. The shape recovery essentially decreases as HSC increases, except in the case of the 15% HSC SMPU sample. This correlation can be explained by the fact that the phase mixing increases as the HSC increases. Furthermore, it is the hard segment that is responsible for the shape recovery, including dipole–dipole interaction, hydrogen bonding and induced dipole–dipole interaction, and the interaction of this segment decreases as the phase mixing increases (Chen *et al.*, 2007a).

Polar MDI was found to reinforce the polarity of the urethane units, due to the strong intermolecular forces expected between the MDI-BDO (MB) hard segments in this PUPy-MB series of BIN-SMPUs. Dynamic mechanical analysis (DMA) testing revealed that the rubber modulus significantly improved as the MB content increased. Thermal-mechanical testing was used to investigate the influence of the hard segment on the SME of BIN-SMPUs, by analyzing the SME of the PUPy-MB series of BIN-SMPUs with different MB contents. Figure 7.7

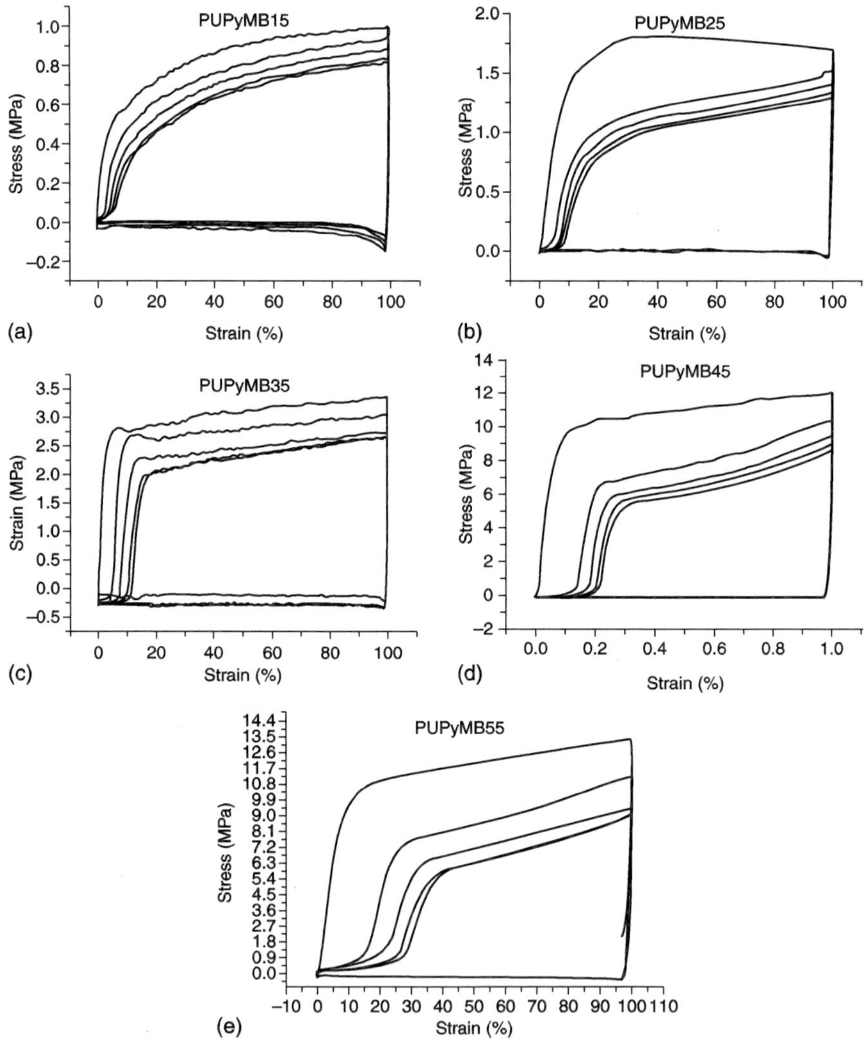

7.7 Cyclic tensile curves of BIN-SMPUs with various MB contents: (a) PUPyMB15; (b) PUPyMB25; (c) PUPyMB35; (d) PUPyMB45; and (e) PUPyMB55.

shows the resulting strain–stress curves, and their shape fixity and shape recovery are summarized in Table 7.2. Figure 7.7 also shows that the addition of MDI-BDO has little influence on the shape fixity of BIN-SMPUs: even the PUPyMB55, which contains only 24.2 wt.% BINA and 55 wt.% MB content, still has a shape fixity of 96.9%. This again confirms that a higher glassy modulus is generated from the strong hydrogen bonds present in both the pyridine ring and the urethane groups, resulting in higher shape fixity in the BIN-SMPUs.

In addition, Table 7.2 shows that the shape recovery decreases as the MB content increases and the BINA content drops. For example, the PUPyMB35 with 35 wt.% BINA and 35% MB content shows a shape recovery of above 90% in the first 3 test cycles. However, when the MB content is increased to 45 wt.%, with the BINA content kept close to 30 wt.%, the shape recovery decreases to only about 79% after the 4th test cycle, showing that the addition of the MDI-BDO segment does not improve the shape recovery. This could be due to the fact that if the urethane unit content of the BIN-SMPU is above 45 wt.%, the excess hard segment will disrupt the hydrogen bonds present in both the pyridine ring and the urethane groups.

Two samples containing a similar amount of BINA were selected, one from the PUPy-BDO series and one the PUPy-MB series, namely PUPyBDO30 with 29 wt.% BINA and PUPyMB45 with about 30 wt.% BINA. The PUPy-BDO series of PUPys were found to still show a high shape recovery: PUPyBDO30 has a shape recovery of 92.54%, while PUPyMB45 has a lower shape recovery of only 86.65%. As mentioned above, this implies that the addition of the MDI-BDO segment decreases the shape recovery as it may destroy the regular structure of the HDI-BDO urethane units. In addition, Fig. 7.8 shows that there is a linear relationship between the increase in BINA content in each test cycle and the improvement in shape recovery. In order to obtain a higher shape recovery of above 90%, the BINA content should clearly reach 30 wt.% in the BINA-based SMPU. It is therefore confirmed that the pyridine content is the key element in determining shape recovery in the BIN-SMPUs.

Table 7.2 Cyclic tensile testing results of BIN-SMPU with various MB contents

Samples	BINA wt.%	MB wt.%	R_f	1st R_r	2nd R_r	3rd R_r	4th R_r
PUPyMB0	53.7	0	98.4	98.54	96.34	94.49	92.57
PUPyMB15	45.5	15	98.6	97.91	96.44	94.45	94.43
PUPyMB25	40.3	25	98.0	96.17	94.08	93.16	92.09
PUPyMB35	34.9	35	98.6	95.70	93.12	90.64	88.99
PUPyMB45	29.5	45	97.4	86.65	82.88	80.41	78.85
PUPyMB55	24.2	55	96.9	86.33	79.81	74.68	73.02

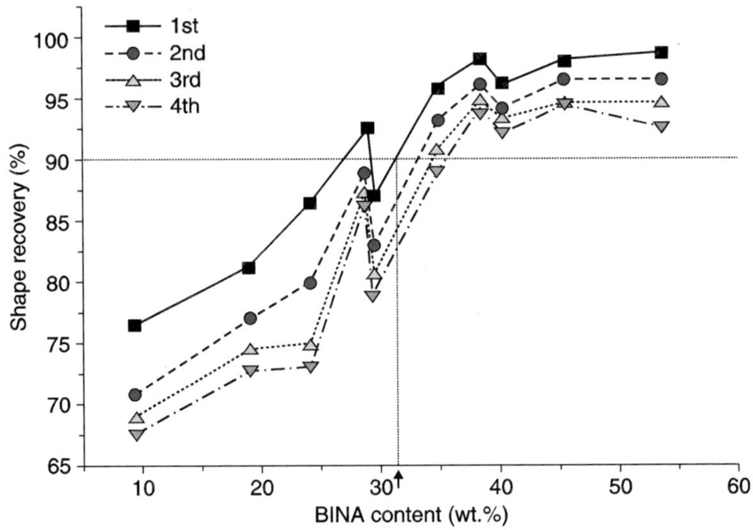

7.8 Dependency of shape recovery with various test cycles on BINA contents.

7.3.3 Shape stability of BIN-SMPUs with different BINA contents

Thermal-mechanical testing was used to determine shape fixity in order to establish the capacity of a material to fix an instant deformation. However, chain relaxation occurs in almost all polymers: over time, the free strain tends to be recovered through the relaxation process, and as a result the shape fixity may decrease if the intermolecular action is weak. Zhu *et al.* (2009), examining supramolecular SMPU grafting with UPy side groups, observed that the shape fixity decreased to only 60% after 24 hours of relaxation, because the supramolecular SMPU contained only a small fraction of UPy side groups.

Li *et al.* (2007) therefore suggested that a side group with a high hydrogen bond fraction might provide a better SME (Li *et al.*, 2007). In BINA-based SMPU, a high fraction of hydrogen bonds are formed in the soft phase and hard phase, and it is therefore expected that BIN-SMPU will have better strain stability than the previous supramolecular SMPs. Figure 7.9 shows the strain stability of PUPyBDO45 with 100% elongation compared to the strain stability of PUPyBDO53 with 100% elongation, 150% elongation and 200% elongation. The shape fixities in all specimens were found to decrease over time. In particular, the strain fixity of PUPyBDO45 decreases significantly from the original value, falling from 100% to only 83% after 96 hours of relaxation. PUPyBDO53 has higher strain stability than PUPyBDO45; in particular, the PUPyBDO53 with

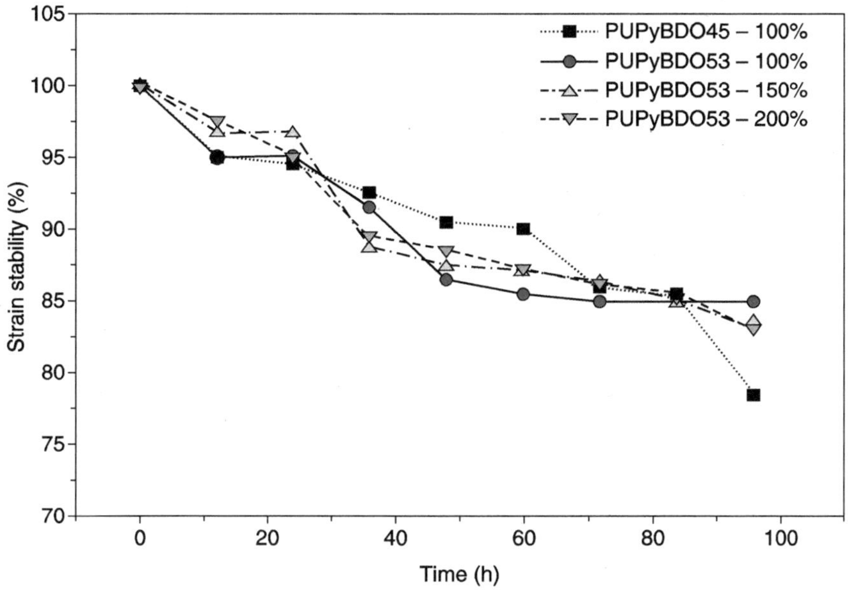

7.9 Strain stability of BIN-SMPU with various BINA contents and elongations.

100% elongation tends to maintain shape fixity of above 85% after 40 hours of relaxation. However, when the elongation is too high, the strain stability becomes worse. For example, the strain stability of PUPyBDO53 with 200% elongation is inversely proportional to relaxation time, particularly during the first 40 hours of relaxation. This is because the PUPyBDO53 contains a large fraction of thermal-reversible hydrogen bonds in the pyridine ring, and these are responsible for the shape fixation.

Higher strain stability is therefore expected in the BIN-SMPUs that have a higher BINA content. However, high elongation will destroy the non-covalent hydrogen bonds, and will also affect the stability of the hydrogen bonds present in the pyridine ring. Higher strain stability will thus result from lower elongation.

7.3.4 Effect of temperature on strain recovery

Thermal recovery testing is used to characterize the shape recovery behavior of a material. The resultant temperature-dependent strain recovery curves, which are usually S-shaped, show the change of recovery rate as a function of temperature upon heating. At lower temperatures, the size of the specimen does not change as the temperature increases. The recovery process usually starts at an initial temperature (T_i). The strain recovery rate changes slowly during the initial stage, but then increases abruptly within a narrow temperature range, before the recovery process finishes at the end temperature (T_e). Finally, several values can be obtained from the recovery curve in order to describe the characteristics of the shape memory behavior of the tested specimens. The final shape recovery rate (R_f) characterizes the recovery ability of the specimen. The temperature corresponding to a recovery rate value of $1/2R_f$ is known as the response temperature or average response temperature (T_r) of SMPs on the thermal recovery curve (Li *et al.*, 1996).

In general, the T_r is principally dependent on the switch temperature of the SMPs, i.e. the T_m, and the R_f is significantly influenced by the SSL and HSC. For example, T_r increases as SSL increases, but decreases as HSC increases in the PHAG/MDI/BDO system. In addition, the R_f is found to decrease as HSC increases. In the T_g-type SMPUs (Lin and Chen, 1998a,b), the T_r increases as HSC increases. A higher HSC tends to result in a higher shape recovery rate (R_r) in both the PTMG250/MDI/BDO system and the PTMG650/MDI/BDO system. The difference between T_m-type SMPU and T_g-type SMPU with regard to their final recovery is the result of their different phase separation structure and crystallization behavior. In the T_m-type SMPU, the increase in HSC prevents the crystallization of the soft segment. A higher HSC results in a continuous hard phase and a separated soft phase, even though the soft segment is dissolved by the hard phase. The recovery temperature therefore decreases as HSC increases. However, in the T_g-type SMPU, T_r is dependent on the T_g, which usually increases as HSC increases. An increase in HSC also leads to improvements in phase separation in segmented polyurethane, causing more perfect hard domains. Hence, the shape fixity rate (R_f) is usually higher in the T_g-type SMPU that have higher HSC.

However, the structure and morphology of BIN-SMPU are different from those of the traditional T_m-type and T_g-type SMPUs. An experiment was therefore conducted to systematically investigate the thermal recovery of BIN-SMPUs. During the test process, each specimen was fixed at 20°C for ten minutes after being elongated to 100% at 80°C. After a further ten minutes of relaxation, the length of each specimen was recorded upon heating, with a heating rate of 3°C/min. Finally, the thermal-recovery curves of BIN-SMPU with different BINA contents and BIN-SMPU with different MB contents were calculated and are plotted in Figs 7.10 and 7.11.

Figure 7.10 shows that PUPyBDO53 and PUPyBDO40 display the typical S-shape strain recovery process. The shape changes slightly below the T_i of 50°C in the PUPyBDO53. However, when the temperature is raised to above T_i, the deformed strain starts to recover immediately, with significant strain recovery being observed between 50 and 70°C.

The T_r is therefore about 60°C in the PUPyBDO53, which is close to the T_g of the soft phase and the dissociation temperature of the hydrogen bonds present in the pyridine ring. This implies that the strain recovery results from the increasing chain movement due to the dissociation of the hydrogen bonds.

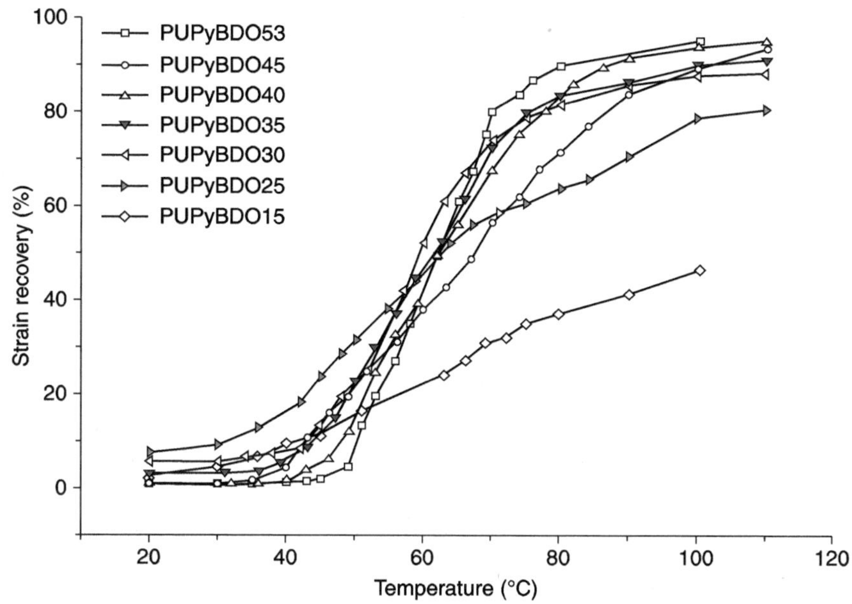

7.10 Strain recovery curves of BIN-SMPU with various BINA contents.

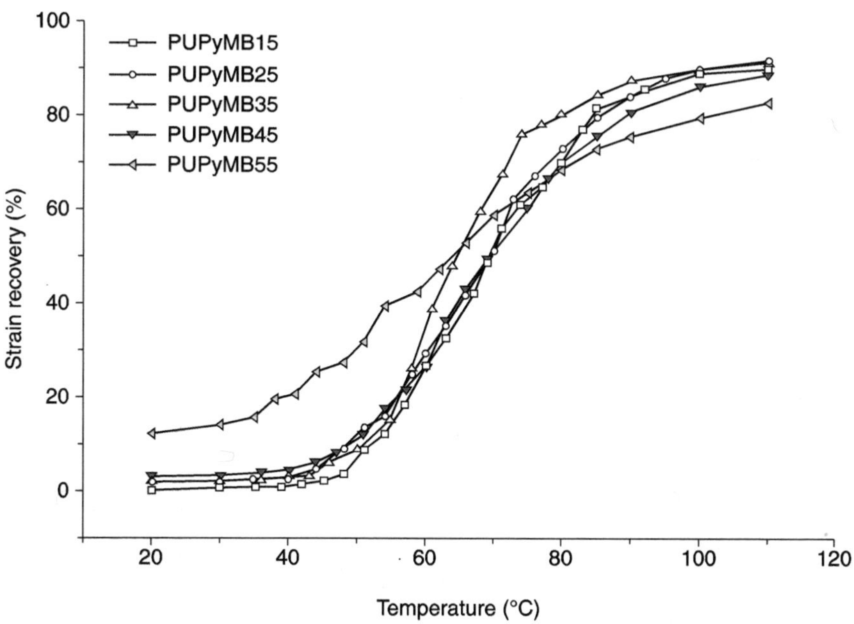

7.11 Strain recovery curves of BIN-SMPU with various MB contents.

When the temperature is raised to above the T_e of 80°C, most of the deformed strains are recovered, although the strain recovery process is slow. This shape recovery process shows that both the PUPyBDO53 and PUPyBDO40 have excellent SMEs. However, an unacceptable final strain recovery is observed if the BINA content drops to below 30 wt.%; this is below 50% in the PUPyBDO10. In addition, the samples with higher BINA content – PUPyBDO53 and PUPyBDO45 – were found to have higher shape fixity. However, the samples with lower BINA content, namely PUPyBDO10 and PUPyBDO20, also have lower shape fixity. These observations are consistent with the results obtained through thermal-mechanical testing. This implies that BIN-SMPUs with a higher BINA content show better shape recovery and better shape fixity. The lowest BINA content that allows good SMEs in BIN-SMPUs is about 30 wt.%.

Figure 7.11 shows that the PUPyMB55 has the lowest shape fixity and the lowest final shape recovery. The other samples containing less than 45 wt.% MB and more than 30 wt.% BINA, i.e. PUPyMB15, PUPyMB25, PUPyMB35 and PUPyMB45, all show the typical S-shaped strain recovery process. Their shape fixity is beyond 95%, and their final shape recovery is approximately 90% at 110°C. This means that the addition of MDI-BDO has no influence on the SME of BIN-SMPUs if the BINA content is kept above 30 wt.%.

Furthermore, the BINA content and MB content have been found to have little influence on the shape recovery temperature of the PUPy-BDO series and the PUPy-MB series of BIN-SMPUs. The result obtained is exactly consistent with the T_g changing tendency that was observed using differential scanning calorimetry (DSC) testing. This is probably due to the fact that the strain recovery is determined by the dissociation of the hydrogen bonds, and the recovery temperature is determined by the closed dissociation temperature of the same hydrogen bonds that exist between the N–H of urethane and the N of the pyridine ring.

7.3.5 Shape recovery force of BIN-SMPUs

The driving force of shape recovery in a polymer is the elastic strain generated during the deformation. It has been proposed that the restoration force of SMPs can be measured as the temperature increases from room temperature to above T_r by fixing the specimen length after a procedure of deformation and fixation (Chen *et al.*, 2009). In an earlier study (Chen *et al.*, 2007a), the restoration stress (F) of the T_m-type SMPU was observed at the beginning of strain recovery, i.e. T_1 = 41°C. It was then increased to its maximum restoration stress (F_{max}) when the temperature was raised to T_2 = 60°C, i.e. near the end temperature of strain recovery. After the deformed strain was entirely released, the recovery stress began to decrease as a result of the increasing temperature.

In the present study, the static stress is recorded in order to determine the restoration force upon heating, by using DMA and controlling the displacement to

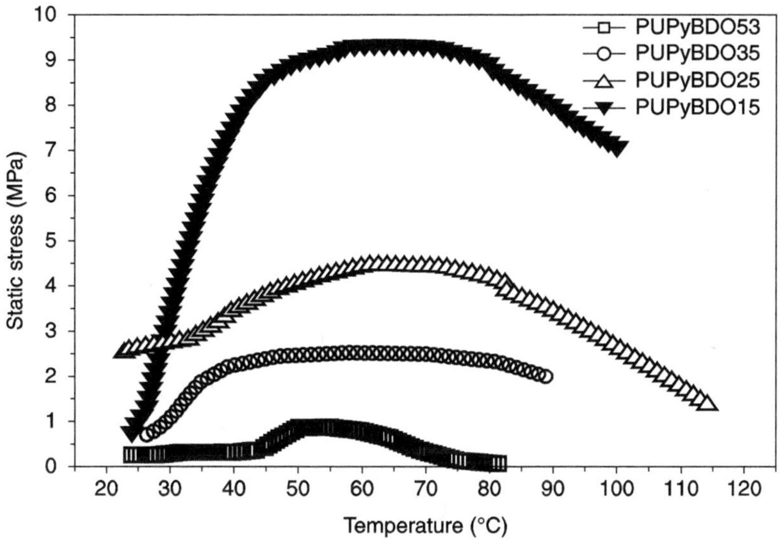

7.12 Static stress curves of BIN-SMPU with various BINA contents.

a small value, specifically 0.001 mm. Figure 7.12 presents the static stress curves of BIN-SMPU with different BINA contents at 100% elongation. It can be observed in Fig. 7.12 that static stress does not appear in the PUPyBDO53 until the temperature is raised to above 43°C. It then increases to its maximum value of about 1.0 MPa as the temperature is raised to T_1 = 50°C. The static stress tends to stay above 1.0 MPa before the temperature is raised to above 58°C. Thereafter, the static stress decreases as the temperature increases. The maximum value of static stress increases significantly as the BINA content decreases; for example, about 2.5 MPa static stress is obtained in the PUPyBDO40, while the PUPyBDO30 shows a maximum static stress of 4.6 MPa at about 60°C.

The PUPyBDO20 shows a much higher static stress of above 9.0 MPa after it is heated to over 60°C, because the rubber modulus increases as the BINA content decreases, as observed in the DMA test. The high rubber modulus causes higher elastic stress when the hydrogen bonds are dissociated at above T_r. This therefore confirms that the recovery energy can be stored in the deformed shape, and the restoration force is considered to be an important factor in determining SMEs.

Figure 7.13 demonstrates that similar tendencies are also found in the PUPy-MB series of BIN-SMPUs, i.e. there is little static stress in the PUPyMB35 before 47°C. The static stress increases to its maximum value of 1.4 MPa when the temperature is raised to above 47°C, but undergoes a significant decrease at a temperature of above 100°C. In addition, Fig. 7.13 shows that BIN-SMPUs

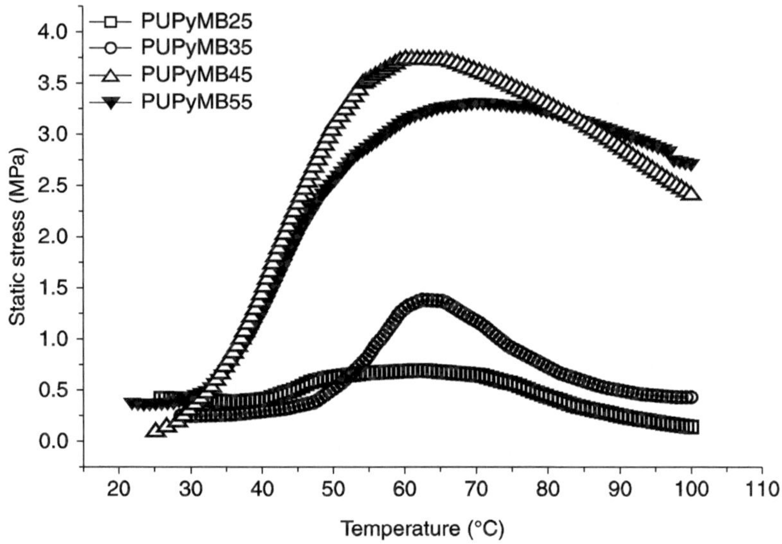

7.13 Static stress curves of BIN-SMPU with various MB contents.

with a higher MB content have a much higher static stress than those with a lower MB content. This is because the former have a higher rubber modulus as a result of the lower BINA content and higher MB content, as shown by the DMA analysis. It can therefore be concluded that BIN-SMPUs with a higher BINA content show a lower restoration force, although they have a better shape recovery. However, BIN-SMPUs with a higher MB content have a lower shape recovery irrespective of their higher restoration force, confirming that the thermally-induced SME is mainly influenced by the pyridine ring. The addition of the MB segment reinforces the hard phase, resulting in a higher shape recovery force.

Furthermore, the elongation also influences the recovery force of BIN-SMPUs. Figure 7.14 presents the static stress curves of PUPyBDO30 with 100% elongation, 150% elongation and 200% elongation. The specimen with the greatest elongation has the highest maximum static stress, while the specimen with the lowest elongation shows the lowest maximum static stress, because the driving force of SMPs comes from the stored elastic strain. According to Hooke's law: $\sigma = k \times \varepsilon$ (σ is the elastic stress, K is the constant parameter and ε is the strain) (Beblo and Mauck Weiland, 2008), higher elongation will result in a greater elastic force.

7.3.6 Effect of deformation temperature

Further investigation into the effects of deformation temperatures was carried out using PUPyBDO30 and PUPyBDO53. Figure 7.15 shows the strain recovery

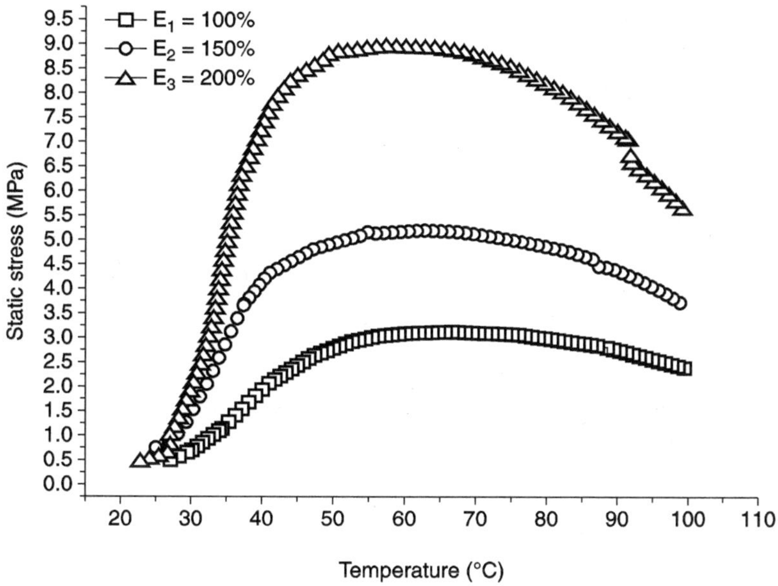

7.14 Static stress curves of PUPyBDO30 with various elongations: E_1 = 100%, E_2 = 150% and E_3 = 200%.

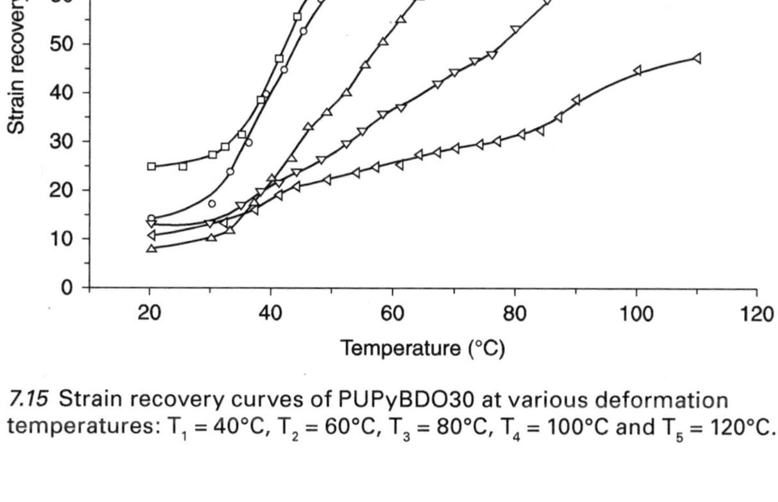

7.15 Strain recovery curves of PUPyBDO30 at various deformation temperatures: T_1 = 40°C, T_2 = 60°C, T_3 = 80°C, T_4 = 100°C and T_5 = 120°C.

curves of PUPyBDO30 at various deformation temperatures: it is clear that PUPyBDO30 shows the highest shape fixity and the highest shape recovery when the sample is deformed at 80°C. As the deformation temperature drops, the shape fixity of PUPyBDO30 also tends to decrease: the lowest shape fixity is observed when the sample is deformed at 40°C, while if the deformation temperature is raised to 100 or 125°C, the shape recovery will be very low. In addition, Fig. 7.16 shows that the shape recovery temperature increases as the deformation temperature increases.

The impact of the deformation temperature on the shape recovery temperature is more obvious in the PUPyBDO53 sample. Figure 7.16 presents the strain recovery curves of PUPyBDO53 at different deformation temperatures. For the sample deformed at 40°C, shape recovery is observed to begin at about 32°C, and its shape recovery temperature is about 50°C. When the deformation temperature is raised to above 80°C, the shape recovery decreases. For example, for a sample deformed at 125°C, only 30% shape recovery is possible before the condition temperature is raised to above 110°C. Although it is possible to fix the strain after deformation at 110°C, most of the deformation cannot be recovered at temperatures lower than the deformation temperature.

Figure 7.17 shows that if the condition temperature is further raised to above 160°C, the strain continues to recover its pre-deformed shape at higher temperatures. In the PUPyBDO30 sample deformed at 120°C with 150% elongation, only 50% strain can be recovered before 100°C. During the raising of

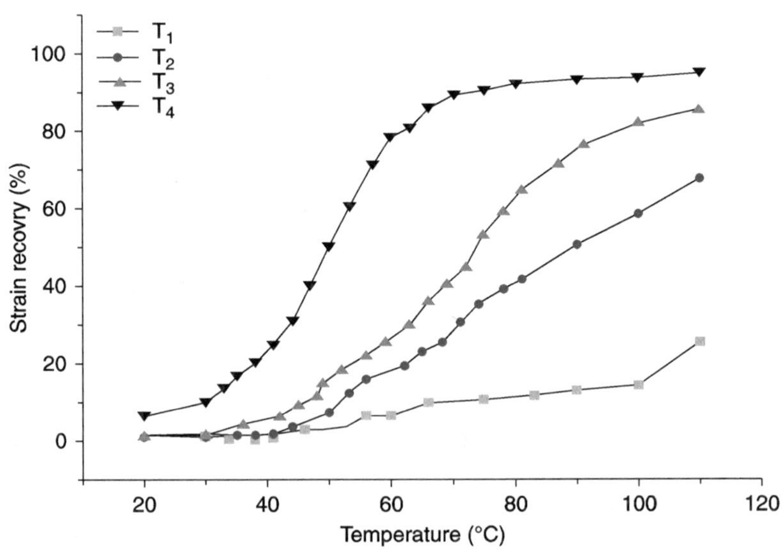

7.16 Strain recovery curves of PUPyBDO53 at various deformation temperatures: T_1 = 125°C, T_2 = 100°C, T_3 = 80°C and T_4 = 40°C.

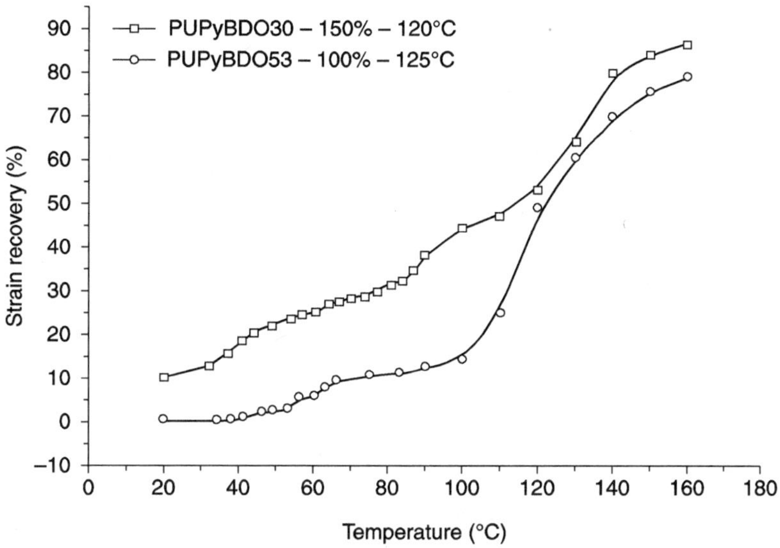

7.17 Strain recovery curves of PUPyBDO53 and PUPyBDO30 at various elongation (150 and 100%) and deformation temperatures (120 and 125°C).

the temperature to 140°C, another 35% strain is recovered continuously, meaning that the final strain recovery can reach 85%. Strain recovery in the PUPyBDO53 sample deformed at 125°C with 100% elongation is slow, and only 15% deformed strain can be recovered before 70°C. The shape then remains relatively stable at temperatures between 70 and 100°C. When the temperature is raised to 160°C, the strain undergoes sudden recovery before reaching a final strain recovery of above 80%. The recovery process of BIN-SMPU therefore clearly has two stages at below 160°C when the deformed temperature is significantly higher. By adjusting the deformation temperature of BIN-SMPUs, a triple shape recovery behavior can be achieved.

BIN-SMPUs contain two kinds of strong hydrogen bonds: one type, those between the N–H of the urethane group and the pyridine ring, is associated at a lower temperature and acts as the molecular switch; the other type, those between the urethane groups, is associated at a higher temperature and acts as the physical net point. The hydrogen bonds with the lower association temperature might be partially destroyed by the application of external force at lower temperatures. Some deformed strain cannot be fixed after BIN-SMPU is cooled to room temperature. Moreover, these unstable hydrogen bonds are dissociated easily upon heating. As a result, deformation at lower temperatures allows a lower strain recovery temperature and a higher strain recovery. However, if the specimen is deformed at a higher temperature, it does not only dissociate the hydrogen bonds with a lower association temperature, but also partly destroys those with a higher association temperature.

When BIN-SMPU is cooled to room temperature, both the hydrogen bonds with a lower association temperature and those with a higher association temperature contribute to the strain fixation. Moreover, most of the strains are fixed by the hydrogen bonds with the higher association temperature since these are first associated when BIN-SMPU is cooled. Hence, the recoverable strain is small at temperatures lower than 100°C. Raising the temperature to the dissociation temperature of hydrogen bonds will result in further strain recovery.

7.4 Thermally-induced SME mechanism of BIN-SMPUs

7.4.1 Molecular thermally-induced SME mechanism of traditional SMPU

Shape memory polymers (SMPs) are elastic polymer networks equipped with suitable stimuli-sensitive switches. The polymer network consists of molecular switches and net points. The net points determine the permanent shape of the polymer network and can be either chemical net points, such as covalent bonds, or physical net points, such as inter-molecular interactions. SMPUs with thermally-induced SMEs are usually composed of a soft segment and a hard segment, which transform into the soft phase and hard domain, respectively. The soft phase has a T_g or T_m above room temperature and can thus be used as the reversible switch segment, while the hard segment has strong inter-molecular interaction (in the form of MDI-BDO units) and acts as the physical net point. When in solution or during melting, where the temperature is usually above the crystal melting temperature of hard domains (T_{mh}), the polyurethane chain is random and free with no inter-molecular interactions. At this point, any kind of initial shape can be fabricated from the polyurethane solutions or melting polyurethane. If the temperature is decreased to T_l ($T_l < T_{ms}$), where T_{ms} is the crystal melting temperature of soft domains, the micro-phase separation structure will result in the formation of a soft phase and a hard domain.

However, when the temperature is raised to the higher dissociation temperature T_h ($T_{ms} < T_h < T_{mh}$), the crystal of the soft phase is melted, but the hard domain is still fixed. The polyurethane can then be deformed into a second shape by applying external force while the polyurethane is in the rubber state, and then cooling to T_l ($T_l < T_{ms}$) under the load, which will fix the deformed shapes. Moreover, when the temperature is again raised to above T_h, the deformed strain will release immediately, resulting in macroscopic shape recovery. Both shape fixation and shape recovery can hence be achieved in segmented SMPU.

The current generation of BIN-SMPUs are synthesized from BINA, HDI with the addition of MDI, and/or BDO, and the BINA used is not a long-chain polyol. The BINA can form strong hydrogen bonds with the N–H of the urethane groups. Their shape memory mechanism is therefore different from that of traditional SMPUs. The following sections present some further investigations into the thermally-induced shape memory mechanism of BIN-SMPU.

7.4.2 Effect of temperature on hydrogen bonding

Temperature-dependent Fourier transform infrared (FT-IR) spectroscopy is widely used to study the association and dissociation of hydrogen bonding. According to the FT-IR spectra taken at different temperatures, the weakening and breaking of hydrogen bonds as the temperature increases can be tracked (Mikhaylova *et al.*, 2006). Figure 7.18 presents the FT-IR spectra of an HDI-BINA copolymer at various temperatures upon heating.

FT-IR analysis of BIN-SMPUs has shown that strong hydrogen bonds are formed:

1. between the pyridine ring and the N–H of urethane groups; and
2. between the C=O of urethane groups and the N–H of urethane groups.

The frequency at 3330 cm^{-1} can be attributed to the stretching vibration of the N–H of the urethane group, and the frequency at 3051 cm^{-1} can be attributed to the stretching vibration of the C–H of the pyridine ring. In addition, the frequency at 1635 cm^{-1} can be attributed to the vibration of the C=O next to the pyridine ring of the BINA unit, and the frequencies at 998.1 and 832.7 cm^{-1} are the result of deformation vibrations caused by C–H movement out of the plane of the pyridine ring. There are also some characteristic frequencies attributed to the pyridine ring, including those at 1600.9, 1462.5, 1434.1, 1410.5, 1136.4, 1093.6 and 1043.8 cm^{-1}. It is also well-established that the associated frequencies such as hydrogen-bonded vibration are different from the non-associated frequencies. Hence, temperature-dependent FT-IR can track the dissociation process of the hydrogen bonds present in both the pyridine ring and urethane groups.

Figure 7.18(a) shows that the frequency of the N–H stretching vibration shifts from 3329 cm^{-1} to a higher frequency of 3349 cm^{-1} as the temperature increases from 20 to 180°C. However, the frequency of the C–H vibration in the pyridine ring at about 3054 cm^{-1} shifts to a lower frequency of about 3041 cm^{-1}. In addition, Fig. 7.18(b) shows that the frequency of the C=O stretching vibration of the urethane groups shifts from 1705 to 1721 cm^{-1}; the frequency of the C=O stretching vibration next to the pyridine ring shifts from 1633 to 1642 cm^{-1}; and the frequency of the C–N–C stretching vibration in the pyridine ring shifts from 1602 to 1596 cm^{-1}. Even the breathing vibration frequency in the pyridine ring can be seen (Fig. 7.18(c)) to shift from 999 to 991 cm^{-1}: in particular, in the temperature range of 40 to 60°C, the breathing vibration frequency of the pyridine ring shifts abruptly from 997.5 to 992.5 cm^{-1}. Furthermore, the frequency at about 3057 cm^{-1} also drops significantly to 3048 cm^{-1} at below 60°C, whereas the frequency of the –N stretching vibration at about 1183 cm^{-1} does not disappear until the temperature

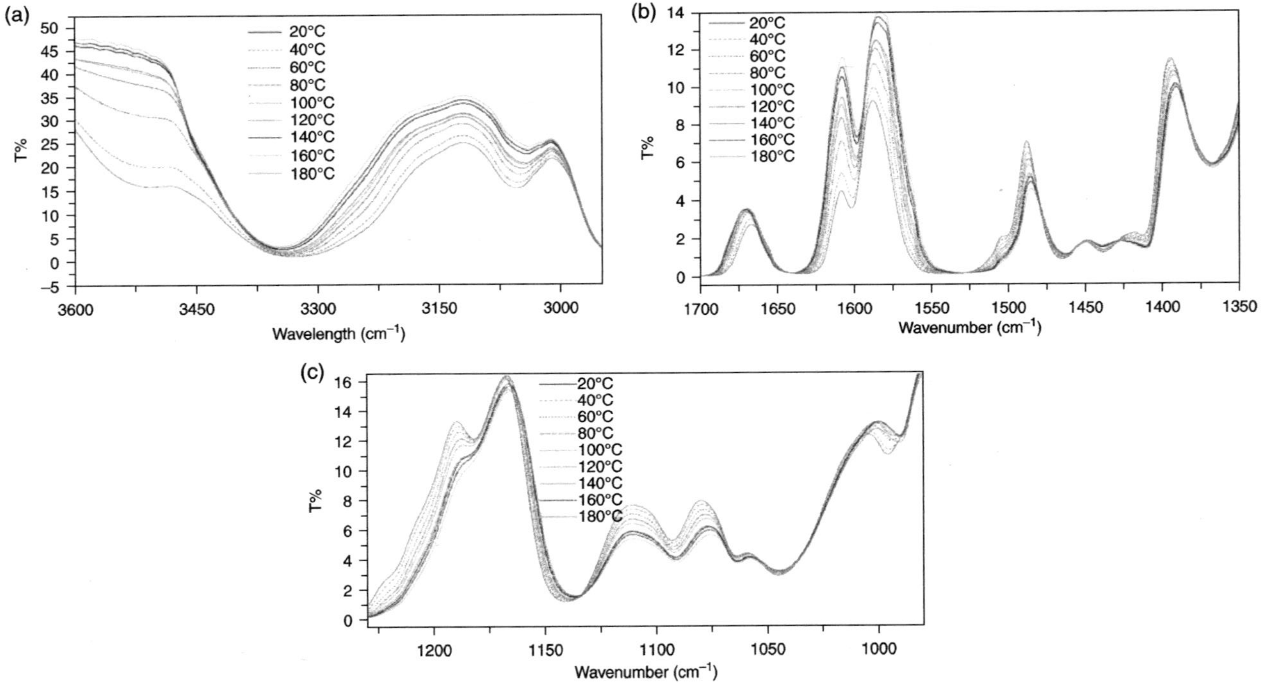

7.18 FT-IR spectra of HDI-BINA at various temperatures upon heating at the frequency range of: (a) 3600–2950 cm^{-1}; (b) 1700–1350 cm^{-1}; and (c) 1230–980 cm^{-1}.

is raised to above 120°C. These observations indicate that most of the hydrogen bonds present in the pyridine ring dissociate at about 40 to 60°C, while the hydrogen bonds present in the urethane group dissociate only slightly until the temperature is raised to above 120°C. The effect of temperature on the frequency of some typical vibrations related to the pyridine ring and the urethane groups upon heating and cooling is shown in Figs 7.19 and 7.20, respectively.

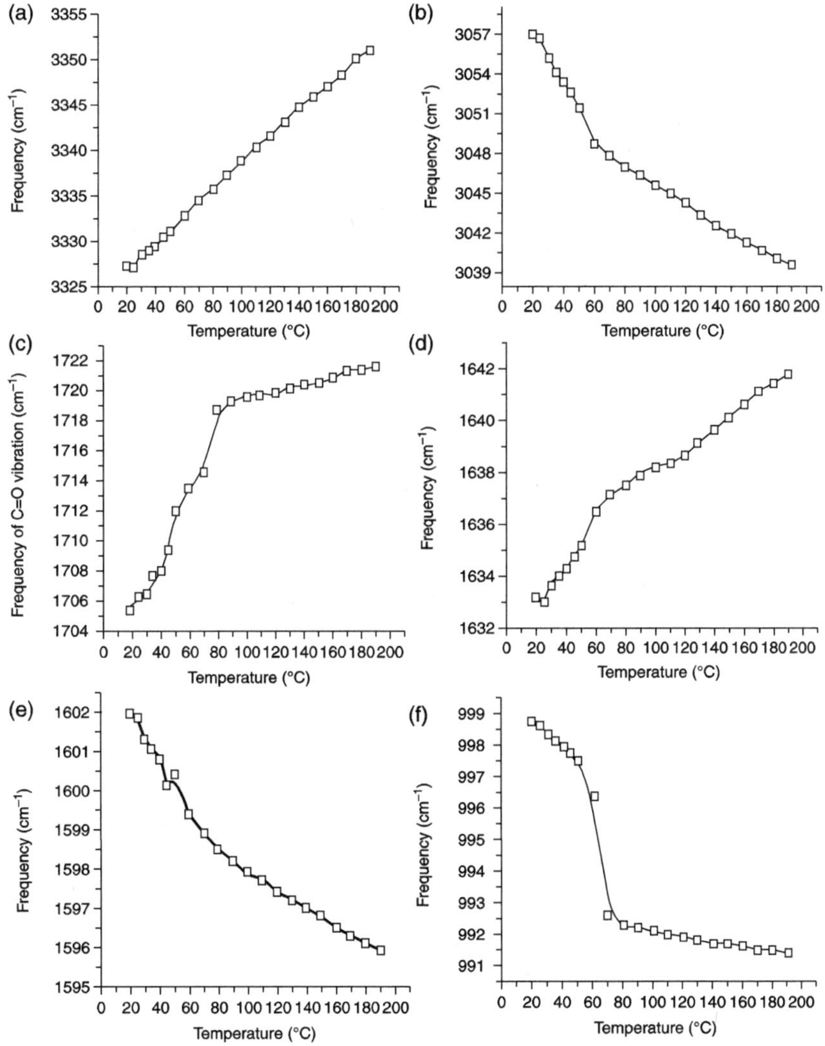

7.19 Dependency of stretching vibrations vs temperature upon heating for: (a) N–H of urethane; (b) C–H of BINA; (c) C=O of urethane; (d) C=O of BINA; (e) stretching vibration of pyridine ring; and (f) breathing vibration of pyridine ring.

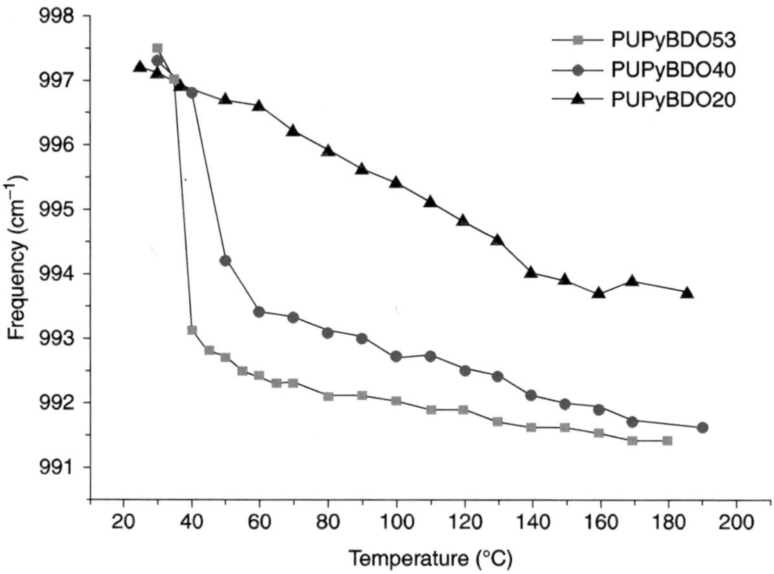

7.20 Dependency of stretching region vs temperature upon cooling for urethane.

In Fig. 7.19, the frequency of the N–H stretching vibration is observed to increase linearly from 3323 cm^{-1} at 20°C to 3353 cm^{-1} at 190°C. It is well-known that the strong N–H hydrogen bonding results in a higher frequency. The shift to high frequency shows that the strength of hydrogen bonds is reduced as the temperature increases. In addition, the frequency of the C–H stretching vibration at about 3057 cm^{-1} also decreases significantly before 60°C and then undergoes a continuous reduction down to below 3040 cm^{-1} at 190°C. Similarly, the frequency of C=O next to the pyridine ring at about 1602 cm^{-1} decreases rapidly from 20 to 70°C down to 1596 cm^{-1}. However, the most obvious frequency shift is observed for the pyridine ring breathing vibration. At the lower temperature range, i.e. before 40°C, the frequency at about 999 cm^{-1} changes slightly and then starts to drop abruptly at 50°C, shifting to 992.5 cm^{-1} at 70°C. Subsequently, the frequency changes slightly within the 70 to 190°C temperature range, implying that the dissociation of the hydrogen bonds present in the pyridine ring is an abrupt process, i.e. the hydrogen bonds are responsive to the temperature stimulus at about 40 to 60°C.

It can also be observed that the frequencies ascribed to the C=O stretching vibration of the urethane group, and the C=O stretching vibration next to the pyridine ring, shift from the lower frequencies of 1702 and 1630 cm^{-1} to higher frequencies, particularly below 80°C for the C=O vibration frequency of the urethane group, and below 60°C for the C=O vibration next to the pyridine ring. The frequency of the C=O vibration next to the pyridine ring then enters into the stable stage before it starts to shift rapidly to a higher frequency of above 120°C. However, the frequency of the C=O vibration of the urethane groups starts to shift again up to the point at which the temperature is raised to above 140°C. This implies that the hydrogen bonds present in the C=O group are partially dissociated below 80°C, while some stronger hydrogen bonds do not dissociate until the temperature is raised to above 140°C.

The majority of the hydrogen bonds present in the pyridine ring are therefore dissociated at below 70°C. However, the hydrogen bonds between N–H and C=O dissociate in two stages, and most do not dissociate until the temperature is raised to above 140°C. Therefore, the modulus of BIN-SMPU decreases significantly within the 40 to 60°C temperature range, as was observed by DMA testing. This is because the dissociation of the hydrogen bonds present in the pyridine ring results in an increase of chain movement. The pre-deformed strain can be recovered due to the weak inter-molecular interaction. However, the whole polymer chain is maintained in an unchanged form due to the presence of the hydrogen bonds in the urethane group at temperatures lower than the dissociation temperature of those hydrogen bonds. It is believed that the hydrogen bonds present in the pyridine ring serve as the molecular switch, whereas the hydrogen bonds present in the urethane groups act as the physical net points for the utilization of BIN-SMPUs as SMMs.

The N–H stretching vibration frequency also drops linearly upon cooling, meaning that the N–H of the urethane group provides the hydrogen donor to form hydrogen bonds continually during the cooling process. However, the frequency of about 992 cm^{-1}, corresponding to the pyridine ring stretching vibration at 160°C, gradually increases to only 992.7 cm^{-1} at 40°C, but then increases abruptly to 999 cm^{-1} within the 40 to 25°C temperature range. This means that the hydrogen bonds present in the pyridine ring are formed quickly as the temperature decreases to below 40°C. As a result, strong inter-molecular interactions are formed at room temperature in the BIN-SMPU, resulting in a high modulus. Good shape fixity is therefore expected in this kind of polymer. Another characteristic frequency of the pyridine ring also shifts from 3037 cm^{-1} at 190°C to 3048 cm^{-1} at 40°C, and then abruptly to 3052 cm^{-1} at 20°C. Similarly, the hydrogen bonds between the N–H and C=O also show a reversible association process. Hence, when the association of hydrogen bonds occurs, the N–H and C=O vibrations both shift to lower frequencies, while the pyridine ring vibrations shift to higher frequencies. These two kinds of hydrogen bonds will therefore influence the morphology of BIN-SMPUs: the hydrogen bonds present in the urethane group limit the movement of the polymer chain at high temperatures, while the association and dissociation of the hydrogen bonds present in the pyridine ring causes changes in the inter-molecular force, significantly influencing the stiffness or modulus.

7.4.3 Influence of pyridine content on the association and dissociation of hydrogen bonds

Changes in the breathing vibration frequency of the pyridine ring have been found to reflect the association–dissociation process of hydrogen bonds. The frequency at about 998 cm^{-1} was therefore chosen in this study to investigate the influence of the pyridine content on the association–dissociation process of hydrogen bonds present in the pyridine ring. Figures 7.21 and 7.22 show the change in the frequency of the breathing vibration of the pyridine ring upon heating and cooling, respectively.

Figure 7.21 shows that when the BINA content is above 40 wt.%, the frequency at about 998 cm^{-1} drops abruptly at a critical temperature (T_c) of 60°C. Moreover, the critical temperature tends to increase as the BINA content decreases. When the BINA content is below 30 wt.%, the breathing vibration frequency of the pyridine ring at room temperature is lower than that of the sample with higher BINA content, but their frequency at the elevated temperature is still higher. This means that the strength of the hydrogen bonds present in the pyridine ring will be reduced at room temperature in PUPys with lower BINA content. At higher temperatures, the hydrogen bonds present in the pyridine ring may become weaker without dissociation, since their dissociation can be affected by the presence of a large hydrogen bond fraction. As a result, the strength of the hydrogen bonds present in

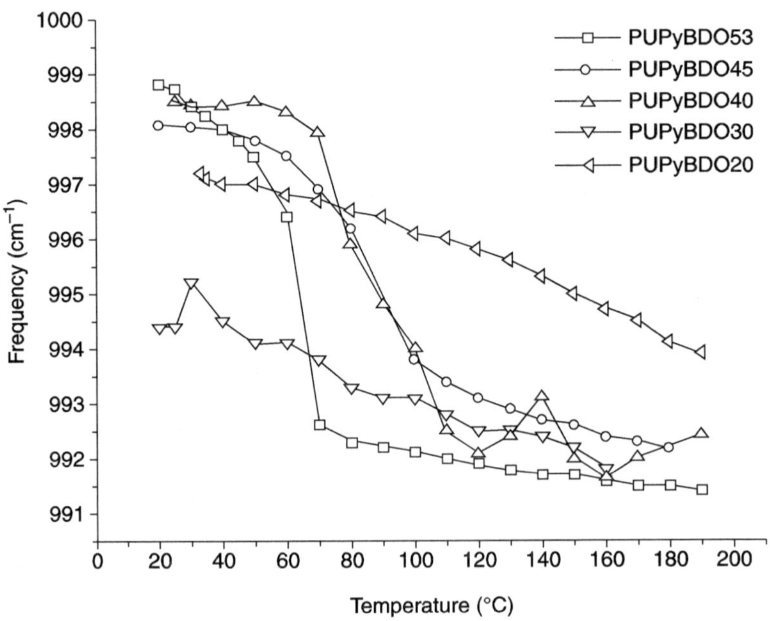

7.21 Frequency of pyridine ring vibration vs temperature upon heating.

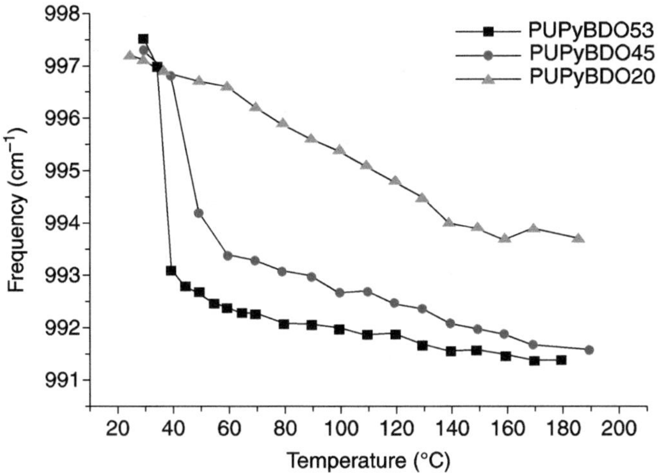

7.22 Frequency of pyridine ring vibration vs temperature upon cooling.

the pyridine ring of BIN-SMPU with lower BINA content gradually reduces as the temperature increases.

In contrast, the breathing vibration frequency of the pyridine ring in PUPys with higher BINA content, namely PUPyBDO53, slowly shifts to a lower frequency upon cooling until the temperature falls below a critical temperature, which is 50°C for the PUPyBDO40 and 40°C for the PUPyBDO53. When the temperature falls below this critical value, the frequency increases abruptly, returning to around 998 cm^{-1}. This means that the strong hydrogen bonds become associated again when they are cooled to a lower temperature, particularly for the BIN-SMPU with higher BINA content. However, when the BINA content is lower than 20 wt.%, no abrupt frequency shift occurs during the cooling process. The hydrogen bonds present in the pyridine ring of BIN-SMPU with lower BINA content become gradually stronger as the temperature decreases: this explains why the BIN-SMPU with higher BINA content has excellent SMEs, while polyurethane with lower BINA content has lower shape recovery and a slow shape recovery process. These results again confirm that the hydrogen bonds present in the pyridine ring play a key role in the shape recovery of the thermally-induced SMEs of BIN-SMPUs.

7.4.4 Model of thermally-induced shape memory mechanism of BIN-SMPUs

In order to explain the mechanism of thermally-induced shape memory behavior, a molecular model of the thermally-induced SME mechanism of BIN-SMPUs is presented in Fig. 7.23.

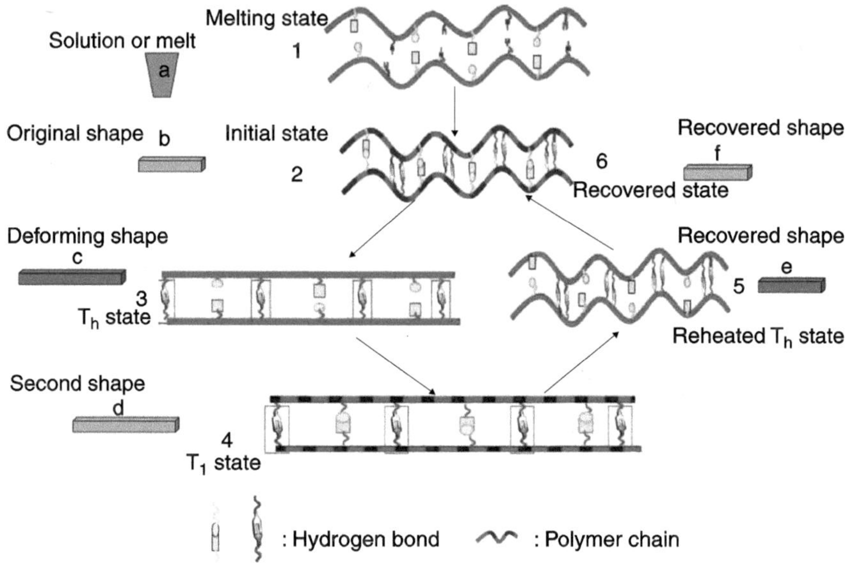

7.23 Shape memory mechanism of thermally-induced SME of BIN-SMPUs.

This figure shows that the large fraction of hydrogen bonds with T_{ah} (high association temperature) acting as the physical net point will determine the permanent shape of the polymer network, whereas the large fraction of hydrogen bonds with T_{al} (low association temperature) acting as the reversible 'switch' will fix the deformed strain upon cooling and release the stored strain upon heating, resulting in the shape recovery. When the BIN-SMPU is in solution or in the melting state ($T > T_{ah}$), referred to as state 1 in Fig. 7.23, the polymer chain is random and free. The BIN-SMPU solutions or melting polymer (shape a) can then be fabricated into any initial shape (shape b) using the conventional process. When the temperature cools to below the T_{al} at state 2, the two kinds of hydrogen bonds are associated and the shape b can be remembered. If the temperature is subsequently raised to a temperature of T_h ($T_{al} < T_h < T_{ah}$) at state 3, the hydrogen bonds with T_{al} are dissociated or become weaker, whereas the hydrogen bonds with T_{ah} are still strong enough to retain the microscopical shapes. The polymer can then be deformed to a temporary shape (shape c) through the application of external force to the rubber state. Cooling to T_l ($T_l < T_{al}$) at state 4 under the load will fix the deformed shape (shape d), because the hydrogen bonds with T_{al} are associated. Moreover, when the temperature is again raised to above T_h at state 5, the deformed strain is immediately released as the hydrogen bonds with T_{al} are dissociated. This results in macroscopic shape recovery (shape e) and in shape fixation again at room temperature (shape f). Shape f is essentially similar to shape b. Hence, both shape fixation and shape recovery can clearly be achieved in the BIN-SMPUs.

7.5 Conclusions

This chapter has provided a thorough investigation and analysis of thermally-induced SMEs, and the following conclusions have been reached:

1. BIN-SMPUs have typical thermally-induced SMEs on an S-shaped curve with a T_r of 45 to 55°C. Relatively high shape fixity and high shape recovery can be achieved in both the PUPy-BDO series and the PUPy-MB series of PUPys.
2. The BINA content or pyridine ring fraction is the key element in determining shape recovery in BIN-SMPUs. As the BINA content drops, the shape fixity decreases slightly, while the shape recovery decreases significantly. In order to achieve a satisfactory shape recovery of above 90%, the PUPy-BDO series of PUPys should contain at least 30 wt.% BINA.
3. In the PUPy-MB series of PUPys, the shape recovery drops as the BINA content decreases and the MDI-BDO content increases. However, the addition of an MDI-BDO segment does not improve the shape recovery, presumably because this segment interrupts the stability of the hard phase of PUPy.
4. Instant shape fixity is good in BIN-SMPUs, but shape retention tends to decrease over time. PUPys with a higher BINA content show a comparatively better shape stability. In addition, lower elongation results in higher strain stability in BIN-SMPUs.
5. Shape recovery force improves as the BINA content drops in the PUPy-BDO series, and as the MDI-BDO content increases in the PUPy-MDI series: the addition of MDI-BDO can thus improve the shape recovery force.
6. The deformation temperature has a significant effect on the SMEs of BIN-SMPUs. The T_r increases with the deformation temperature: a higher deformation temperature results in higher shape fixity, while a lower deformation temperature results in lower shape fixity. Moreover, the final shape recovery is usually higher in the BIN-SMPUs deformed at the lower temperature, whereas the shape recovery process of BIN-SMPUs deformed at higher temperature tends to pass through two stages with a much higher temperature in the second stage of strain recovery.
7. Temperature-dependent FT-IR shows that the hydrogen bonds present in the pyridine ring are responsive to the temperature stimulus at about 40 to 60°C. However, the hydrogen bonds between the C=O and N–H of the urethane groups do not dissociate completely until the temperature increases to above 120°C. It is therefore suggested that the hydrogen bonds present in the pyridine ring serve as the molecular switches, and the hydrogen bonds present in the

urethane groups can provide the PUPys with an elastic polymer network for shape recovery.

7.6 References

Beblo, R., Mauck Weiland, L. (2008), Strain induced anisotropic properties of shape memory polymers, *Smart Materials & Structures*, **17**, 055021 (7).

Chen, S. J., Hu, J. L., Liu, Y. Q., Liem, H. M., Zhu, Y., Liu, Y. J. (2007a), Effect of SSL and HSC on morphology and properties of PHA-based SMPU synthesized by bulk polymerization method, *Polymer Science Part B: Polymer Physics*, **45**, 444–54.

Chen, S. J., Hu, J. L., Liu, Y. Q., Liem, H. M., Zhu, Y., Meng, Q. H. (2007b), Effect of molecular weight on shape memory behavior in polyurethane films, *Polymer International*, **56**(9), 1128–34.

Chen, S. J., Hu, J. L., Yuen, C. W., Chan, L. K. (2009), Supramolecular polyurethane networks containing pyridine moieties for shape memory materials, *Materials Letters*, **63**, 1462–4.

Cho, J. W., Kim, J. W., Jung, Y. C., Goo, N. S. (2005), Electroactive shape-memory polyurethane composites incorporating carbon nanotube, *Macromolecular Rapid Communications*, **26**, 412–16.

Huang, W. M., Yang, B. (2005), Water-driven programmable polyurethane shape memory polymer: Demonstration and mechanism, *Applied Physics Letters*, **86**, 114105.

Jiang, B. H., Zhou, W. M., Liu, Y., Qi, X. (2003), Recent progress of magnetically controlled shape memory materials, *Materials Science Forum*, **3**, 2285–90.

Lee, B. S., Chun, B. C., Chung, Y.-C., Sul, K. I., Cho, J. W. (2001), Structure and thermo-mechanical properties of polyurethane block copolymers with shape memory effect, *Macromolecules*, **34**, 6431–7.

Lendlein, A., Jiang, H., Junger, O., Langer, R. (2005), Light-induced shape-memory polymers, *Nature*, **434**, 879–82.

Li, F. K., Hou, J. N., Zhu, W., Zhang, X., Xu, M., *et al.* (1996), Crystallinity and morphology of segmented polyurethanes with different soft-segment length, *Applied Polymer Science*, **62**, 631–8.

Li, J. H., Viveros, J. A., Wrue, M. H., Anthamatten, M. (2007), Shape-memory effects in polymer networks containing reversibly associating side groups, *Advanced Materials*, **19**, 2851–4.

Lin, J. R., Chen, L. W. (1998a), Study on shape-memory behavior of polyether-based polyurethanes, Part I: Influence of the hard-segment content, *Applied Polymer Science*, **69**, 1563–74.

Lin, J. R., Chen, L. W. (1998b), Study on shape-memory behavior of polyether-based polyurethanes, Part II: Influence of soft-segment molecular weight, *Applied Polymer Science*, **69**, 1575–86.

Liu, G. Q., Ding, X. B., Cao, Y. P., Zheng, Z. H., Peng, Y. X. (2004), Shape memory of hydrogen-bonded polymer network/poly(ethylene glycol) complexes, *Macromolecules*, **37**, 2228–32.

Mikhaylova, Y., Adam, G., Haussler, L., Eichhorn, K. J., Voit, B. (2006), Temperature-dependent FT-IR spectroscopic and thermo-analytic studies of hydrogen bonding of hydroxyl (phenolic group) terminated hyperbranched aromatic polyesters, *Molecular Structure*, **788**(1–3), 80–8.

Ratna, D., Karger-Kocsis, J. (2008), Recent advances in shape memory polymers and composites: A review, *Materials Science*, **43**, 254–69.

Yuan, Z., Ji, B., Wu, L. B. (2009), Synthesis and thermal induced shape memory properties of biodegradable segmented poly(ester-urethane)s, *Acta Polymerica Sinica*, **2**, 153–8.

Zhu, Y., Hu, J. L., Liu, Y. J. (2009), Shape memory effect of thermoplastic segmented polyurethanes with self-complementary quadruple hydrogen bonding in soft segments, *European Physical Journal E*, **28**, 3–10.

8
Moisture-induced properties of supramolecular shape memory polymers containing pyridine

DOI: 10.1533/9780857098542.196

Abstract: This chapter introduces the moisture-induced shape memory effect (SME) observed in supramolecular shape memory polymers, particularly shape memory polyurethane (SMPU), containing pyridine moieties. The moisture absorption of polyurethane networks containing pyridine moeties (PUPy) is discussed followed by an investigation into the effect of relative humidity (RH), temperature, BINA content and MDI-BDO content. The influence of moisture absorption on both the thermal properties and the dynamic mechanical properties of SMPUs is also described, along with the moisture-induced SME mechanism.

Key words: moisture-induced shape memory effect, pyridine, supramolecular shape memory polymer, moisture absorption.

8.1 Introduction

Shape memory polymers (SMPs) are elastic networks equipped with suitable stimuli-sensitive switches, whose shape memory effect (SME) may be induced by temperature, light, pH and other factors (Ratna and Karger-Kocsis, 2008). Many SMPs have thermally-induced SMEs, with a thermal reversible phase that serves as a 'switch' being either amorphous or semi-crystalline (Zhang *et al.*, 2008). In SMPs with light-induced SMEs, the stimulus is achieved through the incorporation of reversibly reacting molecular switches (Lendlein *et al.*, 2005). SMEs are also observed in polymer blends consisting of an immiscible elastomer and a switch polymer (Zhang *et al.*, 2009). The incorporation of suitable switches, such as molecular switches, switch segments and switch polymers, is therefore expected to result in elastic networks with a variety of SMEs, including those induced by moisture.

To date, thermally-induced SMPs have offered the most potential, due to their wide availability and large number of possible applications (Zhang *et al.*, 2009). Their thermally-induced SMEs are generally triggered by heating the specimen above the melting temperature (T_m) of the crystalline soft segment or above the glass transition temperature (T_g) of the amorphous soft segment (Wornyo *et al.*, 2007; Gunes *et al.*, 2008). However, application temperatures are necessarily at the lower end of the scale, particularly in the field of biomedicine, meaning that the pre-stored strain must be recovered without external heating. More recently, water-influenced programmable shape memory behavior has been observed in shape memory polyurethane (SMPU), because the

T_g of SMPU reduces dramatically after immersion in water. Further investigation has shown that hydrogen bonds play a key role in determining water-influenced SMEs in SMPUs, and the water or moisture absorbed by the polymer is a crucial part of the shape recovery process (Huang and Yang, 2005; Yang *et al.*, 2005, 2006). However, the shape recovery of these polymers is only water-influenced; strain recovery is not caused simply by moisture, as the polymers have low moisture absorption.

Earlier studies in this book have discussed the morphology of SMPU and thermally-induced SMEs, and have demonstrated that the hydrogen-bonded supramolecular polyurethane networks containing pyridine moieties (PUPys) can be used as shape memory materials. It was also confirmed that the movement of the polymer chain of BINA-based shape memory polyurethane (BIN-SMPU) was significantly influenced by the hydrogen bonding, and that the dissociation of the hydrogen bonds present in the pyridine ring resulted in strain recovery. Moreover, Jaczewska *et al.* (2007) observed during a study of the humidity and solvent effects of polythiophene and polyvinyl pyridine that the humidity absorption of the polyvinyl pyridine was several times higher than that of the polythiophene.

It was hypothesized that the pyridine ring was responsive to moisture absorption. It should therefore be the case that moisture-induced SMEs are found in SMPU containing pyridine moieties, as the pyridine forms hydrogen bonds with supramolecular 'switches', which have a high moisture response. This chapter discusses and analyses a series of synthesized BIN-SMPUs with moisture-induced SMEs.

8.2 Moisture absorption of BIN-SMPUs

Moisture absorption in polymeric films is an important process used by a variety of industries, ranging from micro-electronics to adhesives and coatings. There have been few studies to date on the moisture absorption of thermally-induced SMPs. Since this plays a fundamental role in determining moisture-induced SMEs, this chapter will begin with an analysis of moisture absorption and moisture diffusion in BIN-SMPUs. Moisture absorption is usually determined by comparing the weight of wet and dry specimens; the percentage of moisture absorption at a time t (M_t) is calculated using the following equation:

$$M_t = [(W_t - W_d)/(W_d)]*100\%$$

where W_d and W_t refer to the weight of the dry specimen and the wet specimen, respectively.

8.2.1 Influence of temperature on moisture absorption

This study uses rectangular specimens 2 mm thick, cut from a sample of PUPyBDO45, for moisture absorption testing. After drying at 100°C for 12 hours, the specimens are placed in a humidity temperature oven with a constant relative humidity (RH) of 65%, but different temperatures, namely 22, 25, 28, 31 and 37°C. In the first 30 minutes, the weight percentage of the specimen is measured every 5 minutes. In the next 30 minutes, the interval time is increased to 10 minutes and then to 20 minutes in the hour after that (i.e. 60 to 120 min). Finally, once the absorption speed has slowed, the interval is increased to 30 minutes. The moisture absorption over time at 65% RH and different conditioning temperatures is shown in Fig. 8.1, while Fig. 8.2 shows the effect of different conditioning temperatures at 65% RH on maximum moisture absorption.

Figure 8.1 shows that PUPyBDO45 absorbs moisture quickly in the first 90 minutes. Moisture absorption then slows, and the sample reaches saturation after 1500 minutes within the temperature range of 22 to 37°C. In the initial stage, this moisture absorption process is in accordance with Fick's second law (Vogt *et al.*, 2005). The Shen–Springer equation determines that the diffusion coefficient increases as the conditioning temperature rises (Lin *et al.*, 2002). The moisture content of a polymer at any one time similarly increases with the rise in conditioning temperature, as shown in Fig. 8.1. For example, the moisture absorption is only 0.7% at 22°C but increases to 2.1% at 37°C within 90 minutes. Finally, Fig. 8.2 shows that the maximum moisture absorption also increases as the conditioning temperature increases, particularly below 28°C. The maximum

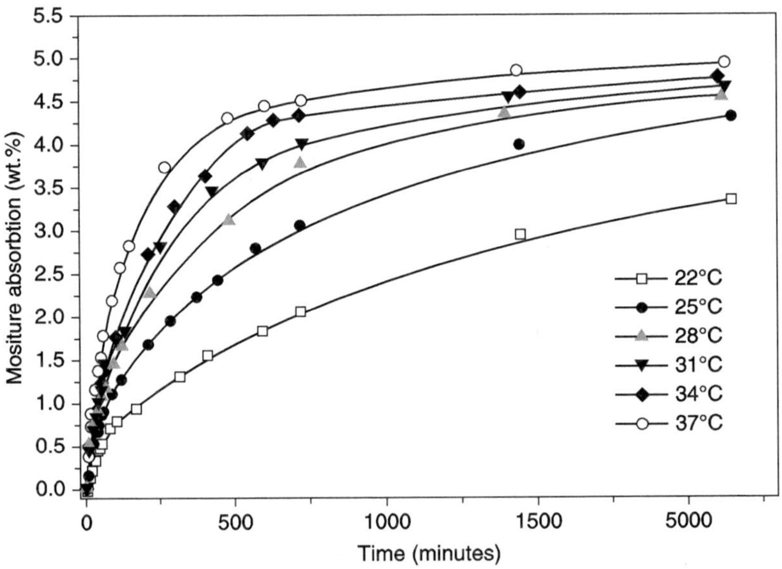

8.1 Moisture absorption over time under the condition of RH = 65% with various conditioning temperatures: T_1 = 22°C; T_2 = 25°C; T_3 = 28°C; T_4 = 31°C; T_5 = 34°C; and T_6 = 37°C.

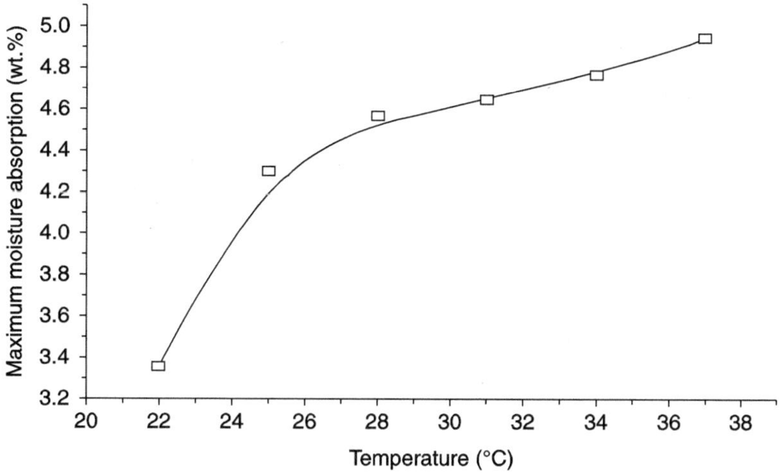

8.2 Dependency of maximum moisture absorption on temperature at RH = 65%.

moisture absorption increases slightly when the temperature is raised to above 28°C, probably because moisture absorption is mainly determined by the difference in the moisture content between the conditioning environment and the materials (Vogt *et al.*, 2005). In addition, Fig. 8.1 shows that the slope of moisture absorption at 37°C is much higher than that at 22°C, i.e. moisture absorption is faster at the higher temperature. Although the moisture content may be lower at higher temperature, the diffusion speed is still the main driving force for moisture absorption, due to its high diffusion coefficient at higher temperatures.

8.2.2 Influence of relative humidity on moisture absorption

Figure 8.3 shows the absorption over time at different RH and at a constant temperature of 34°C, while the effect of RH on maximum moisture absorption is shown in Fig. 8.4. The moisture absorption of PUPyBDO45 rises continually with the increase in time until the specimen is saturated. The moisture absorption of a specimen tends to vary depending on the length of time for which it is exposed to moisture conditions, for example 300 minutes at 90% RH, over 450 minutes at 70% RH, and over 700 minutes at 60% RH. Figure 8.4 also shows that the maximum moisture absorption increases as RH increases, particularly at over 70% RH. For example, the maximum moisture absorption is only 4.5% at 70% RH, while it reaches 7.23% at 80% RH and 10.51% at 90% RH. This confirms that the moisture absorption at equilibrium is dependent on the RH, as higher RH

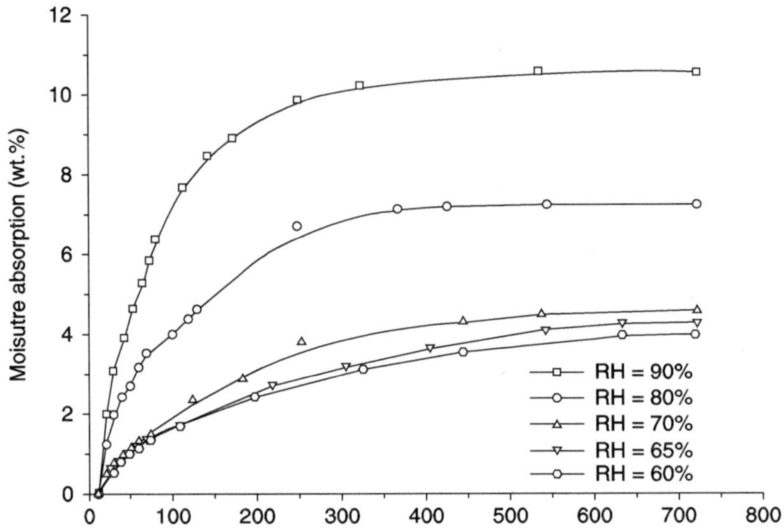

8.3 Moisture absorption over time with various RH and at a constant temperature of 34°C.

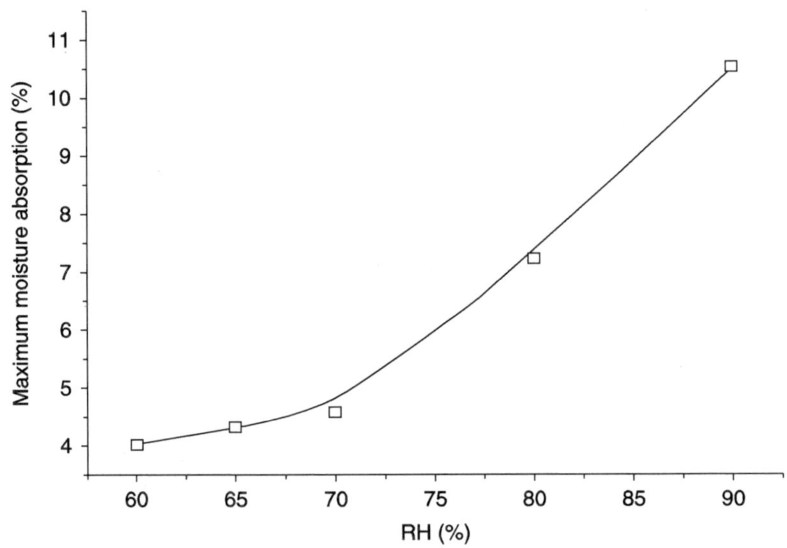

8.4 Effect of maximum moisture absorption on RH at T = 34°C.

results in a higher moisture absorption speed. BIN-SMPU must therefore be sensitive to moisture. As both temperature and RH can have a significant influence on the moisture absorption speed and maximum moisture absorption, higher moisture absorption can be achieved by controlling the RH and temperature.

8.2.3 Influence of BINA content on moisture absorption

Under the same conditions of 32.5°C and 85% RH, BIN-SMPUs containing different amounts of BINA show different levels of moisture absorption. Figure 8.5 shows the moisture absorption over time of BIN-SMPUs containing different amounts of BINA, and Fig. 8.6 presents the effect of the BINA content on moisture absorption over time. Figure 8.5 shows that all BIN-SMPU samples have a similar moisture absorption process; the main difference between them is that after the saturation point, the moisture absorption tends to decrease slightly over time, particularly for the BIN-SMPUs with high BINA content, namely PUPyBDO53 and PUPyBDO45. More importantly, the BIN-SMPU samples with higher BINA content have higher moisture absorption than those with lower BINA content after any length of time. The maximum value of moisture absorption and absorption speed are also observed to be higher at 24, 48 and 65 hours in the samples with higher BINA content.

To test the influence of the thickness of the contact surfaces on moisture absorption, Fig. 8.7 presents the moisture absorption per unit over time for BIN-SMPUs containing different amounts of BINA. The initial slope of moisture absorption is observed to increase in close correlation with the increase in BINA content, i.e. the BIN-SMPU sample with higher BINA content has a higher moisture absorption speed. The BINA content reflects the pyridine content in the

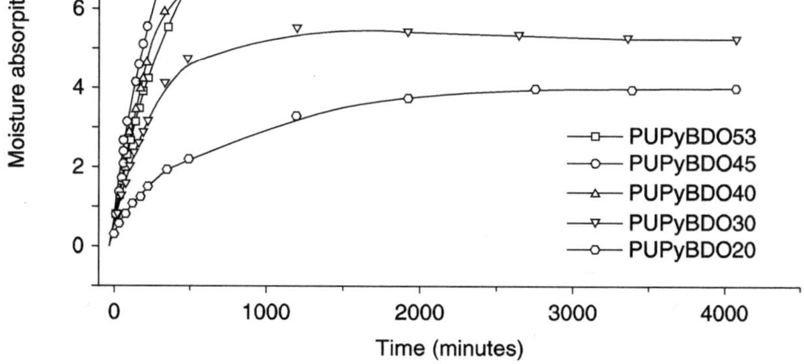

8.5 Dependency of moisture absorption on time for BIN-SMPU with various BINA contents.

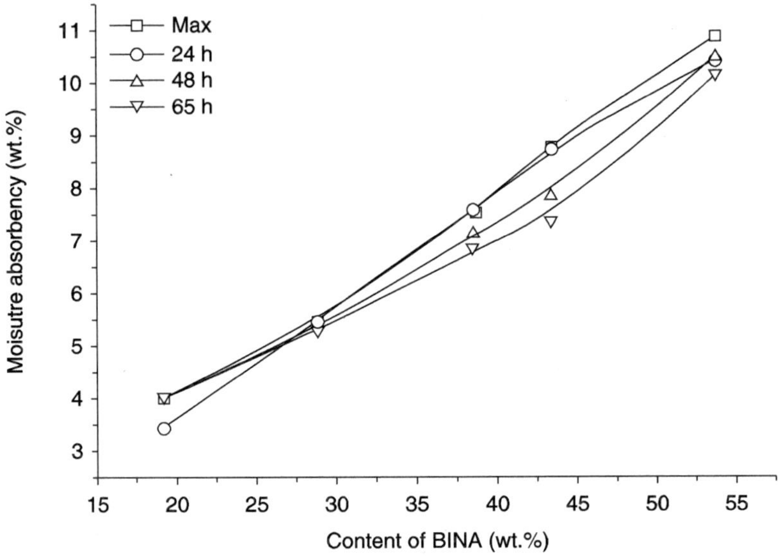

8.6 Effect of the BINA content on moisture absorption over time.

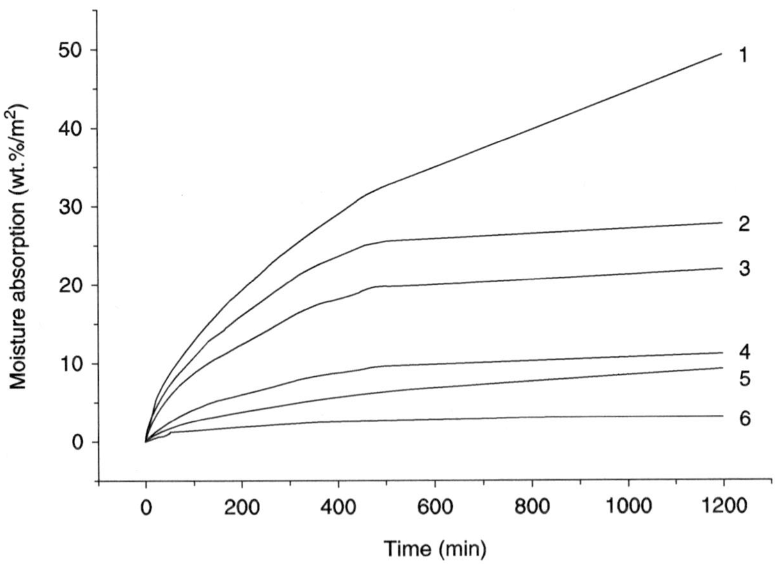

8.7 Moisture absorption per unit vs time for BIN-SMPU with various BINA contents.

BIN-SMPUs. It can therefore be concluded that a higher pyridine content results in higher moisture absorption at higher speed.

8.2.4 Influence of MDI-BDO content on moisture absorption

Figure 8.8 shows the moisture absorption over time for BIN-SMPUs containing different amounts of MDI-BDO. The PUPyMB0 sample without an MDI-BDO segment reaches its maximum moisture absorption of 10.2 wt.% within 354 minutes. The maximum moisture absorption of the samples is 9.12 wt.% for PUPyMB15 (achieved after 1238 min); 8.0 wt.% for PUPyMB25 (1188 min); 6.78 wt.% for PUPyMB35 (2000 min); and 5.45 wt.% for PUPyMB55 (2600 min). However, it should be noted that these are not their saturated values. This is due to the fact that the absorbed moisture molecule can be divided into two parts: the free water molecule and the hydrogen-bonded water molecule. Although the specimens can reach their maximum moisture absorption, the free water molecule still continues to exchange with the environment. As a result, the moisture absorption decreases slightly after reaching its maximum moisture absorption.

More importantly, a drop in the maximum moisture absorption and the final moisture absorption is observed as the MDI-BDO content increases and the BINA content decreases in the PUPy-MB series of BIN-SMPUs. This confirms that the

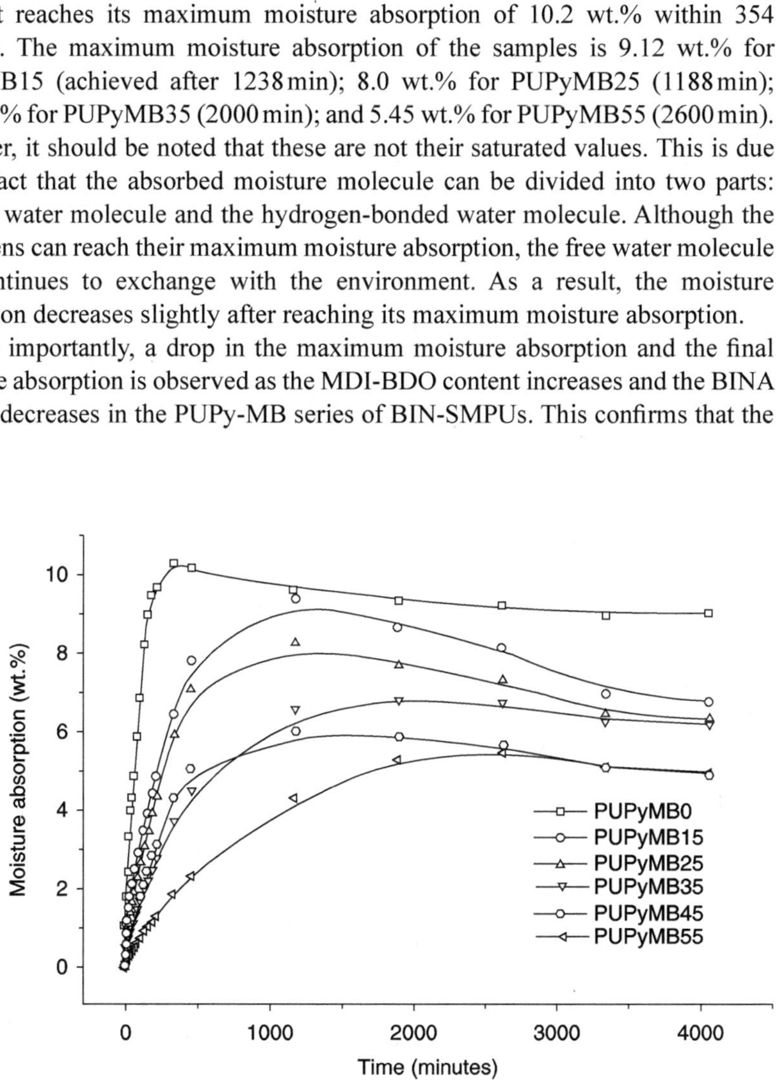

8.8 Moisture absorption over time for BIN-SMPUs containing various amounts of MDI-BDO.

moisture absorption is mainly dependent on the BINA content, i.e. the pyridine ring plays an important role in moisture absorption in the BIN-SMPUs.

8.2.5 Thermal gravimetric analysis

Moisture absorption is traditionally determined by measuring the weight of specimens before and after exposure to moisture. In the present study, thermal gravimetric analysis (TGA) is also performed to investigate the moisture absorption at equilibrium. Specimens are placed in a humidity temperature oven at 85% RH and at 37°C in order to reach the saturated state before weight loss is measured. The TGA curves of BIN-SMPUs containing different amounts of BINA are shown in Fig. 8.9, and the weight loss at different temperature ranges is summarized in Table 8.1. Figure 8.9 shows that the first stage of weight loss in BIN-SMPUs starts at about 60°C, and then stops temporarily at about 150°C. The second stage starts at 220°C and then stops after the temperature is raised to above 500°C.

The same figure also shows that weight loss in the first stage decreases as the BINA content drops. Moreover, the data summarized in Table 8.1 show that the weight loss below 200°C is similar to the value of moisture absorption at equilibrium as tested using the weight method. For example, the moisture absorption at equilibrium tested by the weight method is 7.31 wt.% for the PUPyBDO45 and 6.80 wt.% for the PUPyBDO40. However, the TGA results show that the weight loss in the first stage is 7.18 and 60 wt.% for PUPyBDO45

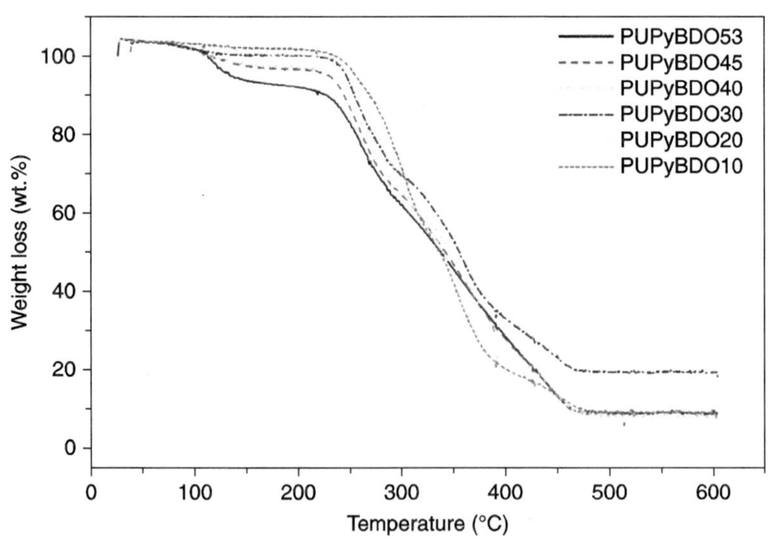

8.9 TGA curves of BIN-SMPU with various BINA contents.

Table 8.1 TGA data of BIN-SMPU with various BINA contents

Samples	Weight loss at 50–200°C (wt.%)	Weight loss at 200–300°C (wt.%)	Weight loss at 300–420°C (wt.%)	Weight loss at 300–500°C (wt.%)
PUPyBDO53	11.08	30.84	53.06	53.06
PUPyBDO45	7.18	31.22	56.61	56.61
PUPyBDO40	6.80	30.99	58.08	58.08
PUPyBDO30	3.66	30.39	50.19	50.19
PUPyBDO20	3.60	41.82	39.86	9.98
PUPyBDO10	2.39	46.66	37.24	8.56

and PUPyBDO40, respectively. This therefore confirms that the weight loss occurring between 20 and 200°C is the result of moisture desorption in the polymers, and the moisture absorption at equilibrium enhances as the BINA content increases.

Figure 8.10 shows the TGA curves of BIN-SMPU samples containing different amounts of MDI-BDO, and their TGA data are summarized in Table 8.2. In the PUPy-MB series of BIN-SMPUs, the weight loss in the first stage is lower when MDI-BDO content is higher. This is consistent with the moisture absorption tendency observed using the weight method. These results show that the moisture absorption in both the PUPy-BDO series and the PUPy-MB series of PUPys is dependent on the BINA content. The moisture absorption at equilibrium is increased in samples containing larger amounts of BINA.

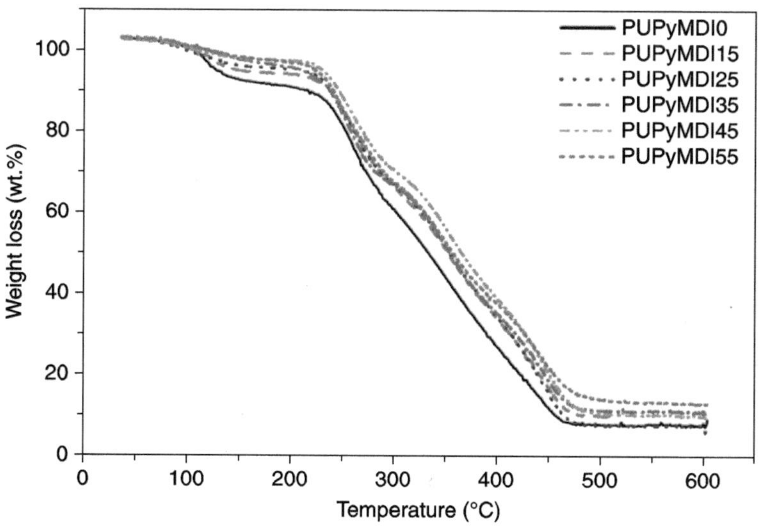

8.10 TGA curves of BIN-SMPU with various MDI-BDO contents.

Table 8.2 TGA data of BIN-SMPUs with various MDI-BDO contents

Samples	50–200°C	200–300°C	300–420°C	420–500°C
PUPyMB0	11.08	30.84	53.06	53.06
PUPyMB15	8.87	30.07	53.92	53.92
PUPyMB25	7.02	32.93	54.78	54.78
PUPyMB35	7.46	33.11	27.57	23.74
PUPyMB45	5.10	31.28	32.82	23.03
PUPyMB55	5.83	34.13	30.28	19.00

8.3 Effect of moisture absorption on the thermal properties of polyurethane

A large fraction of water molecules may be absorbed into BIN-SMPU in the form of moisture. According to Couchman (1978) and Couchman and Karasz (1978), moisture influences the thermal properties of polymers. As the moisture content increases from 0 to 20 wt.%, the T_g of collagen, elastin and cellulose decreases by approximately 200, 145 and 260°C, respectively (Agarwal *et al.*, 1997). Recently, Yang *et al.* (2005) found that a decrease in T_g of more than 35°C was obtained when the polymer absorbed about 4 wt.% water. Shape recovery could consequently be achieved through immersion in water instead of heating.

The hydrogen-bonded supramolecular structure of BIN-SMPU has been described in previous chapters. Compared to the water-driven SMPUs investigated in earlier studies (Huang *et al.*, 2006; Yang *et al.*, 2005, 2006), this kind of SMPU is more responsive to moisture. For example, more than 7 wt.% moisture absorption can be obtained by conditioning the PUPyBDO45 specimens at 85% RH and 32.5°C. Their thermal properties are thus significantly affected by the absorbed moisture. Figures 8.11 to 8.13 respectively show the differential scanning calorimetry (DSC) curves of PUPyBDO45 after conditioning for different lengths of time at 90% RH and 34°C; 80% RH and 34°C; and 65% RH and 22°C.

Figure 8.11 displays obvious T_g at about 40.7°C on the first heating curves of dry PUPyBDO45. However, after it has been conditioned at 90% RH and 34°C for 30 minutes, the glass transition process takes place over a broader temperature range, 3 to 77°C, and the T_g tends to shift to a lower temperature of 34°C. This implies that the amorphous phase is significantly influenced by the absorbed moisture. Moreover, no difference is observed at the higher temperature range after 30 minutes of conditioning time. After the sample has been exposed to moisture conditions at higher temperature for more than 3.5 hours, the T_g moves to an even lower temperature of 13°C. An abrupt phase transition with a large enthalpy change (Cp) accompanied by a number of small exothermic peaks is measured at above 130°C. The high temperature phase transition with a large

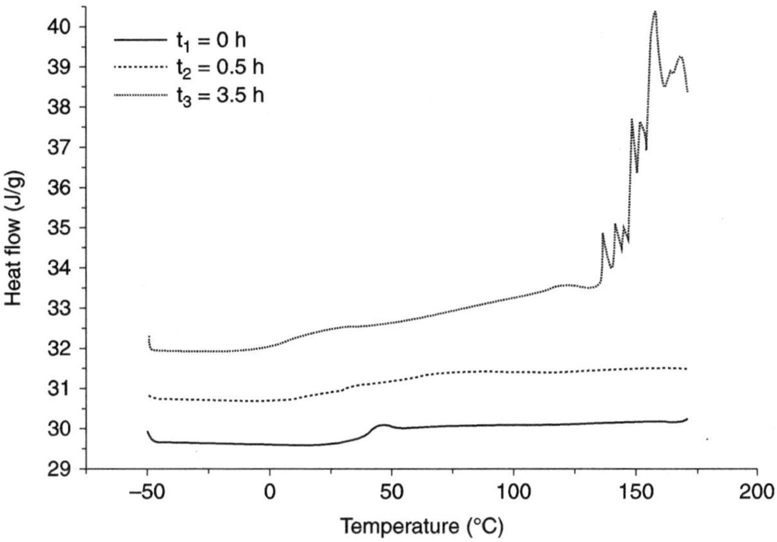

8.11 DSC curves of PUPyBDO45 after conditioning at RH = 90% and T = 34°C for various lengths of time.

enthalpy change can be attributed to the glass phase transition of the hard phase, while the exothermic peaks are presumably the result of the dissociation of bonded water.

When RH is lowered to 80% without changing the temperature, the T_g of PUPyBDO45 moves to a significantly lower temperature over time (Fig. 8.12). Within 6.5 hours, the T_g decreases by approximately 39°C, i.e. from 41 to 2°C. A large enthalpy change also appears on the heating curves at about 133°C after conditioning at 80% RH for two hours, and a small exothermal peak appears during the high temperature phase change process. However, as the moisture absorption of the polymer increases, the enthalpy change becomes larger and takes place at a lower temperature. For example, a high temperature phase transition appears at 127°C after conditioning for three hours, and then moves to 124°C after 6.5 hours. Furthermore, a number of exothermic peaks appear in the enthalpy changing process after conditioning for more than 4.5 hours. This implies that both the RH and conditioning time affect the morphology of BIN-SMPU.

As described above, higher RH results in higher moisture absorption for the BIN-SMPU, while PUPyBDO45 tends to have lower moisture absorption at lower RH. Since higher temperature can increase the moisture absorption speed, the specimens cut from the same PUPyBDO45 sample were conditioned at a lower RH of 65%, and a lower temperature of 22°C. The DSC curves of PUPyBDO45 after conditioning at 65% RH and 22°C for different lengths of time are presented in Fig. 8.13. The T_g can be observed to undergo a slight change

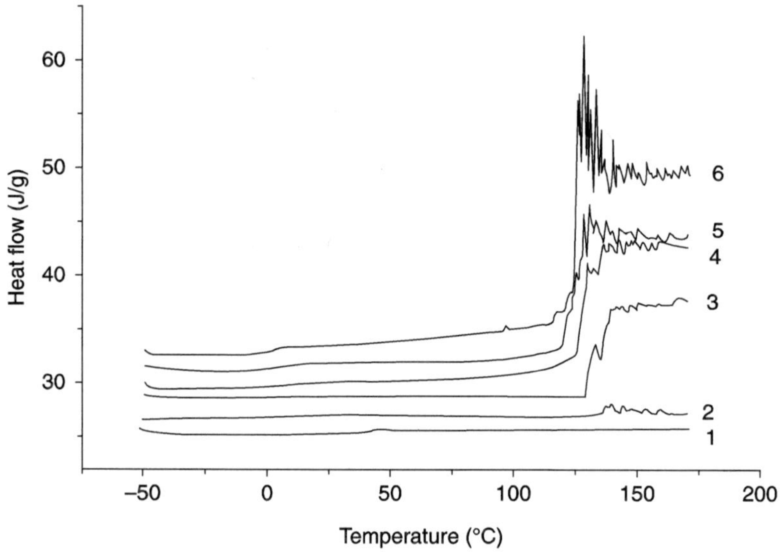

8.12 DSC curves of PUPyBDO45 after conditioning at RH = 80% and T = 34°C for: (1) 0 hours; (2) 1 hour; (3) 2 hours; (4) 3 hours; (5) 4 hours; and (6) 6.5 hours.

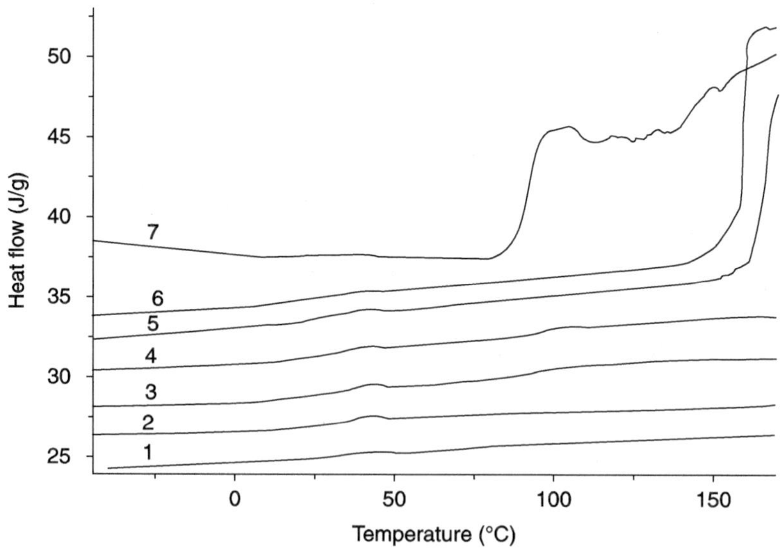

8.13 DSC curves of PUPyBDO45 after conditioning at RH = 65% and T = 22°C for: (1) 0 hours; (2) 1 hour, (3) 2 hours; (4) 4 hours; (5) 8 hours; (6) 10 hours; and (7) 11 hours.

within the first four hours, and the large enthalpy change in the higher temperature range also does not appear until the sample is conditioned at lower RH and lower temperature for more than 7.5 hours. The T_g of the hard phase is then observed to move to a lower temperature range as the conditioning time increases. The phase transition even starts at 82°C after conditioning for more than 11 hours. More importantly, the DSC analysis did not reveal any small exothermic peaks at above 100°C under these conditions. It can thus be concluded that the high temperature phase transition is not the result of the evaporation of water molecules, but of the phase transition of a long-range disordered structure.

The DSC results suggest that during conditioning, the moisture absorption of the polymer increases over time. As mentioned earlier, the absorbed moisture is principally composed of two parts, free moisture and bonded moisture, particularly at higher RH. The free moisture molecule lubricates the polymer chain, thereby causing a significant decrease in the T_g. The bonded moisture molecule between the polymer chains also reduces the T_g, in this case due to the decrease of intermolecular interaction. The bonded moisture molecule is thought to promote the aggregation of the ordered urethane units via the water bridge hydrogen bonding, i.e. –N–H. . .H–O–H. . .O=C–. Finally, the long-range disordered structure tends to form a hard phase with high Cp.

8.4 Effect of moisture absorption on the dynamic mechanical properties of polyurethane

The analysis of PUPyBDO45 found that its glassy modulus is high, at 4.0 GPa, but the rubber modulus is low, at 6.5 MPa. A significant decrease in modulus is observed during the glass transition process due to the dissociation of hydrogen bonds, which is responsive to the temperature stimulus. It is also observed that during conditioning at a particular RH and room temperature, the polymer becomes softer over time. This implies that the moisture sorption also influences the dynamic mechanical properties of BIN-SMPUs.

Figures 8.14 and 8.15 show the dynamic mechanical properties of the PUPyBDO45 sample after conditioning at 65% RH and 20°C for different lengths of time. In Fig. 8.14, it can be observed that the modulus starts to decrease at 45°C, and the polymer enters into its rubber modulus plateau after 75°C. In the rubber state, the modulus drops continuously as the temperature rises, followed by an abrupt decrease in modulus at about 90°C. This means that the dry PUPyBDO45 sample has an imperfect phase separation structure. In addition to the amorphous soft phase and hard phase, unstable mixing phases can also be observed. The first stage of the abrupt decrease in modulus is therefore observed during the glass transition process of the soft phase. As the temperature increases, the dissociation of hydrogen bonds in the mixing phase results in a continuous decrease in modulus. However, the second stage of the abrupt decrease in modulus should be understood as the result of the phase transition of the hard phase.

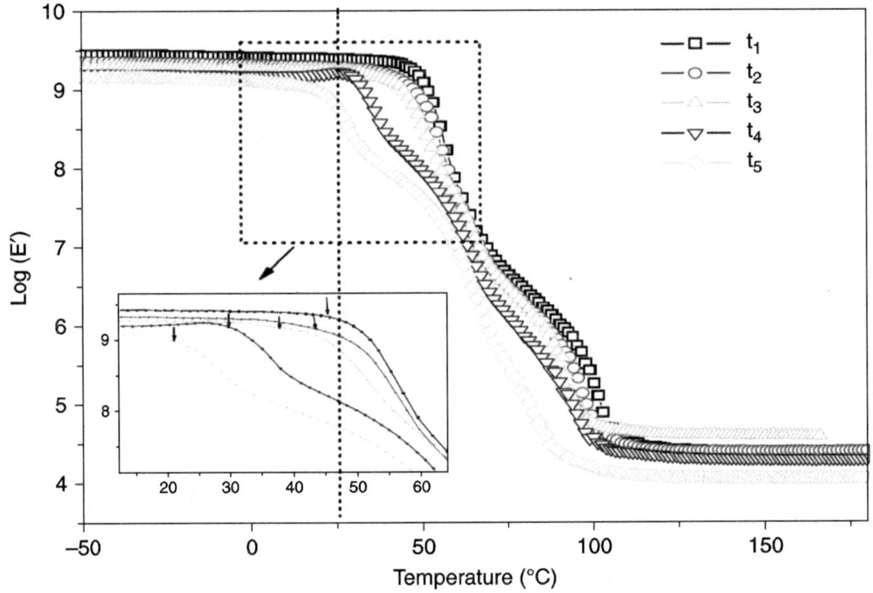

8.14 Storage modulus curves of PUPyBDO45 after conditioning at RH = 65% and T = 20°C for conditioning time: t_1 = 0; t_2 = 2; t_3 = 5; t_4 = 24; and t_5 = 36 hours.

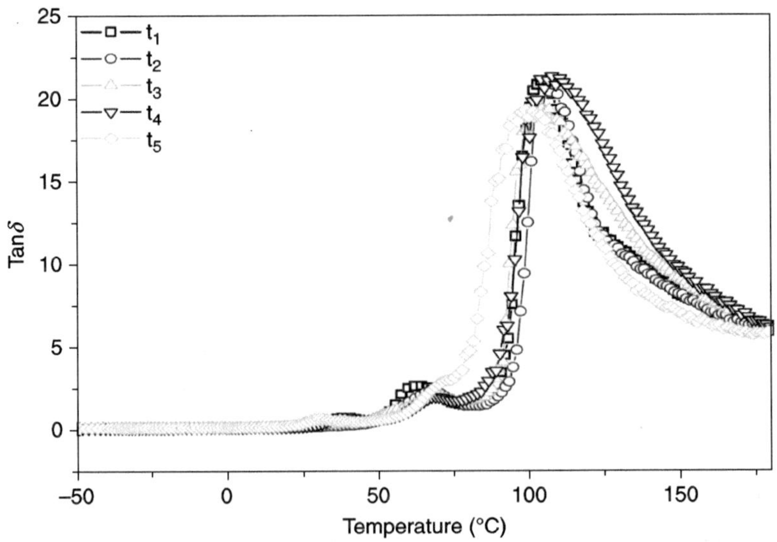

8.15 Tanδ of PUPyBDO45 under the condition of RH = 65% and T = 20°C for: t_1 = 0; t_2 = 2; t_3 = 5; t_4 = 24; and t_5 = 36 hours.

When the PUPyBDO45 sample is conditioned at 65% RH at 20°C, moisture is also absorbed into the polymer. Since the moisture absorption speed is slow at lower temperatures, the absorbed moisture has only a slight effect on the dynamic mechanical properties over the first five hours. Over time, the onset temperature for the first stage of the decrease in modulus falls, shifting from 45°C before conditioning to 41°C after conditioning for five hours; to 29°C after conditioning for 24 hours; and finally to 20°C after conditioning for 36 hours. The glassy modulus plateau similarly decreases to a lower modulus level over time. Polymer softening therefore occurs due to the significant decrease in modulus after conditioning.

The onset temperature of the second stage decrease in modulus is also observed to decrease. During conditioning, the decrease in modulus becomes slower over time, but no abrupt decrease can be measured after five hours. This means that the mixing phase fraction increases while the hard phase fraction decreases. Finally, the rubber modulus plateau disappears and the polymer passes through a third stage of decrease in modulus after 24 hours of conditioning time, causing the polymer to enter into a viscosity modulus plateau. The amorphous soft phase fraction also becomes smaller; the first stage of decrease in modulus appears only within the temperature range of 35 to 55°C after conditioning for 36 hours, while the second stage of decrease in modulus starts at about 60°C, and the polymer enters the visco-elastic state at about 90°C.

In addition, $\tan\delta$ analysis demonstrates the influence of moisture absorption on the polymer chain. In Fig. 8.15, a large peak with a maximum $\tan\delta$ of more than 20 is observed within the higher temperature range of 90 to 180°C. This substantial energy loss can be taken as the result of the phase transition of the hard phase, and again confirms that a large hydrogen bond fraction is present in the hard domain. After longer conditioning periods, the peak range becomes broader and the right shoulder shifts to a higher temperature within 24 hours. This implies that the fraction of hydrogen-bonded hard domains become richer as the moisture absorption increases. However, after conditioning for more than 36 hours, the $\tan\delta$ peak shifts to a lower temperature, since the high moisture absorption of hard domains renders them unstable. Moisture absorption therefore not only promotes the aggregation of hard domains, but also destroys the stability of hard domains. As a result, the original hydrogen bonds present in the urethane groups become weaker after the moisture molecules enter into the polymer chain.

A magnified image of the $\tan\delta$ curve at the lower temperature range is presented in Fig. 8.16. A clear glass transition can be observed at about 66°C for the PUPyBDO45 sample after conditioning for 140 minutes, while a small phase transition appears at about 54°C after conditioning for 280 minutes, which gradually grows and shifts to a lower temperature range over time. For example, the peak appears at 38, 32 and 29°C after conditioning for 24, 25 and 26 hours, respectively. This implies that the moisture absorption results in a new transition which can be attributed to the relaxation of the pendant pyridine side group. The pyridine ring is thought to be bonded in the dry state through the strong hydrogen

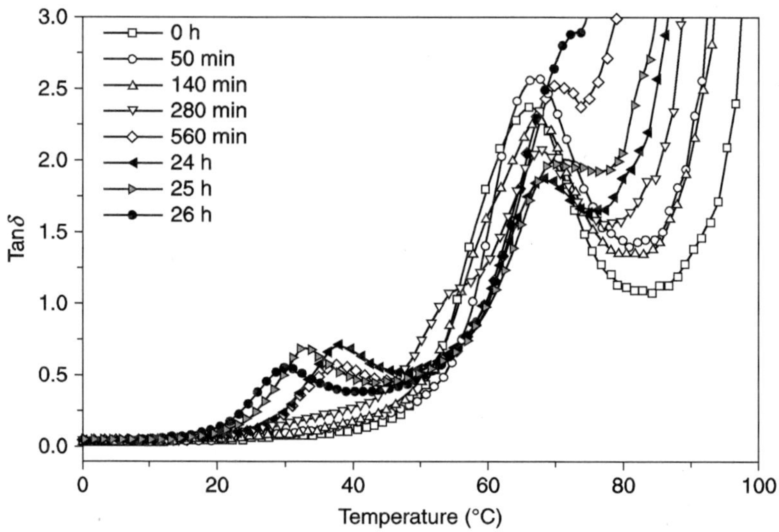

8.16 Tanδ of PUPyBDO45 after conditioning at RH = 65% and T = 20°C for various lengths of time.

bonds between the N–H of the urethane groups and the N of the pyridine ring. However, the absorbed moisture interrupts this kind of hydrogen bonding, resulting in the formation of a large water-bonded pyridine ring fraction as a result of the new hydrogen bonds between the pyridine ring and the moisture molecules. The tanδ curve therefore shows that a clear new transition appears at the lower temperature range after conditioning for 24 hours.

When the RH is raised to 80% and the temperature is raised to 34°C, the influence of moisture absorption on the dynamic mechanical properties of the polymer becomes clearer. Figures 8.17 and 8.18 show the dynamic mechanical properties of PUPyBDO45 at 80% RH and 34°C for different lengths of time. These two figures show that the onset temperature (T_1) for the first stage of decrease in modulus shifts to 16°C within the first hour of conditioning time, and the onset temperature (T_2) for the second stage of decrease in modulus also moves to 68°C. After two hours of conditioning, the first stage of decrease in modulus starts at 12°C, i.e. T_1 = 12°C, while T_2 remains constant. Moreover, after conditioning for four hours, the T_1 also stops at 9°C. This implies that the influence of moisture on both the soft phase and the hard phase remains constant once moisture absorption reaches saturation.

In addition, the tanδ curves shown in Fig. 8.18 demonstrate that during conditioning the transition temperature (T_r) shifts quickly when both RH and temperature are high. These results suggest that the higher moisture absorption has a rapid effect on the dynamic mechanical properties of the polymer.

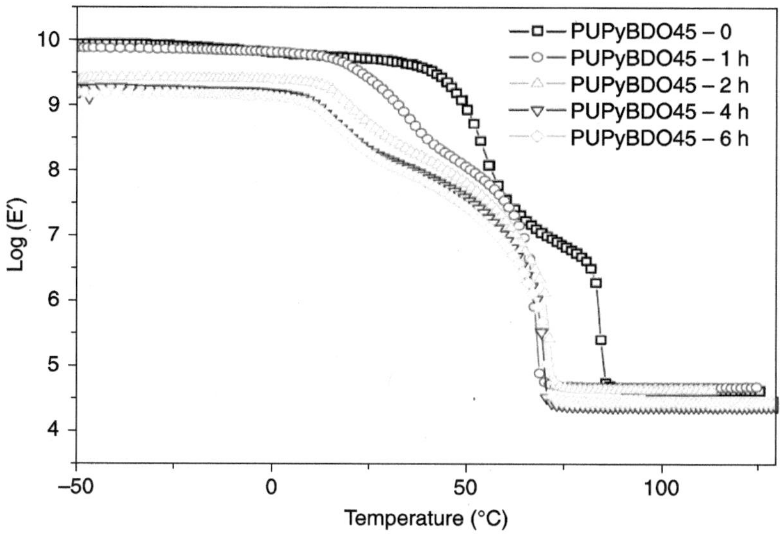

8.17 Storage modulus of PUPyBDO45 after conditioning at RH = 80% and T = 34°C for various lengths of time.

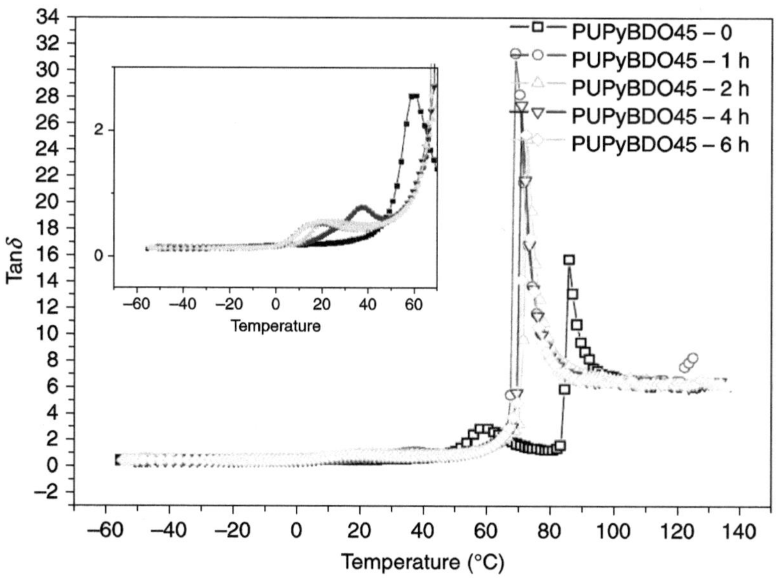

8.18 Tanδ of PUPyBDO45 after conditioning at RH = 80% and T = 34°C for various lengths of time.

Water immersion has a more rapid and more severe effect than moisture conditioning on the dynamic mechanical properties of the polymer. Since the key component responsible for moisture absorption is the BINA unit, samples with a higher BINA content, such as PUPyBDO53, are more sensitive to moisture and water. Figure 8.19 shows the dynamic mechanical analysis (DMA) curves of PUPyBDO53 after different periods of water immersion. Similar to that observed for PUPyBDO45, the decrease in modulus of dry PUPyBDO53 is also divided into two stages, as shown in Fig. 8.19. However, these two stages occur close together; after only ten minutes of immersion, the onset temperature (T_1) for the first stage of decrease in modulus decreases, and continues to fall over the following 20 and 60 minutes. However, the onset temperature (T_2) for the second stage of decrease in modulus moves only slightly over 60 minutes. The tanδ curves in Fig. 8.20 also show that the tanδ peak reflecting the glass transition of the soft phase moves to the lower temperature the longer the immersion time. However, the tanδ peak in the higher temperature range is relatively stable within 60 minutes. This implies that the absorbed moisture only enters into the soft phase, causing the hydrogen bonds present in the hard phase to become relatively stable within the first 60 minutes of water immersion.

The tanδ curves of PUPyBDO53 shown in Fig. 8.20 also show that the β relaxation, reflecting the relaxation of the side chain (Gomez-Carracedo *et al.*, 2003), appears at −70°C after the polymer is immersed in water for more than 60 minutes. As the lower temperature β transition is not observed in moisture

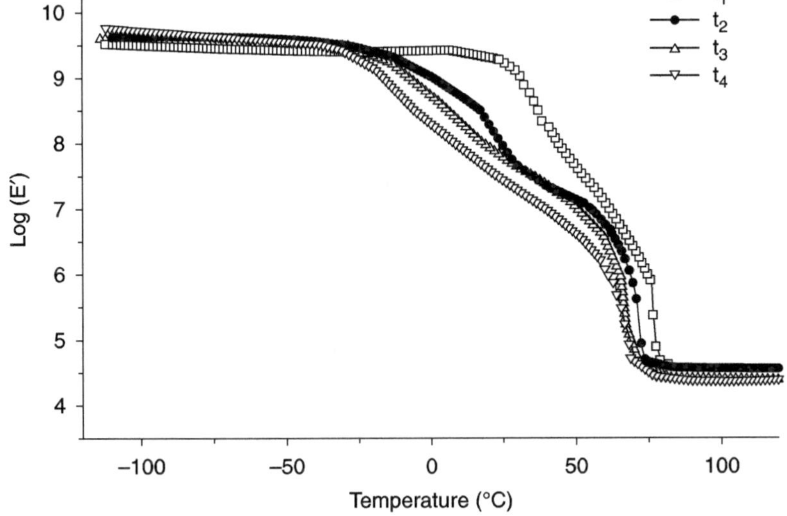

8.19 Storage modulus of PUPyBDO53 after immersing in water for time: t_1 = 0; t_2 = 10; t_3 = 20; and t_4 = 60 minutes.

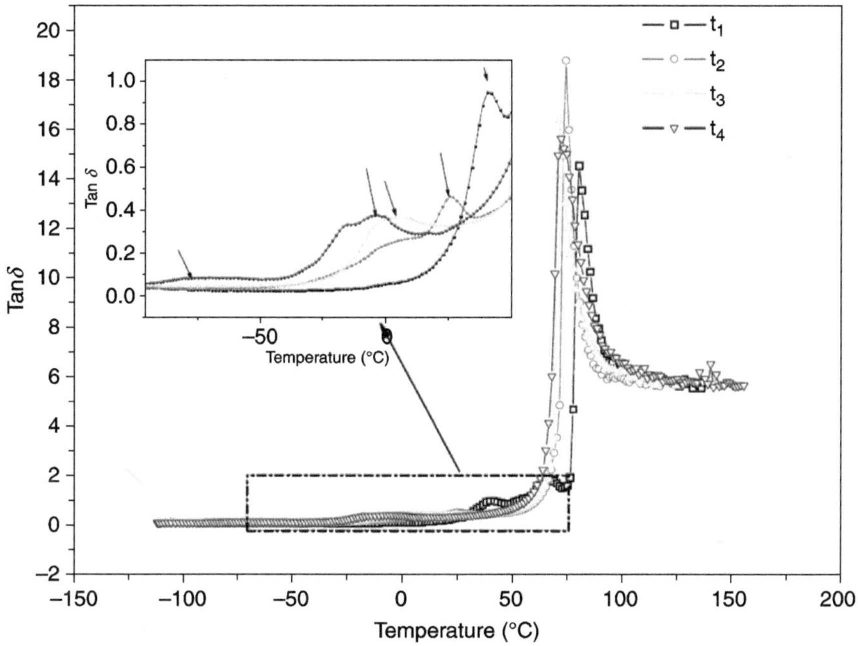

8.20 Tanδ of PUPyBDO53 after immersing in water for time: $t_1 = 0$; $t_2 = 10$; $t_3 = 20$; and $t_4 = 60$ minutes.

conditions, it can be concluded that the pendant pyridine ring is free after 60 minutes of immersion in water. This can be explained by the fact that after a short period of water immersion, the pyridine ring tends to physically link together via the formation of new hydrogen bonds between the pyridine ring and the water molecules (.>N. . . .O–H–O. . . .N<), which replace the original hydrogen bonds between the pyridine ring and the N–H of the urethane group (>N. . .H–N–). However, when the absorbed water molecule content is too high in the polymer following a long period of immersion, the pyridine ring is also protonized by the water molecules due to the formation of new hydrogen bonds (>N. . .O–H–O).

8.5 Moisture-induced and water-influenced shape memory effects (SMEs)

Previous studies (Yang *et al.*, 2004 2005; Huang and Yang, 2005) have reported that the T_g of some SMPUs, such as MM3520 and MM5520, could be reduced dramatically by water immersion. For example, the T_g of MM3520 was reduced by about 35°C after absorbing the moisture fraction of 4.5 wt.%. A water-influenced actuation and recovery were hence observed in sequence in this kind of SMPU (Yang *et al.*, 2004, 2005; Huang and Yang, 2005). In the present study, a higher level of moisture absorption of 10% can be achieved by conditioning BIN-SMPUs at a high RH of 90% and a high temperature of 34°C. As the moisture fraction changes, the T_g of BIN-SMPU also moves to the lower temperature. Hence, shape recovery is expected to occur by applying the stimulus of moisture without external heating.

8.5.1 Moisture-induced SMEs

Figure 8.21 shows the moisture-induced shape recovery process of a flower made of PUPyBDO45 film at 65% RH and 20°C. The shape memory flower opens its petals in sequence at a temperature above T_r (80°C). The present study also found that the petals opened step by step. The difference is that in this experiment the petals open at 65% RH and 20°C without external heating. As shown in Fig. 8.21, a closed flower starts to open its outside petals after being exposed to moisture conditions for 24 hours, and the flower relaxes during the conditioning time. After conditioning for 35 hours, the petals open fully by themselves, thus confirming that the shape recovery of BIN-SMPU can be triggered by the moisture stimulus.

Compared to the water-influenced SMPs investigated in previous studies, such as MM3520 (Huang and Yang, 2005), BIN-SMPU is triggered by moisture at a much lower RH of 65% and a lower temperature of 20°C. However, the shape recovery of MM3520 is triggered by water, and requires a long immersion time. Experimental results confirm that BIN-SMPU should be more sensitive to moisture; shape recovery can therefore be achieved in BIN-SMPUs through the

8.21 Moisture-sensitive shape recovery process of flower under the condition of RH = 65% and T = 20°C.

reduction of the transition temperature, also known as the recovery temperature, to below ambient temperature, i.e. T_r less than 20°C. This discovery is a highly significant and a promising development for the application of SMPs.

8.5.2 Water-influenced SMEs

In order to further investigate the water-influenced shape memory behavior, specimens of PUPyBDO53, PUPyBDO35 and PUPyMB30 were immersed in water for different lengths of time, and their length after different immersion times was recorded. The results are compared in Table 8.3 and show that the PUPyBDO53 sample containing about 53 wt.% BINA starts to recover its deformed strain after only 10 minutes of immersion, and then reaches its equilibrium state within only 20 minutes.

The PUPyBDO35 sample containing 35 wt.% BINA starts to recover after 40 minutes of water immersion, while the PUPyMB30 containing 35 wt.% BINA and 30 wt.% MDI-BDO takes more than 70 minutes to begin shape recovery. The PUPyBDO35 sample reaches its final equilibrium state within 85 minutes, while the PUPyMB30 sample takes more than 180 minutes to complete the shape recovery process. The PUPyMB30 sample shows the highest final shape recovery of 77%, while the PUPyBDO53 sample shows the lowest shape recovery of 7.7%. This is because the samples with a higher pyridine ring content are more sensitive to water absorption. The high moisture absorption will destroy the physical net points of BIN-SMPUs. In addition, the polar aromatic structure influences the polymer's sensitivity to water, and the hard phase of the MDI-BDO hard segment of PUPyMB30 shows greater stability to water.

Finally, it should be noted that although shape recovery can be achieved either by moisture or by water, the final shape recovery of PUPys triggered by water and moisture is still less than that triggered by heat. This is due to the fact that water interrupts the hydrogen bonds both in the reversible phase and physical net points, resulting in significant reduction in the shape recovery force. Water-driven or moisture-induced switches, as well as stable net points or domains, are the two necessary conditions for SMPs that display water-influenced or moisture-induced SMEs

Table 8.3 Results of water-sensitive shape recovery behavior of PUPyBDO53, PUPyBDO35 and PUPyMB30

Sample	L_{start} (mm)	L_{final} (mm)	T_{start} (min)	T_{final} (min)	R_r (%)
PUPyBD053	52.0	50.0	10	20	7.7
PUPyBD035	50.0	39.1	50	85	44.0
PUPyMB30	58.0	35.0	60	180	79.0

8.6 Effect of immersion time on strain recovery

8.6.1 Quantitative analysis of moisture-induced SMEs

Thermally-induced SMEs are usually quantified by thermal-mechanical investigations. A typical test protocol is as follows. First, the test specimen is heated to a temperature (T_h) above the switching temperature (T_{trans}), and stretched to a strain (ε_{max}). The sample is then cooled to below the transition temperature under a constant strain (ε_{max}), thereby fixing the temporary shape. Retracting the clamps to its original position may cause partial strain relaxation. When T_h is raised to above T_{trans}, the sample contracts and the permanent shape is recovered. The fixed strain (ε_f), recovered strain (ε_r), shape fixity (R_f) and shape recovery (R_r) can all be calculated on the basis of the maximum strain (ε_{max}):

$$R_f = \varepsilon_f/\varepsilon_{max} * 100\%$$

$$R_r = \varepsilon_r/\varepsilon_{max} * 100\%$$

In previous studies, a microscope (Leitz Wetzlar) with a hot stage (Mettler Toledo FP90 Central Processor & FP82 Hot Stage) and a camera (Pixera PVC 100C) has also been used to record the strain recovery process upon heating, allowing the curves of temperature-dependent shape recovery to be obtained. The curve also allows the calculation of shape fixity and shape recovery at any temperature.

In the present study, moisture-induced SMEs are quantitatively investigated using specimens with a thickness of 1.0 mm, width of 5.0 mm, and length (L_0) of 20.0 mm, which are stretched to a certain maximum elongation (L_{max}) by raising the temperature to above T_{trans}. After the deformed strains (L_f) are fixed at lower temperature, the deformed specimen is subjected to either moisture or water conditions, with the resulting length at certain time points (L_t) recorded. Finally, the strain recovery can also be calculated at any time (R_t) using the following equation:

$$R_t = (L_f - L_t)/(L_{max} - L_0) * 100\%$$

Hence, the relationship between R_t and time can be established. Figure 8.22 shows a typical curve for strain recovery vs time for moisture-induced SMEs.

The final balanced strain recovery (R_r) is defined as the maximum strain recovery (R_{max}); the strain recovery onset time (t_s) is conditioning time at which the strain recovery is 10% R_{max}; the strain recovery time (t_r) is the conditioning time at which the strain recovery is 50% R_{max}; and the strain recovery end time (t_e) is the conditioning time at which the strain recovery is the 90% R_{max}. The time length ($t = t_e - t_s$) is also used to characterize the moisture recovery speed, i.e. when t is shorter, the recovery speed is quicker.

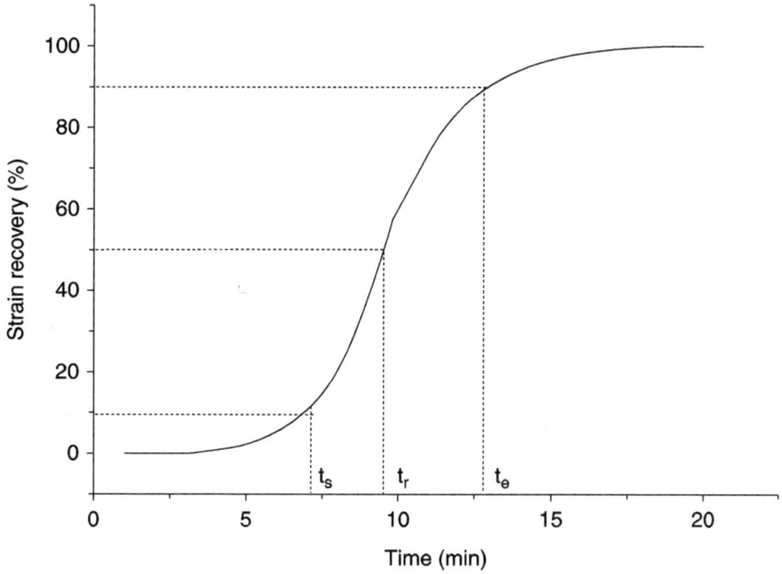

8.22 Typical strain recovery vs time for moisture-sensitive SME.

8.6.2 Effect of relative humidity on moisture-induced strain recovery

Figures 8.23 and 8.24 show the dependency of strain recovery on the time of exposure to different RH conditions, but a constant temperature 30°C for the PUPyBDO53 sample and the PUPyBDO45 sample, respectively. Their strain recovery onset time, strain recovery time, strain recovery end time, and their final strain recovery are summarized in Table 8.4. The recovery curves of the PUPyBDO53 sample and the PUPyBDO45 sample at 65% RH show that the shapes tend to remain unchanged at the first stage, and the deformed strain starts to recover after the polymer is conditioned for a critical time of 77 minutes for the PUPyBDO53 and 200 minutes for the PUPyBDO45. Abrupt strain recovery then occurs, before strain recovery slows after the t_e. Since this pattern of strain recovery is similar to that observed with thermally-induced SMEs, this process can be assumed to represent the typical pattern for moisture-induced SMEs.

In addition, Table 8.4 shows that t_s in both the PUPyBDO53 and PUPyBDO45 samples tends to be short at higher RH, and both the t_r and t_e decrease as the RH increases. Moreover, the final strain recovery drops as RH decreases. However, the PUPyBDO53 shows poor strain recovery at 85% RH, due to the destruction not only of the switch phase, but also of the physical net points at higher RH. According to the _t, PUPyBDO53 recovers most of its deformed strain within only 18 minutes at 85% RH. However, it takes about 120 minutes to achieve the

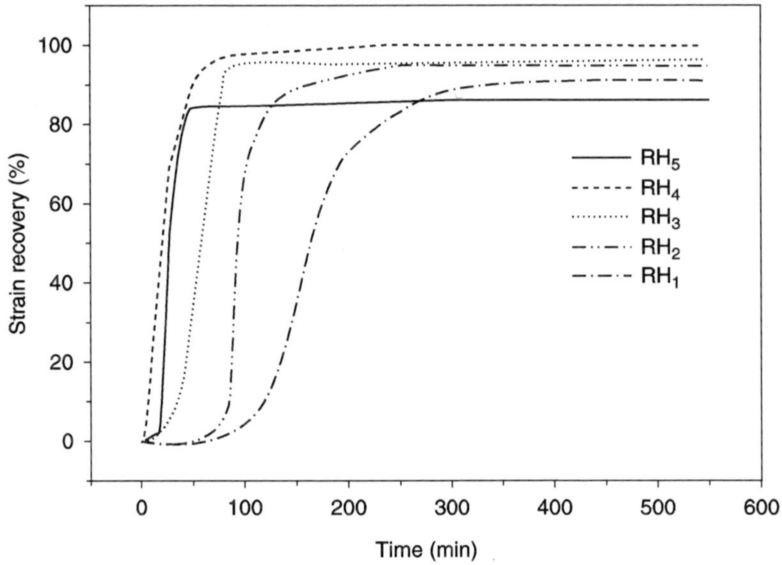

8.23 Dependency of strain recovery of PUPyBDO53 on time at various RH: $RH_1 = 65\%$; $RH_2 = 70\%$; $RH_3 = 75\%$; $RH_4 = 80\%$; and $RH_5 = 85\%$, at the constant temperature of T = 30°C.

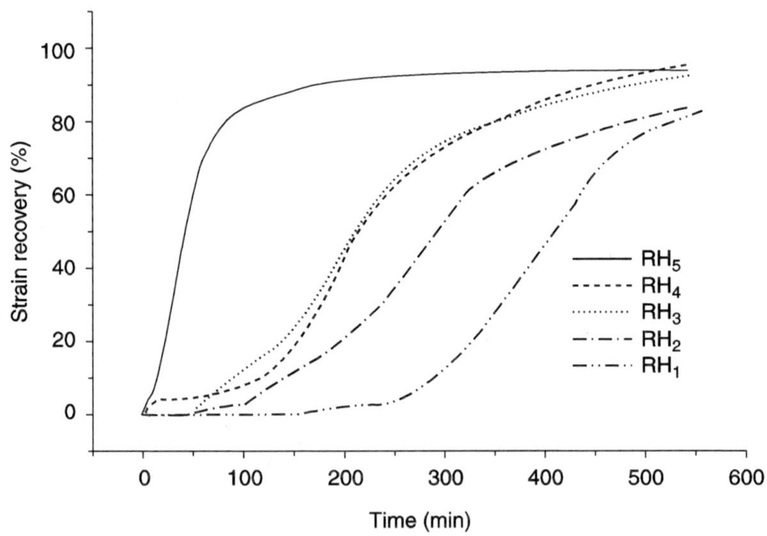

8.24 Dependency of strain recovery on time for PUPyBDO45 at various RH: $RH_1 = 65\%$; $RH_2 = 70\%$; $RH_3 = 75\%$; $RH_4 = 80\%$; and $RH_5 = 85\%$, at the constant temperature of T = 30°C.

Table 8.4 Moisture-sensitive SME results of BIN-SMPU at various RH

Samples	RH (%)	T (°C)	t_s (mins)	t_r (mins)	t_e (mins)	R_r (%)	Δt
PUPyBDO53	85	30	21.2	28.0	39.54	86.1	18.34
	80	30	5.1	17.9	50.4	99.4	45.3
	75	30	38.2	60.9	79.5	95.9	41.3
	70	30	83.7	94.7	124.8	94.7	41.1
	65	30	115	158.8	234.9	90.6	119.9
PUPyBDO45	85	30	13.7	40.8	106.3	93.9	92.6
	80	30	114.1	207.6	388.1	93.8	274
	75	30	92.1	203.1	388.2	92.2	296.1
	70	30	128.4	271.6	428.8	83.2	300.4
	65	30	280.2	387.0	473.5	81.1	193.3

same strain recovery at 65% RH. This means that the strain recovery speed at higher RH is better than that at lower RH.

8.6.3 Effect of temperature on moisture-induced strain recovery

Figures 8.25 and 8.26 show the dependency of strain recovery on time exposed to different temperatures at a constant RH of 80% for PUPyBDO53 and PUPyBDO45, respectively. Their strain recovery onset time, strain recovery time, strain recovery

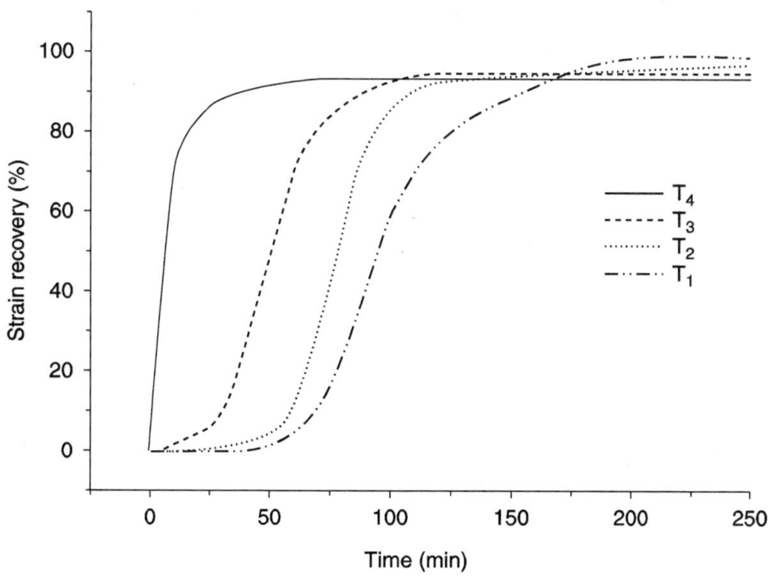

8.25 Dependency of strain recovery on time for PUPyBDO53 at various temperatures: T_1 = 26°C; T_2 = 28°C; T_3 = 30°C; and T_4 = 32°C at RH = 80%.

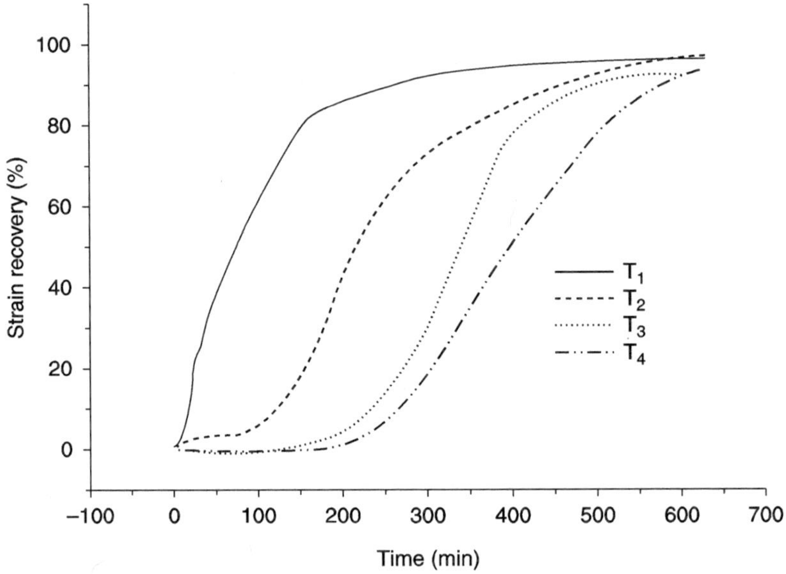

8.26 Dependency of strain recovery on time for PUPyBDO45 at various temperatures: T_1 = 32°C; T_2 = 30°C; T_3 = 28°C; and T_4 = 26°C at RH = 80%.

end time and final strain recovery are summarized in Table 8.5. It can be observed that the temperature has a significant effect on the strain recovery time. Both the PUPyBDO53 and PUPyBDO45 show a high strain recovery at 80% RH, and as the temperature increases, the strain-time recovery curves are observed to shift over a longer time period.

Table 8.5 also demonstrates that all the critical times, including t_s, t_r, and t_e increase as the temperature increases. This is because moisture absorption is faster at higher temperatures due to the higher diffusion coefficient. The t reveals that

Table 8.5 Moisture-sensitive SME results of BIN-SMPU at various temperatures

Samples	RH (%)	T (°C)	t_{10} (mins)	t_r (mins)	t_{90} (mins)	R_r (%)	Δt (mins)
PUPyBDO53	80	32	1.0	5.0	19.8	**93.11**	18.8
	80	30	**29.8**	**49.3**	74.5	**94.64**	44.6
	80	28	59.2	77.2	101.5	**96.65**	42.3
	80	26	67.9	93.9	147.1	**98.32**	79.2
PUPyBDO45	80	32	17.0	69.7	208.6	96.23	**19.2**
	80	30	125.1	212.6	423.0	97.15	**29.8**
	80	28	234.0	334.8	436.5	93.10	**20.3**
	80	26	267.3	384.2	534.0	94.26	26.7

both the PUPyBDO53 and PUPyBDO45 have a higher strain recovery speed at higher temperatures, for example in the PUPyBDO53 sample, the t is 18.8 minutes at 32°C but 79.2 minutes at 26°C. PUPyBDO45 also has a high recovery speed than PUPyBDO53 within the temperature range of 26 to 32°C. Moreover, it is observed that the maximum strain recovery of the PUPyBDO53 sample decreases as the temperature increases, suggesting that high temperature may affect the physical net points at 80% RH.

8.6.4 Effect of BINA content on moisture-induced strain recovery

The effect of the pyridine (i.e. BINA) content on strain recovery is shown in Fig. 8.27, and Table 8.6 summarizes the results. It is clear that PUPyBDO53, PUPyBDO40 and PUPyBDO30 all have a higher strain recovery at 34°C, while the PUPyBDO20 and PUPyBDO10 show poor strain recovery at 85% RH. This demonstrates that BIN-SMPU tends to show lower strain recovery, even at high RH and high temperature when the BINA content is below 20%. BIN-SMPU with a low BINA content usually has poor shape recovery. BINA therefore plays an important role in the shape recovery, not only for thermally-induced SME, but also for moisture-induced SMEs in BIN-SMPUs. The data presented in Table 8.6 demonstrate that the t_s, t_r and t_e of PUPyBDO53 are generally shorter than those

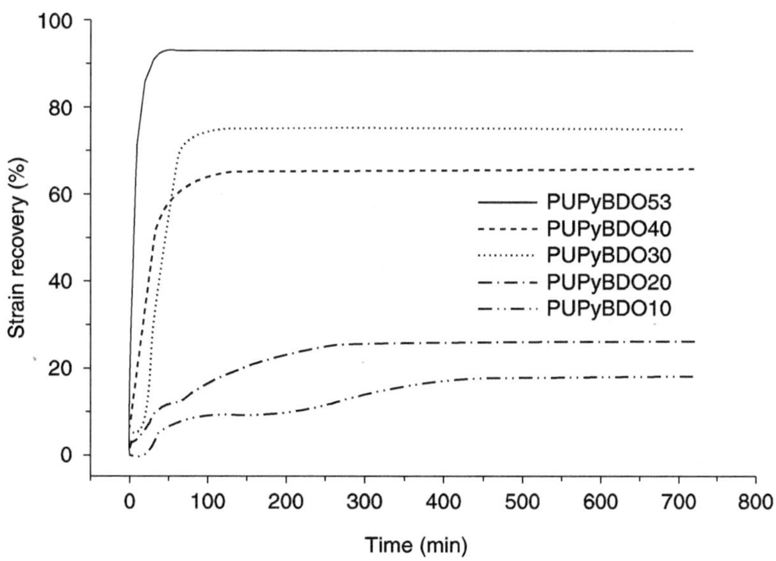

8.27 Effect of strain recovery on time for BIN-SMPU with various BINA contents under the conditions of RH = 85% and T = 34°C.

Table 8.6 Moisture-sensitive SME results of PUPys with various BINA contents

Samples	R (%)	T (°C)	t_s (mins)	t_r (mins)	t_e (mins)	R_r (%)	Δt (mins)
PUPyBD053	85	34	0.1	5.0	17.4	93.42	**17.3**
PUPyBDO40	85	34	4.5	**21.95**	59.1	65.82	**21.5**
PUPyBDO30	85	34	16.9	36.3	54.1	75.35	**37.2**
PUPyBDO20	85	34	–	–	–	25.53	–
PUPyBDOIO	85	34	–	–	–	17.86	

of PUPyBDO45 within the RH range of 65 to 85% and within the temperature range of 26 to 32°C. Table 8.6 also shows that in BIN-SMPUs with higher BINA content, the t_s and t_r become shorter and the speed of strain recovery increases.

8.6.5 Effect of MDI-BDO content on moisture-induced strain recovery

The effect of the MDI-BDO content on the strain recovery of BIN-SMPU at 85% RH and 34°C is shown in Fig. 8.28. PUPyMB0, PUPyMB15 and PUPyMB25 show a higher strain recovery with a fast strain recovery process. However, the BIN-SMPUs with a higher MDI-BDO content, namely PUPyMB45 and PUPyMB55, tend to show a lower strain recovery with a slow strain recovery process. The data presented in Table 8.7 demonstrate that the final strain recovery decreases as the MDI-BDO content increases. Furthermore, the time t_s, t_r and t_e all increase as the MDI-BDO content increases. This pattern is consistent with the results observed in the study of the influence of BINA content on strain recovery. An increase in MDI-BDO content results in a decrease of BINA content for the PUPyMB series of BIN-SMPUs; hence, increasing the MDI-BDO content does not result in an increase in strain recovery or a reduction of recovery time. Instead,

Table 8.7 Results of moisture-sensitive SME of BIN-SMPU with various MDI-BDO contents

Samples	RH (%)	T (°C)	t_{10} (mins)	t_r (mins)	t_w (mins)	R_r (%)	Δt (mins)
PUPyMB0	85	34	6.0	17.0	49.9	99.20	43.9
PUPyMB15	85	34	**12.4**	**61.5**	134.8	93.66	122.4
PUPyMB25	85	34	10.0	88.2	214.4	90.67	204.4
PUPyMB45	85	34	34.4	116.0	555.0	58.70	520.6
PUPyMB55	85	34	117.0	336.8	609.3	58.70	492.3

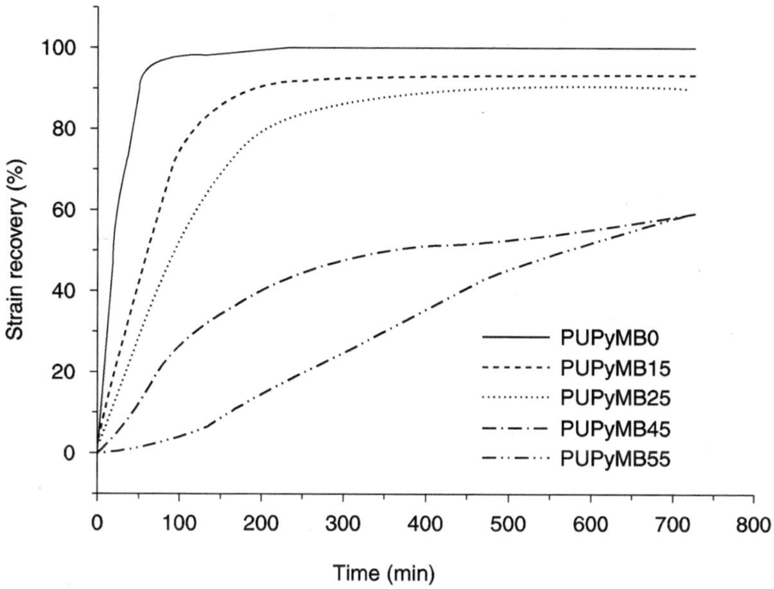

8.28 Effect of strain recovery on time for BIN-SMPU with various MDI-BDO contents at RH = 85% and T = 34°C.

as noted above, it is an increase in the BINA content that leads to improvements in strain recovery.

It can therefore be concluded that under certain conditions, the strain recovery and recovery time are mainly dependent on the BINA content. However, the PUPy-MB series of BIN-SMPUs show a relatively higher final strain recovery than the PUPy-BDO series. This is due to the fact that among the samples with the same BINA content, those containing an MDI-BDO unit show a much higher strain recovery than those without an MDI-BDO unit. For example, PUPyMB25, which contains about 40 wt.% BINA, has a final strain recovery of 90.7%, while the PUPyBDO40, which also contains 40 wt.% BINA, has a final strain recovery of 65.8%. This means that the addition of the MDI-BDO unit can significantly improve the moisture-induced strain recovery. This is because the MDI-BDO unit will reinforce the physical net points or the hard phase, which can provide a stable polymer network for moisture-induced SMEs.

8.7 Mechanism of the moisture-induced shape memory effect in BIN-SMPUs

Previous studies (Papanicolaou *et al.*, 2006; Premkumar and Srikanth, 2006) have carried out research to investigate the moisture/water absorption mechanism in the polymer system. Agarwal *et al.* (1997) listed three mechanisms by which moisture could be held in a polymer system:

1. as mono-molecular layers bound to the surface, namely adsorbed moisture;
2. as isolated molecules distributed internally, either bound or free, namely absorbed moisture; and
3. as multi-molecular layers, namely condensed moisture.

In order to investigate the molecular mechanism of moisture absorption and the influence of moisture on the structure, the Fourier transform infrared (FT-IR) spectra of PUPyBDO53 after different conditioning times were recorded. Figure 8.29 shows some typical vibration frequencies at different times at 80% RH and 30°C. Figure 8.29(a) shows that the density of the frequency corresponding to water vibration increases quickly within the first 20 minutes until it reaches a stable density, which indicates that the PUPyBDO53 reaches its equilibrium state within 20 minutes under these conditions. Since the moisture absorption time is close to the strain recovery time of PUPyBDO53, it may be concluded that moisture absorption results in strain recovery.

Furthermore, the frequencies corresponding to the vibration of the pyridine ring are also observed to shift to a higher frequency. Figure 8.29(c) shows that moisture absorption results in a shift in the breathing vibration frequency of the pyridine ring from 994 to 999 cm^{-1}. In the equilibrium state, the frequencies of the pyridine ring appear at 3060, 1603 and 999 cm^{-1}, which are close to those of the pyridine ring observed at higher temperature, as shown in the temperature-dependent FT-IR spectra (Chen *et al.*, 2009). These results indicate that the hydrogen bonds present in the pyridine ring are free or weak after moisture absorption, i.e. moisture causes the dissociation of the hydrogen bonds between the pyridine ring and the N–H of the urethane group.

In previous models of the interaction of water with hydrophilic polymers, it is generally hypothesized that the water molecules are either bonded to specific polymer chain sites or are freely dispersed homogeneously throughout the amorphous polymer matrix (Huang and Yang, 2005). However, based on the FT-IR analysis of the BIN-SMPU, the moisture absorption mechanism can be explained by the theory of dynamic combinatorial chemistry or constitutional dynamic chemistry proposed by Lehn (2007), which relies on the selection of the thermodynamically most stable product from an equilibrating mixture.

An illustration of the moisture absorption mechanism in the pyridine ring is presented in Fig. 8.30. Under the stimulus of moisture, the pre-formed hydrogen bonds, N–H. . .N–Py, in the dry state is replaced by a more adaptive hydrogen bond. The pyridine ring tends to be free due to its protonization by water (O–H. . .N–Py), or else is linked via water molecules in the form of bridged hydrogen bonding (Py–N–H. . .O–H–O. . .H–N–Py) with a long bond distance (Papai and Jancso, 2000). As a result, the interaction becomes weaker, and the stiffness or modulus of the polymer becomes softer as the moisture absorption increases. Similar to the dissociation of hydrogen bonds induced by temperature, the replacements of hydrogen bonds present in the pyridine ring also results in shape recovery. Therefore, the hydrogen bonds between the pyridine ring and the

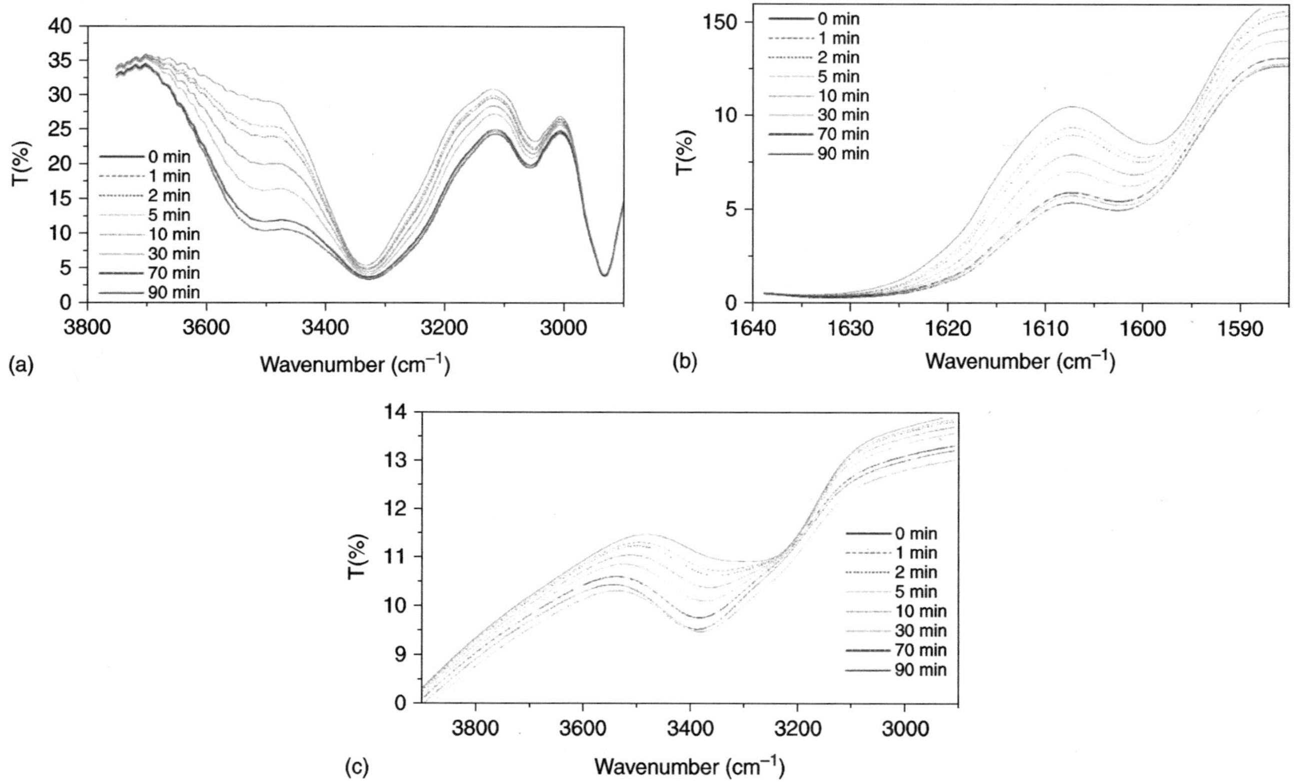

8.29 FT-IR spectra of BIN-SMPUA moisture absorption for various times in the frequency range of: (a) 3800–2900 cm^{-1}; (b) 1640–1585 cm^{-1}; and (c) 1015–985 cm^{-1}.

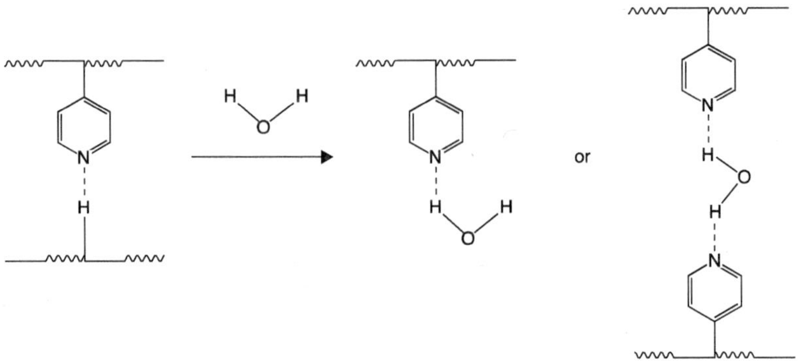

8.30 Moisture absorption mechanism at the hydrogen-bonded pyridine ring.

N–H of the urethane group are used as the moisture-induced 'switch', while the hard phase formed via hydrogen bonds among the urethane groups are the physical net points in moisture-induced SMPUs.

8.8 Summary

By means of an examination of the structure and morphology of PUPys, this chapter has provided a systematic study of moisture-induced SMEs, allowing the following conclusions to be drawn:

1. BIN-SMPUs have high moisture absorption which increases with temperature as well as with RH. The moisture absorption also increases as the BINA content increases and as the MDI-BDO content decreases.
2. Moisture absorption influences the morphology of BIN-SMPUs. The T_g of the soft segment decreases as conditioning time increases, particularly at higher RH and higher temperature. The phase transition of the hard phase moves to a lower temperature as the conditioning time increases, leading to a softening of the polymer at room temperature.
3. Moisture-induced SMEs and water-influenced SMEs can both be obtained in PUPys. Moreover, the RH, temperature, BINA content and MDI-BDO content all play a significant role in determining moisture-induced SMEs. The deformed strain generally recovers quickly at high RH and high temperature. The strain recovery start time (t_s), strain recovery time (t_r) and strain recovery end time (t_e) all increase as RH and temperature decrease, while the recovery speed increases at high RH and high temperature. The final strain recovery generally decreases as RH falls and temperature rises. If the RH is too high, a lower final strain recovery will be observed in the PUPy-MDI series of PUPys.
4. The main component of PUPys that affects the moisture-induced SMEs is the BINA unit. The addition of MDI-BDO enhances moisture-induced shape

recovery. In the PUPy-BDO series of PUPys, the final shape recovery decreases as the BINA content drops. To achieve a satisfying shape recovery of above 90%, the PUPy-BDO series must contain at least 30 wt.% BINA. In addition, the t_s, t_r, t_e and _t also decrease in PUPys with a higher BINA content. In the PUPy-MDI series, the final shape recovery decreases as BDO content increases and the BINA content decreases. The t_s, t_r, t_e and _t also decrease in PUPys with a lower MDI-BDO content.

5. Finally, it is proposed that the hydrogen bonds present in the pyridine ring serve as molecular 'switches', while the hydrogen bonds present in the urethane groups can provide the PUPys with an elastic polymer network for shape recovery as part of the moisture-induced SMEs of PUPys. Moisture-induced shape recovery mainly results from the dissociation of hydrogen bonds present in the pyridine ring by means of the replacement of new hydrogen bonds with absorbed water molecules.

8.9 References

Agarwal, N., Hoagland, D. A., Farris, R. J. (1997), Effect of moisture absorption on the thermal properties of *Bombyx mori* silk fibroin films, *J. Appl. Polym. Sci.*, **63**, 401–10.

Chen, S. J., Hu, J. L., Yuen, C. W., Chan, L. K. (2009), Supramolecular polyurethane networks containing pyridine moieties for shape memory materials, *Mater. Lett.*, **63**, 1462–4.

Couchman, P. R. (1978), Compositional variation of glass-transition temperatures, Part II: Application of thermodynamic theory to compatible polymer blends, *Macromolecules*, **11**, 1156–61.

Couchman, P. R., Karasz, F. E. (1978), Classical thermodynamic discussion of effect of composition on glass-transition temperatures, *Macromolecules*, **11**, 117–19.

Gomez-Carracedo, A., Alvarez-Lorenzo, C., Gomez-Amoza, J. L., Concheiro, A. (2003), Chemical structure and glass transition temperature of non-ionic cellulose ethers DSC, TMDSC (r) – oscillatory rheometry study, *J. Therm. Anal. Calorim.*, **73**, 587–96.

Gunes, I. S., Cao, F., Jana, S. C. (2008), Evaluation of nanoparticulate fillers for development of shape memory polyurethane nanocomposites, *Polymer*, **49**, 2223–34.

Huang, W. M., Yang, B. (2005), Water-driven programmable polyurethane shape memory polymer: Demonstration and mechanism, *Appl. Phys. Lett.*, **86**, 114105.

Huang, W. M., Lee, C. W., Teo, H. P. (2006), Thermo-mechanical behavior of a polyurethane shape memory polymer foam., *J. Intell. Mater. Syst. Struct.*, **17**, 753–60.

Jaczewska, J., Budkowski, A., Bernasik, A., Raptis, I., Raczkowska, J., *et al.* (2007), Humidity and solvent effects in spin-coated polythiophene-polystyrene blends, *J. Appl. Polym. Sci.*, **105**, 67–79.

Lehn, J. M. (2007), From supramolecular chemistry towards constitutional dynamic chemistry and adaptive chemistry, *Chemical Society Reviews*, **36**(2), 151–60.

Lendlein, A., Jiang, H., Junger, O., Langer, R. (2005), Light-induced shape-memory polymers, *Nature*, **434**, 879–82.

Lin, Q. F., Zhou, X. D., Dai, G. (2002), Effect of hydrothermal environment on moisture absorption and mechanical properties of wood flour-filled polypropylene composites, *J. Appl. Polym. Sci.*, **85**, 2824–32.

Papai, I., Jancso, G. (2000), Hydrogen bonding in methyl-substituted pyridine-water complexes: A theoretical study, *Journal of Physical Chemistry A*, **104**, 2132–7.

Papanicolaou, G. C., Kosmidou, T. V., Vatalis, A. S., Delides, C. G. (2006), Water absorption mechanism and some anomalous effects on the mechanical and visco-elastic behavior of an epoxy system, *J. Appl. Polym. Sci.*, **99**, 1328–39.

Premkumar, J., Srikanth, N. (2006), Moisture absorption mechanism in epoxy molding compounds used in IC encapsulation. *Eighth Electronic Packaging Technology Conference 2006*, **1**, 60–2.

Ratna, D., Karger-Kocsis, J. (2008), Recent advances in shape memory polymers and composites: A review, *J. Mater. Sci.*, **43**, 254–69.

Vogt, B. D., Soles, C. L., Lee, H. J., Lin, E. K., Wu, W. (2005), Moisture absorption into ultrathin hydrophilic polymer films on different substrate surfaces, *Polymer*, **46**, 1635–42.

Wornyo, E., Gall, K., Yang, F. Z., King, W. (2007), Nano-indentation of shape memory polymer networks, *Polymer*, **48**, 3213–25.

Yang, B., Huang, W. M., Li, C., Lee, C. M., Li, L. (2004), On the effects of moisture in a polyurethane shape memory polymer, *Smart Materials and Structures*, **13**, 191–5.

Yang, B., Huang, W. M., Li, C., Chor, J. H. (2005), Effects of moisture on the glass transition temperature of polyurethane shape memory polymer filled with nano-carbon powder, *Eur. Polym. J.*, **41**, 1123–8.

Yang, B., Huang, W. M., Li, C., Li, L. (2006), Effects of moisture on the thermo-mechanical properties of a polyurethane shape memory polymer, *Polymer*, **47**, 1348–56.

Zhang, H., Wang, H., Zhong, W., Du, Q. (2009), A novel type of shape memory polymer blend and the shape memory mechanism, *Polymer*, **50**, 1596–601.

Zhang, S., Yu, Z., Govender, T., Luo, H., Li, B. (2008), A novel supramolecular shape memory material based on partial A-CD-PEG inclusion complex, *Polymer*, **49**, 3205–10.

9
Shape memory polymers with novel functions: electro-active, magnetically-active, light-adaptive and phase change materials

DOI: 10.1533/9780857098542.231

Abstract: This chapter introduces the theory of inducing the shape memory effect through indirect heating, and current and potential applications. Shape memory polymers with temperature regulating effects are also discussed. The chapter also summarizes the different mechanisms of phase change materials and the fabrication of phase change textiles and fabrics.

Key words: shape memory effect through indirect heating; electro-active shape memory effect; magnetic-active shape memory effect; light-active shape memory polymer; phase change material.

9.1 Introduction

Shape memory polymers (SMPs) can display shape memory effect (SME) if, when deformed, they are heated to a temperature above the switching temperature. Normally, this is induced by direct heating of the polymer. However, some researchers have achieved SME through indirect heating, using agents such as an electrical field (Cho *et al.*, 2005), a magnetic field (Buckley *et al.*, 2006) or a solution (Yang *et al.*, 2006). The SME of SMPs induced through indirect heating further facilitates the expansion of the technique to a much wider range of applications.

When electrical conductive materials are incorporated into thermal-active SMPs, the current passing through them can induce Joule heating and trigger shape recovery. When ferromagnetic fillers are employed, heat produced by the magnetic field can be used to create magnetic-active SMEs. Light absorbed by a deformed SMP can also increase its temperature to the switch transition temperature, which in turn triggers shape recovery. As well as electrical and magnetic fields, light, moisture or solvent can all be used as stimuli for a SME. This chapter introduces the theory of inducing the SME through indirect heating, and explores current and potential applications.

SMPs with temperature regulating effects are also discussed. Through phase change materials (PCMs), these can interactively respond to an individual human's unique physiological condition, and thus absorb, store and release heat to help the body remain comfortable. They store or release heat energy over a narrow temperature range, thereby helping the body to maintain its natural temperature across hot and cold environments. This chapter also summarizes the different mechanisms of PCMs and the fabrication of phase change textiles and fabrics.

9.2 Electro-active shape memory polymers

Conventionally, the SME of SMPs is induced by directly heating the polymer to a temperature above the switching temperature. However, it has been demonstrated that the shape recovery of SMPs can be achieved by Joule heating after they have been filled with conductive fillers. A certain level of conductivity can be achieved after conductive ingredients are incorporated into thermally-adaptive SMPs. When a current passes through, the resulting Joule heating may raise the internal temperature to above the switching transition temperature of the polymer, thus triggering shape recovery, as shown in Fig. 9.1 (Koerner *et al.*, 2004; Cho *et al.*, 2005; Paik *et al.*, 2005, 2006; Goo *et al.*, 2007).

The added electrical-adaptive effect allows even more applications of thermally-adaptive SMPs, such as biomedical devices, self-deployable aerospace structures, actuators and sensors, while indirect induction heating of SMPs using electricity can also alleviate some conventional heat transfer problems. The conductive ingredients, which have been used as conductive fillers, include carbon nanotubes (CNTs), polypyrrole (PPy) (Sahoo *et al.*, 2005, 2007a,b), carbon black and short carbon fiber (Leng *et al.*, 2007, 2008a,b; Lv *et al.*, 2008a). To obtain a good, electro-responsive SME, good conductivity is necessary (Lan *et al.*, 2008; Leng *et al.* 2008c,d). Multiwalled carbon nanotube (MWCNT)/SMPU composite fiber with electro-responsive SME has also been developed by Meng *et al.* (2007a).

One disadvantage of shape memory fibers (SMFs) compared with shape memory alloys (SMAs) is the lower recovery stress of the former. However, SMFs do have the advantages of low cost, good processability, high deformability, high shape recoverability, soft mechanical properties and tailor-able switching temperature (Buckley *et al.*, 2007; Zhang *et al.*, 2008, 2009a,b; Xie and Rousseau, 2009).

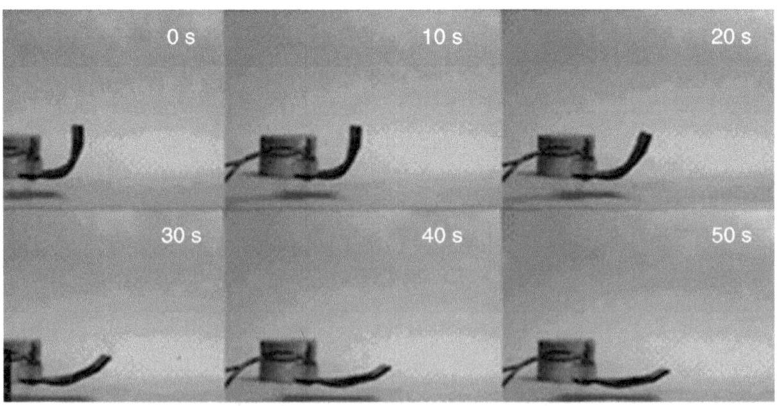

9.1 The electro-responsive shape-recovery behavior of carbon nanotube/SMPU composites (Liu *et al.*, 2008).

Furthermore, because a glass transition or melting transition is employed as the 'switch', the mechanical properties of the SMFs decrease severely at the switching temperature. One of the most obvious methods of improving recovery stress in SMFs is to reinforce them with different kinds of fillers. It has been shown, for example, that CNTs are excellent nano-fillers, particularly when used in a polymer matrix to prepare nanocomposites with advanced properties (Wei *et al.*, 1998; Jia *et al.*, 1999; Lourie, 1999; Gommans *et al.*, 2000; Gong *et al.*, 2000; Haggenmueller *et al.*, 2000; Sholz, 2000; Thostenson and Chou, 2001; Biercuk *et al.*, 2002; Cooper *et al.*, 2002; Demczyk *et al.*, 2002; Lau Kin-Tak, 2002a,b; Ounaies *et al.*, 2003; Andrews and Weisenberger, 2004; Cadek *et al.*, 2004; Huang *et al.*, 2004; Rahul *et al.*, 2004; Zhang *et al.*, 2004a; Jung *et al.*, 2006).

Many research groups have studied polymer reinforcement by employing CNTs (Gommans *et al.*, 2000; Haggenmueller *et al.*, 2000; Biercuk *et al.*, 2002; Demczyk *et al.*, 2002; Andrews and Weisenberger, 2004; Cadek *et al.*, 2004; Huang *et al.*, 2004; Rahul *et al.*, 2004; Zhang *et al.*, 2004b; Jung *et al.*, 2006). Thostenson and Chou (2002) utilized a micro-scale twin-screw extruder to obtain the high shear mixing necessary to disentangle MWCNTs and disperse them uniformly in a polystyrene thermoplastic matrix. Hou *et al.* (2005) prepared polyacrylonitrile nanofibers containing a high concentration of MWCNTs. The spinning solution of polyacrylonitrile and MWCNTs in dimethyl formamide (DMF) proved very stable. They ascribed the high stability of the polyacrylonitrile-MWCNTs-DMF to the high interaction between the molecules of solvent DMF and the MWCNT surface. For conventional elastic CNT/polyurethane systems, Kwon and Kim (2005) and Jung *et al.* (2006) improved the compatibility of MWCNTs with an elastic polyurethane matrix by treating them with 60% nitric acid. Chen and Tao (2005) prepared polyurethane composites with aligned single-walled CNTs by dispersing CNTs in polyurethane solution with tetrahydrofuran as a solvent. They reported that the solvent–polymer interactions served as a driving force for the macroscopic alignment of the CNTs (Hu *et al.*, 2007).

Meng *et al.* (2007a) meanwhile prepared an MWNT/SMPU composite fiber by melt spinning. They found that MWNTs are axially aligned in the polyurethane fiber, which is helpful in improving the electrical conductivity of the fiber. The prepared fibers show electro-active shape memory behavior.

Finally, a CNT/SMPU system has been reported, where CNTs were distributed in a shape memory polyurethane (SMPU) solution using an ultrasonic process, to study their influence on the shape fixity and recovery ratios. The composite showed excellent shape recoverability (Tao, 2001; Mattila, 2006).

The scanning electron microscope (SEM) images of the SMFs with different MWCNT contents are shown in Fig. 9.2. The surface of the MWCNT/SMPU fiber becomes rough and coarse with increasing MWCNT content of 3 to 5 wt.%. This may be due to a high MWCNT content, as some MWCNTs begin to spread and become exposed at the fiber surface. Figure 9.3 shows the SEM image of the MWCNTs used in the study. These were entangled together with long nanosized

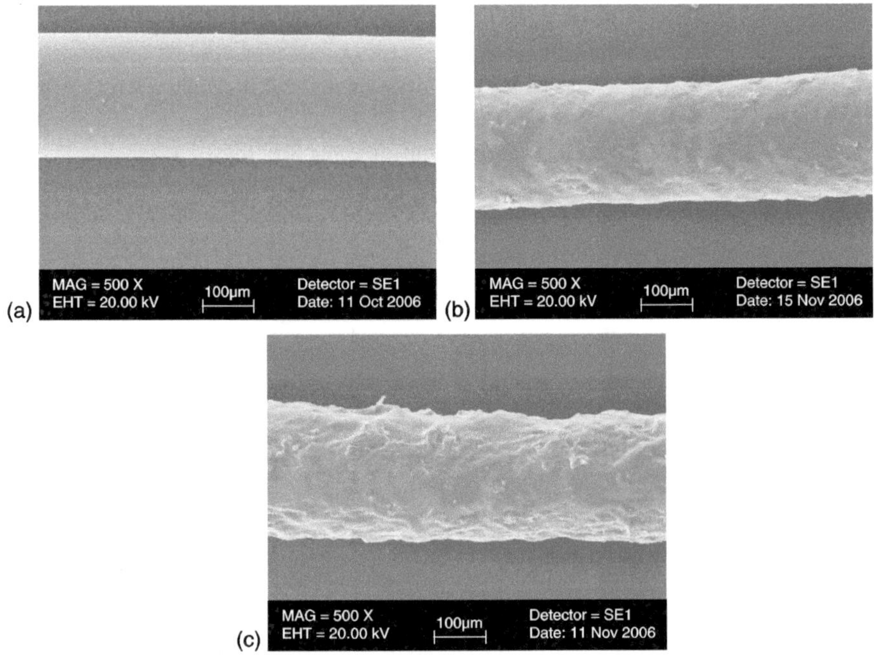

9.2 SEM images of the SMFs with various MWCNT contents. (The scales are 100 μm for each figure.)

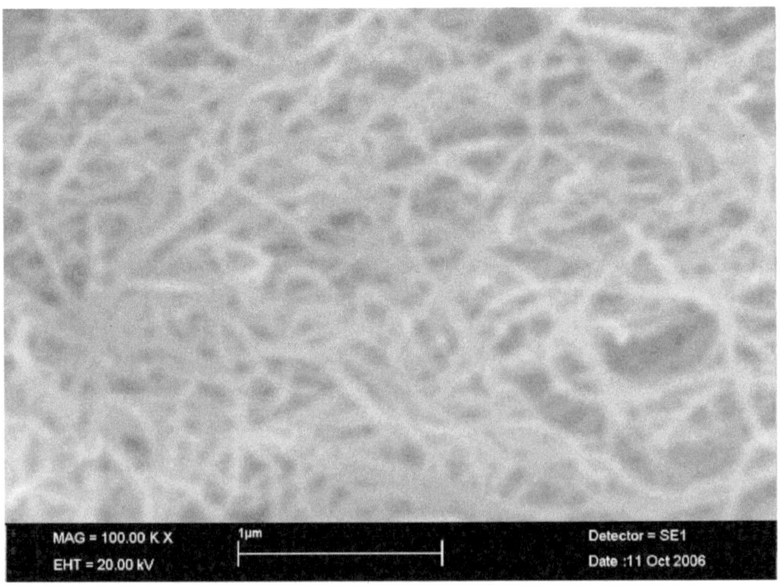

9.3 SEM image of MWCNTs. (The scale is 1 μm.)

dimensions, the outer diameter being about 10 to 20 nm. Figure 9.4 shows the cross-sectional SEM images of MWCNT/SMPU fibers. In the MWCNT/SMF, most MWCNTs are perpendicular to the fiber cross-section, which suggests that most of them are preferentially aligned in the fiber axial.

Many researchers have developed polymeric composites with aligned CNTs by using different methods (Gommans *et al.*, 2000; Haggenmueller *et al.*, 2000; Hwang *et al.*, 2000; Cooper *et al.*, 2002; Thostenson and Chou, 2002; Chen and Tao, 2005; Hou *et al.*, 2005; Miaudet *et al.*, 2005; Borca-Tasciuc *et al.*, 2007; Zamora-Ledezma *et al.*, 2009). Haggenmueller *et al.* (2000) prepared CNT/poly(methyl methacrylate) fibers with aligned CNTs by solvent casting and

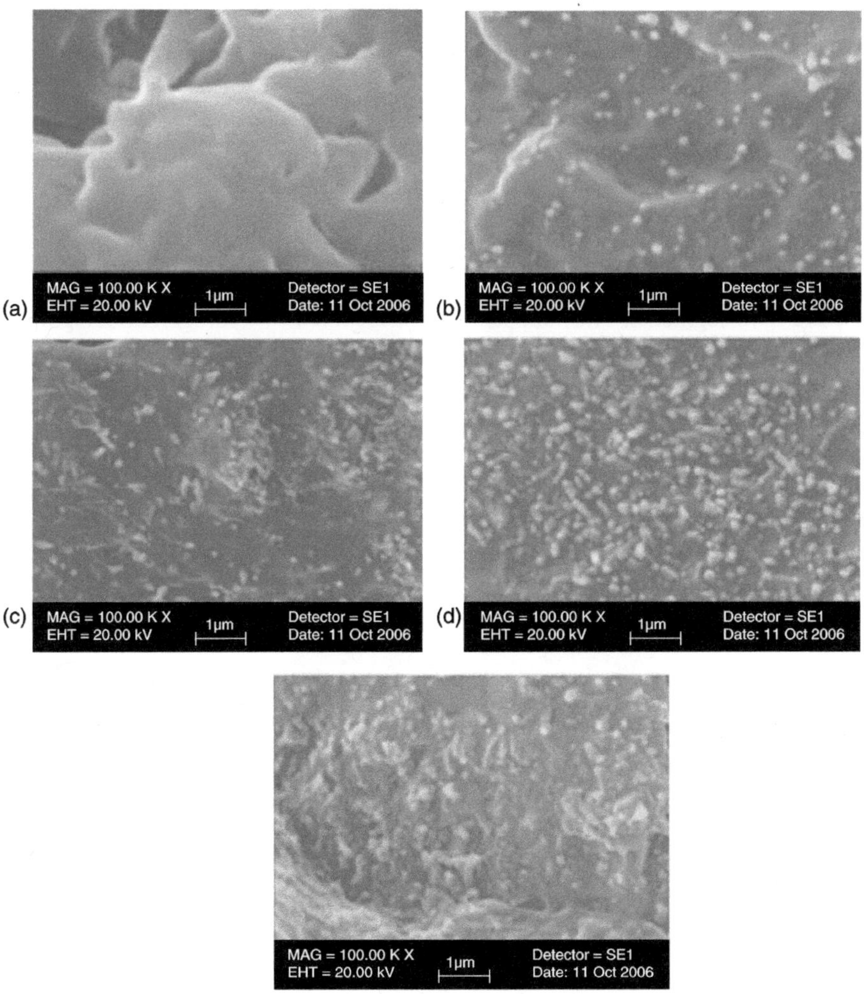

9.4 SEM images of the fracture surface of the fibers with: (a) 0 wt.%; (b) 1.0 wt.%; (c) 3.0 wt.%; (d) 5.0 wt.%; and (e) 7.0 wt.% MWCNT content. (The scales are 1 μm for each figure.)

melt spinning. Cooper *et al.* (2002) developed CNT/poly(methyl methacrylate) composites with aligned CNT by melting, blending and extrusion. Miaudet *et al.* (2005) prepared CNT/polyvinyl alcohol (PVA) fibers with aligned CNT by a solution spinning and post drawing method.

The model in Fig. 9.5 can be employed to explain the distribution and alignment effect of the MWCNTs in the SMFs induced by melt extrusion and melt spinning processes. In Fig. 9.5, the soft segments of polyester are shown as being coiled or folded on themselves. The schematic section length of the zig-zag line corresponds to one repeating unit within the polyol. The isocyanate is shown as a series of rigid circles. The hard segments are fixed, and have a tendency to adhere to each other through strong hydrogen bonding. The MWCNTs are entangled with polyurethane at a molecular level in the polymer matrix. During *in-situ* polymerization, the treated MWCNTs are predistributed in the MDI by the ultrasonic process, and the SMPU molecular chains grow on the treated MWCNT surface. As a result, the MWCNTs adhere to the SMPU matrix, and in particular to the hard segments (diisocyanates). When the MWCNT/SMPU composites are heated above the transition temperature, the soft segment phases are melted. During melt blending, extrusion and the spinning process, the higher shear force and drawing ratio contribute to the improved homogenous distribution and axial alignment of the MWCNTs. This alignment effect is more pronounced as the SMP is composed of soft and hard segments. The hard segment phase fixed by hydrogen bonding may stretch the curled MWCNTs to align axially.

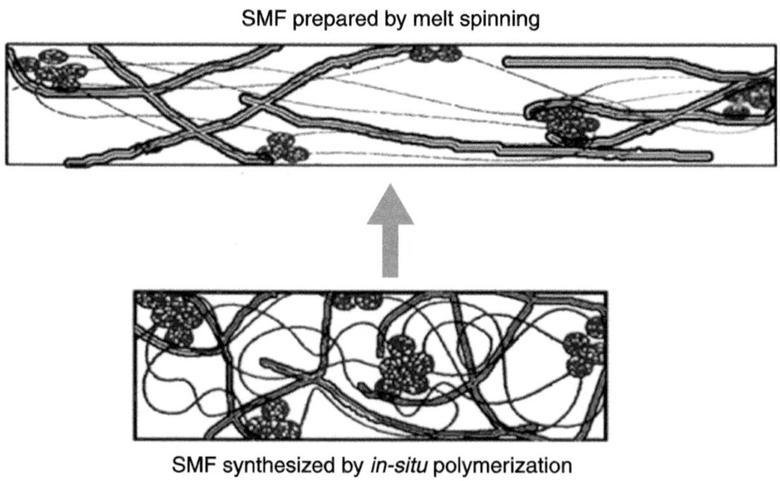

9.5 Schematic representation of MWCNT alignment in the SMF during blending and spinning (the thin lines represent chains of polyols, the circles represent aggregated hard segment groups, and the thick hollow lines represent MWCNTs).

The homogenously and axially aligned MWCNTs in the SMP matrix can contribute to the conductivity of MWCNT/SMF. However, due to the aggregation of MWCNTs and the difficulty of fabricating MWCNT/SMFs with a high MWCNT content, the MWCNT/SMF conductivity cannot reach a high level. Similarly, when the MWCNT content reaches 8.0 wt.%, the MWCNT/SMF cannot be fabricated.

With low MWCNT content, the MWCNT/SMF electrical resistance was too high and no electro-responsive SME was observed. When the MWCNT content reached 7.0 wt.%, the prepared MWCNT/SMF maximum strain was low and large stretching deformation could not be performed. In this study, MWCNT/SMFs with an MWCNT content of 6.0 wt.% and a 90% elongation at break were used. The electro-active shape recovery of the SMPU rod with an MWCNT of 6.0 wt.% content is presented in Fig. 9.6(a–e). The SMPU rod was obtained from the melt extrusion immediately before pelleting, and the temporary shape was obtained by bending the rod in hot water. The ends of the rods were connected to a power source by conductive wires. The shape recovery process finished within one minute.

When studying the electro-active SME of MWCNT/SMFs, 20 pieces of fibers were employed together, because one fiber's resistance was too high to produce enough heat for the sample temperature to be raised above the switching temperature (~42°C). The fibers' original length was 60 mm (the original length in Fig. 9.7(a)), although they were then stretched to 90 mm (50% strain) in a temperature controlled chamber at 60°C. After being cooled to the ambient temperature and alleviated of external stress, the fibers' length reduced to 80 mm (the length after elongation in Fig. 9.7(b)). The MWCNT/SMFs were then connected to a power source with a voltage of 210 v, and the fibers began to shrink, finally recovering their original length of about 60 mm.

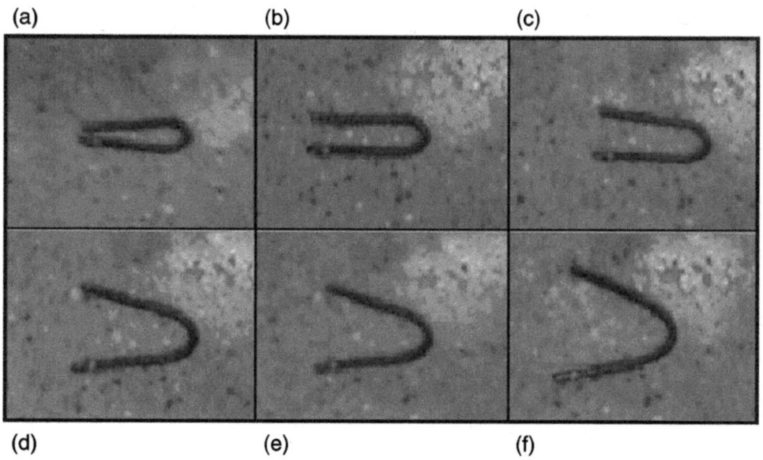

9.6 The electro-active shape recovery effect of MWCNT/SMFs rod.

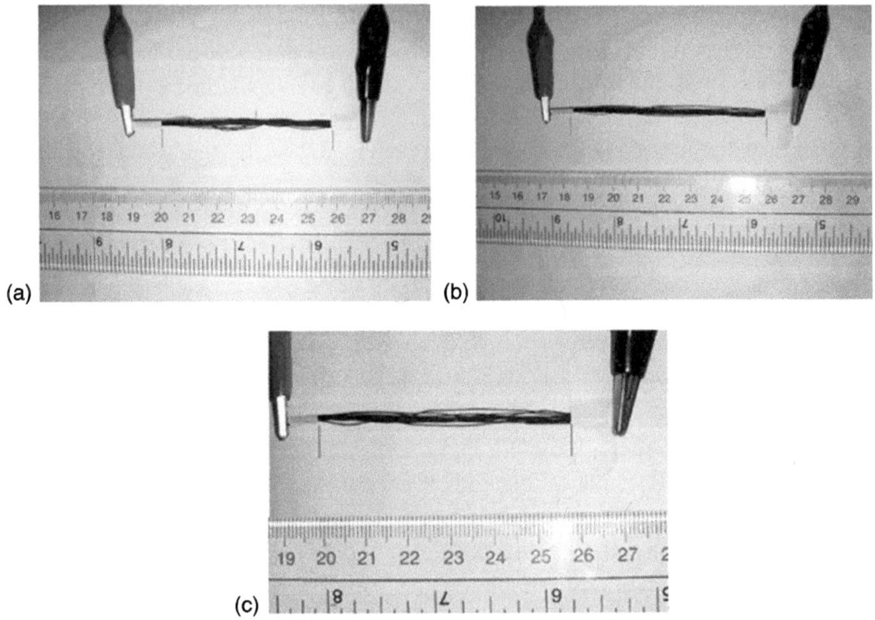

9.7 The electro-active shape memory effect of MWCNT/SMFs: (a) the original length; (b) the length after elongation; and (c) the length after recovery.

9.3 Magnetically-active shape memory polymers

For thermally-adaptive SMPs, energy from direct heating is used to trigger the shape change. However, a second approach, whereby Joule heating after the SMP is incorporated with ferromagnetic fillers, can be used to induce indirect heating and recovery of SMPs. Compared with direct and electro-heating to induce shape change, the magnetic filled adaptive SME is significantly more advantageous. First, through the selection of a ferromagnetic particle material, the Curie temperature is kept within safe medical limits, while Curie thermo-regulation eliminates the danger of overheating. Second, the need for power transmission lines leading to an SMP device is eliminated. Third, more complex device shapes are possible because consistent heating is expected to be achievable for any type of device geometry. Fourth, selective heating of specific device areas is made possible, and finally, remote actuation allows for the possibility of embedded devices that can be later activated by an externally applied magnetic field.

Magnetic-responsive SMPs have also been prepared by employing ferromagnetic fillers. The SMP recovered its original shape as a result of heating

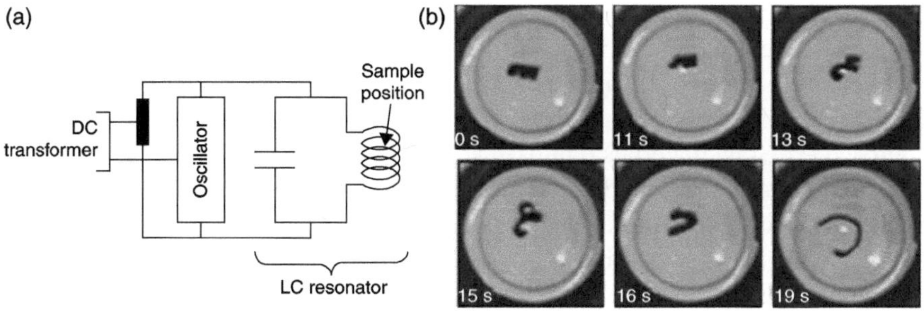

9.8 Magnetically-induced shape memory effect of a SMPU filled with magnetic nanoparticles inside a magnetic field of an inductor ring (Behl, 2008).

induced through a magnetic field. Many researchers employed a magnetic field to enable remote actuation of SMP composites by incorporating nano iron(II, III) oxide nanoparticles and Ni-Zn ferrite particles into SMPs (Buckley *et al.*, 2006; Mohr *et al.*, 2006; Schmidt, 2006; Cuevas *et al.*, 2009; Weigel *et al.*, 2009; Yakacki *et al.*, 2009). The magnetic-responsive SME of an SMPU filled with magnetic nanoparticles is shown in Fig. 9.8(b). Magnetic-responsive SMPs also proved to have good potential biomedical applications (Buckley *et al.*, 2006; Varga *et al.*, 2006; Razzaq *et al.*, 2007a,b).

9.4 Moisture-active shape memory polymers

Moisture-active SMPs can be developed by one of two strategies. First, the shape recovery is achieved by decreasing the switch temperature of the thermally-adaptive SMPs through water that is absorbed by the materials. Second, a highly hydrophilic segment is employed in a segmented polymer so that the shape recovery can be triggered by water/moisture. If T_g-type SMPUs are exposed to high humidity at room temperature for several days, their glass transition temperature drops significantly due to the plasticizing effect of the water. A straight SMPU wire is bent into a circular shape at 40°C and the circular shape is fixed after the wire is cooled to an ambient temperature. In the following week, at an ambient temperature and in a dry cabinet with a humidity of 30 RH, the shape memory wire does not show significant shape recovery. However, after being immersed in water at room temperature for about 30 minutes, the SMPU circular wire starts to recover gradually.

Another strategy for moisture-adaptive shape-memory polymer preparation is to use water soluble segments as a soft segment to prepare SMPUs. Preparation is by polyhedral oligomeric silsesquioxane (POSS) molecules and poly(ethylene glycol) that act as the hydrophobic and hydrophilic groups. The POSS groups aggregate, and the separated group of inorganic moieties is due to the physical interactions between the groups hindering the irreversible sliding of the organic main chain. Similarly, a water sensitive, biodegradable stent, made of chitosan films cross-linked with an epoxy compound, is fabricated by following this approach. The raw materials (chitosan and PEO400) for developing the stent are relatively hydrophilic, and the polymer prepared is thus responsive to hydration. It has been noted that the ether bonds (–O–) in the epoxy compound serve as flexible joints in the cross-linking bridges. The biodegradable stent can be used as an alternative to metallic stents and for local drug delivery.

The pyridine unit, which is responsive to moisture, can be used to improve the moisture absorption in polyurethane. Chen *et al.* (2009) introduced a pyridine unit into SMPU by N-bis(2-hydroxylethyl) isonicotinamine (BINA) and prepared a moisture-responsive SMPU with high strain recovery. The synthesis route of the SMPU with the pyridine unit is presented in Fig. 9.9. First, 1,6-hexamethylene diisocyanate (HDI) reacts with BINA to form a prepolymer, before the pre-polymer is extended with 1,4-butanediol (BDO). It can be deduced that SMPs that are sensitive to their suitable solvents, and similarly hydrophilic SMPs that are sensitive to water/moisture, can be obtained in this way. Lv *et al.* (2008b) observed the DMF (N,N′-dimethylformamide) (a good solvent of SMPU) responsive SME of SMPU.

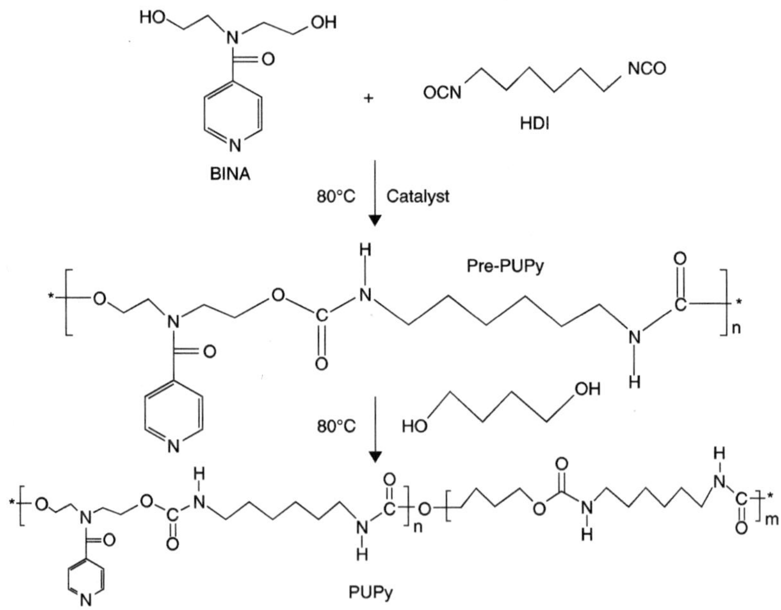

9.9 Synthesis routine of shape memory polyurethane with pyridine unit (reprinted with kind permission of Elsevier Ltd).

9.5 Light-adaptive composites from thermally-adaptive shape memory polymers

Another type of indirect heating SME in SMPs is the light-active SME. Light can be used to increase the material temperature to above the switch transition temperature. The heat absorbance can be enhanced by the incorporation of carbon black and CNTs. Heat absorbed by the light-adaptive polymers raises the internal temperature and melts polymer crystallites or changes the material from a glassy to a rubbery state to trigger the release of stored strain energy. This process is shown in Fig. 9.10.

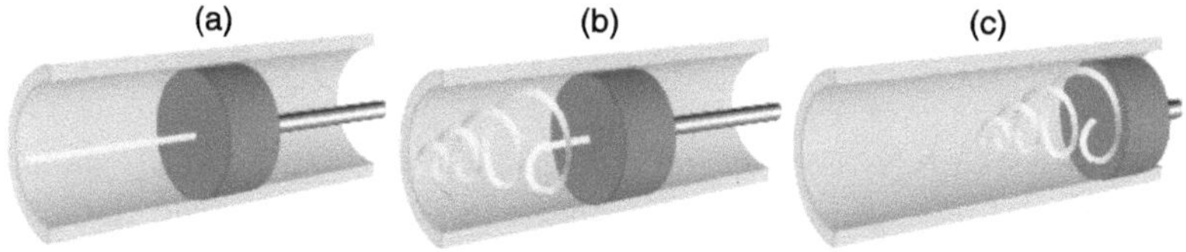

9.10 Light-responsive shape recovery of a SMP micro-actuator coupled to: (a) an optical fiber: (b) temporary straight rod; and (c) permanent corkscrew form (Small *et al.*, 2005).

A two-way SME was observed on a multi-walled CNT (MWNT)-elastomer composite (Ahir and Terentjev, 2005; Vaia, 2005). The expansion and contraction of the composite depends on the extent to which the composite is strained. If the material is slightly pulled, it will expand when exposed to infrared light. Conversely, if the material is pulled to a strain of higher than 10%, it will contract under identical exposure to infrared light. This process is completely reversible and persists after numerous cycles. The reason for this bimodal response has yet to be completely identified.

9.6 Introduction to phase change materials

PCMs, also known as latent heat-storage materials, are a series of functional materials with storing and releasing large heat energy capability under slight or no temperature change. They can be applied in solar energy storage, smart housing, agricultural greenhouses, temperature-regulating textiles, heat management of electronics, telecommunications and microprocessor equipment, and biomedical and biological carrying systems (Vigo and Frost, 1983, 1985; Bruno and Vigo, 1994; Zhang *et al.*, 1999). In recent years, as the energy crisis has become increasingly serious, PCMs are attracting much attention as a potentially energy saving development.

The pioneering research into PCM was conducted by Triangle Research and Development Corporation in the early 1980s. The initial objective was to protect astronauts and instruments from extreme temperature fluctuations in space. The corporation used a variety of different methods to incorporate versatile PCM into fibers or fabrics, a project from which the commercialized technology known as Outlast product technologies eventually grew. Dincer and Rosen (2002) give a comprehensive review of the different types of PCMs and their respective advantages and disadvantages.

In fact, a great number of organic, inorganic, polymeric and eutectic PCMs have latent heat-storage properties. According to their phase change states, they fall into three groups: solid–solid PCMs, solid–liquid PCMs and liquid–gas PCMs. For solid–liquid PCMs, energy is absorbed by the materials to break down the bonding responsible for the solid structure. A large amount of heat is absorbed during the phase change, without the temperature rising. When the materials cool down, the latent heat will be released to the surroundings. Generally, solid–liquid PCMs have the disadvantage of a lack of reliability and poor thermal behavior on repeated thermal-cool cycles (Bryant, 1999; Tao, 2001). Liquid–gas PCMs are not yet practical for use due to the large increase in volume they incur during the phase change from liquid to gas. Solid–solid PCMs absorb and release heat in the same manner as solid–liquid PCMs. However, these materials do not change into a liquid state above the phase change temperature. They merely soften or harden with temperature change above and below the phase change temperature.

Compared with solid–liquid PCMs, solid–solid PCMs have several notable advantages, such as a lack of liquid or gas generation, small volume change and the lack of the necessity for a seal (Liang and Guo, 1995; Ye and Ge, 2000; He and Zhang, 2001). In addition, they can be processed into desired shapes and do not need to be encapsulated like most solid–liquid PCMs do prior to use (Wang *et al.*, 2000; Zalba *et al.*, 2003; Mattila, 2006).

9.7 Phase change materials for textiles applications

In the human body, each area of skin retains an optimum temperature for comfort. The most comfortable skin temperature is reportedly 33.4°C, though this can fluctuate a little according to the particular part of the body and the time of day. When the temperature of the skin differs more than 3.0°C from this ideal temperature, the human body feels uncomfortable. In physical exercise the temperature of the muscles can rise to 39 to 40°C. In addition, a core body temperature of 36.5°C is required, and a rise or fall of 1.5°C can be fatal (Tao, 2001; Mattila, 2006).

One of the primary objectives in clothing design is therefore to protect the wearer's skin temperature from rising or falling too far. The performance of cold insulation in conventional garments relies on the air trapped in the clothing layers. If the air trapped in the textiles layers is low, the thermal insulation will be reduced, while if the air trapped is high, the thermal insulation is increased. The near infrared textiles developed in the late 1980s can absorb solar rays and use them to heat the textiles. The far infrared textiles developed at the same time can absorb the human body's far infrared irradiation and turn it into heat. However, the ultra-violet absorbing textiles developed in the 1990s can absorb the ultra-violet and reflect the near infrared rays from the sun, which can make the body feel cooler (Tao, 2001).

The ideal system in terms of textile insulation would be for the excess heat a person produces to be stored in the clothing system and then released again when they begin to cool. PCMs that have been used to regulate temperature fluctuations have this function (Han and Hubbell (1997); Langer, 1990; Son *et al.*, 1991; Veronese *et al.*, 1996). When the temperature is above the phase change temperature, PCMs absorb and store heat energy with no temperature increase. When the material cools down, the latent heat will be released to the body.

To be suitable for use in textiles, a PCM should meet several requirements. It should have a melting point between 15 and 35°C, large fusion heat, limited temperature fluctuations between the phase change processes and should be environmentally friendly. Similarly, it should have low or no toxicity, stability and repetition, effective heat transfer, ease of availability; and a low cost (Pause, 1995; Colvin, 1998, 1999; Bryant, 1999; Mondal, 2008).

9.8 Solid–liquid phase change materials

A great number of organic, inorganic, polymeric and eutectic PCMs have latent heat-storage properties with a phase change temperature at this interval (Liang and Guo, 1995; Pause, 1995; Colvin, 1998, 1999; Bryant, 1999; Wang *et al.*, 2000; Ye and Ge, 2000; He and Zhang, 2001; Tao, 2001; Dincer and Rosen, 2002; Zalba *et al.*, 2003; Mattila, 2006). The most popular of these include hydrated inorganic salts, polyhydric alcohol mixture (water solution), polyethylene glycol (PEG), polytetramethylene glycol, aliphatic polyester, linear chain hydrocarbon, hydrocarbon alcohol and hydrocarbon acid (Vigo and Frost, 1983, 1985; Bruno and Vigo, 1994; Zhang *et al.*, 1999). Of the above PCMs, paraffin hydrocarbons and PEG are the most widely used in textiles because they can better meet the above standards. Paraffin hydrocarbons are organic compounds consisting entirely of hydrogen and carbon as by-products of oil refining processes. The phase change temperature of hydrocarbons with 16–21 carbons is in the range of 10–40°C. The heat capacity and crystallization temperature of linear hydrocarbons with different numbers of carbon atoms, are reported by various researchers (Pause, 2001; Zuckerman *et al.*, 2003; Zhang *et al.*, 2005a) is recorded in Table 9.1 (Mondal, 2008). The popularity of parrafinic hydrocarbons is because they are non-corrosive, chemically and thermally stable, inexpensive, and contain an extensive supply of raw materials (Erkan, 2004; Mondal, 2008).

PEG is another PCM which has been intensively studied (Table 9.2). It has a melting temperature from around 3.2 to 68.7°C and a high phase change enthalpy due to its molecular weight (Son *et al.*, 1991). In addition, PEG has good biocompatibility due to its intrinsic molecular structure (Han and Hubbell (1997); Langer, 1990; Veronese *et al.*, 1996). Mondal (2008) presents the melting points of the PEG with

Table 9.1 The heat capacity and crystallization temperature of hydrocarbons with various numbers of carbon atoms

Hydrocarbons	No. of C atoms	Latent heat of adsorption (ΔH)(J/g)	Crystallization temperature (T_c,°C)
n-Hexadecane	16	235.2	12.2
n-Heptadecane	17	176.4	16.5
n-Octadecane	18	244.8	22.0
n-Nonadecane	19	177.6	26.4
n-Eiscosane	20	242	30.4

Table 9.2 Molecular weight and melting point of PEG

Materials	Molecular weight	Melting point (°C)
PEG1000	1000	35
PEG1500	1500	50
PEG3400	3400	59
PEG8000	8000	60
PEG10000	10,000	62
PEG20000	20,000	63

different molecular weights. With a molecular structure similar to PEG, poly(tetramethylene glycol) (PTMG) is another kind of polymeric PCM. However, its application is limited because it needs intensive supercooling for crystallization due to the weak interaction between the molecular chains (Mattila, 2006).

9.9 Shape memory polymers with phase change properties

Despite the advantages of PEG listed above, as a solid–liquid PCM, it is prone to exhibit unreliable and poor thermal behavior on repeated thermally-cool cycles (Cox, 1998, 2001; Zhang *et al.*, 2005b). Solid–solid PCMs are generally preferable to solid–liquid PCMs, as they generate no liquid or gas, incur only limited volume change, and do not require a seal (Liang and Guo, 1995; Ye and Ge, 2000; He and Zhang, 2001).

Several research groups have made great efforts to prepare solid–solid PCMs by employing PEG as a phase change ingredient while another 'rigid' ingredient acts as a skeleton to keep the material in solid state after the melting of PEG. Hu *et al.* (2006) prepared a solid–solid PCM by copolymerizing poly(ethylene terephthalate)-PEG. The phase change enthalpy increased to a maximum of 26.85 J/g at 45 wt.% PEG proportion. However, even the maximum enthalpy of 26.85 J/g was still low in comparison with that of pure PEG. Jiang *et al.* (2002) prepared a network solid–solid PCM with rigid polymer cellulose diacetate (CDA) serving as a skeleton, and the PEG as a functional branch chain. However, because of the covalent network structure of the PEG-grafted CDA, the material proved unfit for melt processing.

More recently, Li and Ding (2007) prepared a novel PCM using PEG/MDI (diphenylmethane diisocyanate)/PE (pentaerythritol), cross-linking the copolymer via the condensation reaction of PEG with tetrafunctional pentaerythritol isocyanate. In this research, PEG10000 was used as an energy-storing, working ingredient and tetrafunctional pentaerythritol isocyanate as a multifunctional reactant. First, the tetrafunctional pentaerythritol was end-caped with isocyanate moieties, before the reaction product reacted with PEG to form the star-like copolymer. The reaction mechanism is shown in Fig. 9.11. The phase transition enthalpy was more than 100 J/g, with a transition point at 58.68°C.

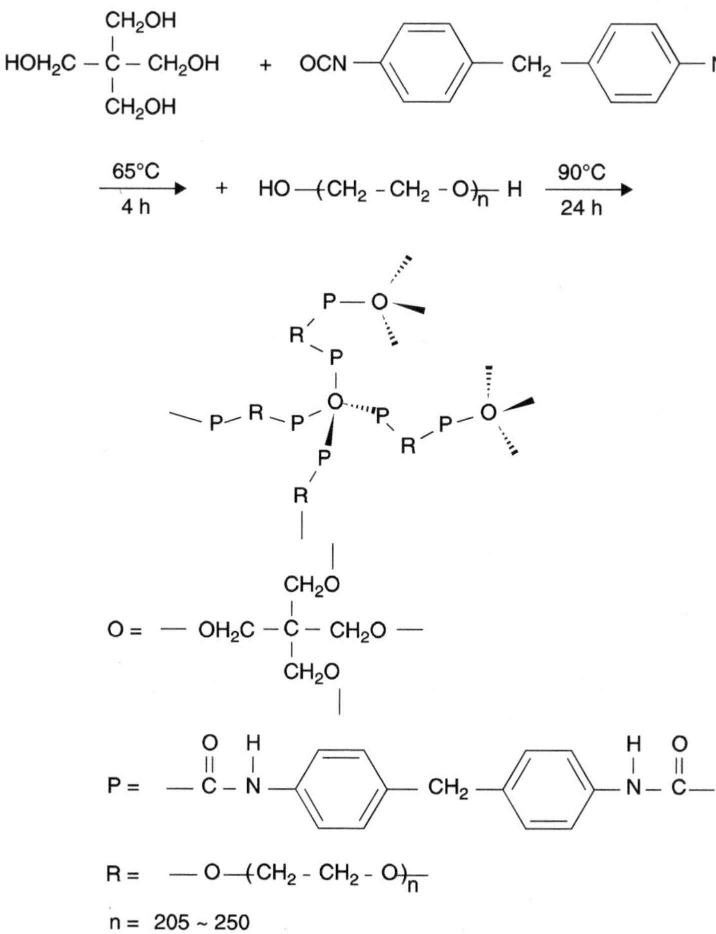

9.11 Reaction mechanism to star-like cross-linking PEG/MDI/PE copolymer.

9.10 Fabrication strategies for phase change materials

9.10.1 Solid–liquid phase change materials

Currently, the most widely-used PCMs in textiles are solid–liquid PCMs, because of their high thermal capacity and low volume change during the phase change. During applications, PCMs have to be encapsulated by specialist technologies because, once the phase change temperature is surpassed, they will become liquid. An alternative method is to fill hollow synthetic fibers with PCMs.

9.10.2 Microcapsules of phase change materials

The technology for encapsulating microcapsules for textile applications dates back to the early 1980s, and the work of Nelson (2001) of NASA. Microcapsules usually have walls of less than 2 μm in thickness and 20 to 40 μm in diameter, as shown in Fig. 9.12. The microcapsules can be reduced by depositing a thin polymer coating onto small solid particles or on dispersions of solids in liquids.

In their application in textiles, the paraffins are either in a solid or liquid state. In order to prevent dissolution of the paraffin while in the liquid state, it is enclosed in a small plastic sphere with a diameter of only a few micrometers. The shell polymer should be resistant to mechanical action such as friction, heat and most types of chemicals. When the temperature rises, the microcapsules absorb heat from the surroundings by melting the PCM in the microcapsules, thus allowing them to store heat energy. When the temperature falls, the microcapsules release the heat by crystallizing the PCM in the microcapsules (Pause, 2000).

The micro-encapsulated PCMs used in thermal regulating textiles are normally low due to the processing difficulties incurred by the microcapsules at a

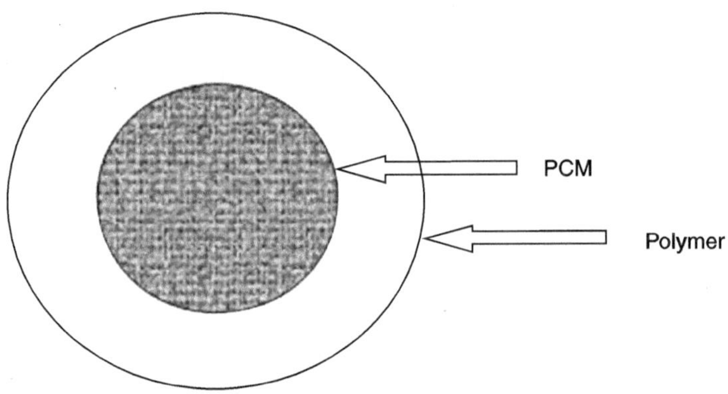

9.12 Microcapsule structure with a paraffinic PCM core and a polymeric shell (Nelson, 2001).

high content. The large particle size and bad distribution of PCM microcapsules in textiles such as fibers may lead the spinnability of the fiber to deteriorate. The most recent research is *in-situ* micro-encapsulation of PCMs by special treatment, which means that the micro-encapsulation happens after the polymer processing. For example, PCM and membrane-forming reagents may be added to polymer melts or solutions to form lacteal spinning dopes. The PCM fibers are then spun by conventional spinning techniques. At or after the final step of the spinning, the PCM microcapsules in the fiber are formed through heat treatment.

9.10.3 Coating methods

The employment of the coating method to make PCM fiber and textile is both convenient and practical. Microspheres containing PCM are dispersed in water solution containing a surfactant, a dispersant, an antifoam agent and a polymer mixture, before the coating is applied to a textile substrate.

PCM microcapsules can be coated onto textiles by traditional coating methods such as the knife-over-roll, knife-over-air, pad-dry-cure, gravure, dip coating and transfer coating. First, the microspheres containing PCM solution or emulsion are made. Then the PCMs are attached to textile surfaces by the coating. The efficiency of encapsulation and stability of the PCM microcapsules during applications are the key challenges in preparing PCM textiles by coating. To maximize the longevity of the PCM textile products, the polymeric shells of the PCM microcapsules have to be resistant to abrasion and pressure. Microcapsules may also be first embedded in a coating compound such as acrylic, polyurethane and rubber latex, and then applied to a fabric.

9.10.4 Filling of hollow fibers

PCMs may also be filled into synthetic hollow fibers to fabricate PCM fibers and fabrics. Researchers have fabricated different kinds of hollow fibers filled with both inorganic and organic PCMs such as $LiNO_3.3H_2O$, Zn $(NO_3)2.6H_2O$ and PEG. The hollow fibers employed include rayon fibers and polypropylene fibers (Vigo and Frost, 1983). For application in textiles, the wash-resistance, durability and handling of the treated PCM have to be taken into consideration.

9.10.5 Laminating methods

Microcapsules can also be blended into a polymer matrix to fabricate PCM films and then develop PCM textiles by laminating the films with the fabrics. Pause (2003) incorporated PCM directly into a polymer film, which was then laminated into the non-woven fabric system. In comparison to the methods listed above, the laminating approach has the significant advantage that a high PCM concentration can be obtained without the cost of the micro-encapsulation procedure. Laminating also minimizes the amount of surplus weight added to the garment.

9.10.6 Spinning methods

Multi-component PCM fibers can be prepared using a PCM as the core and other polymer materials as the sheath. The PCMs are often micro-encapsulated and blended with other fiber-forming polymers prior to spinning, using the melt spinning or wet spinning process.

Zhang *et al.* (2005b) prepared sheath/core composite fibers with 4 to 24 wt.% of micro-encapsulated n-octadecane by melt spinning with a spinning speed of 720 m/min. Polyethylene blended with n-octadecane microcapsules was used as the core and polypropylene was employed as the sheath material. With 20 wt.% microcapsules, the enthalpy, tensile strength and strain of the fiber were 11.0 J/g, 1.8 cN/dtex and 30.2%, respectively. Zhang *et al.* (1999, 2004) also studied the phase change properties of PEG blends with polypropylene and polyester or ethylene-vinyl acetate to prepare PCM fibers. By controlling the relevant spinning parameters and component contents, they managed to prepare temperature-regulating fibers. Magill *et al.* (2006) invented a multi-component fiber with enhanced reversible thermal properties. The multi-component fibers are made of at least two polymers, one of which is a temperature-regulating material with a latent heat of at least 40 J/g and a transition temperature in the range of 22 to 40°C.

Many research groups have made great efforts to prepare temperature-regulating fibers by incorporating micro-encapsulated PCMs into fibers via wet or melt spinning (Bryant and Colvin, 1988; Leskovsek *et al.*, 2004; Zhang *et al.*, 2005b, 2006). This method has proven to be impractical on several grounds. First, microcapsules are prone to form clusters and cause discontinuity in the spinning process; second, the thermal capacity of the prepared fiber is rather modest (~8–12 J/g) because the micro-encapsulated PCMs have an upper loading limit (Zhang *et al.*, 2005b). Finally, it is more complex to incorporate PCMs into melt-spun fibers, because PCMs are mostly volatile materials, and are thus unfit for high temperature extrusion and pumping.

Taking advantages of the *in-situ* micro-encapsulation, Jiang *et al.* (2008) prepared temperature-regulating fibers by wet spinning, using polyvinyl alcohol as the polymer matrix, with ethyl orthosilicate (TEOS) as the membrane-forming reagent and paraffin as the PCM. The microcapsules were not added to the spinning dopes before spinning but generated themselves *in situ*, indicating that this method can be used to develop temperature-regulating fibers with good processability, high latent heat and low cost.

SMPUs are segmented copolymers which may be synthesized from three starting materials: long-chain polyol, diisocyanate and chain extender. Diisocyanate and chain extender form the hard segment. The long-chain polyol forms the soft segment (Hu *et al.*, 2007; Meng *et al.*, 2007b; Zhu *et al.*, 2007). Meng and Hu (2008) synthesized a shape memory thermoplastic polyurethane as a PCM by employing PEG as the soft segment, while isophorone diisocyanate (IPDI) and 1,4-butanediol (BDO) were used as the hard segment. This solid–solid PCM possessing SME may find applications in such fields as smart textiles, smart housing and smart packaging in the near future.

The PCM polyurethanes with PEG as the soft segment show SME as a result of the phase separation of the segmented polyurethanes. Then temperature-regulating fibers were prepared from the PCM SMPUs. PEG-based SMPU fibers are not ideal for temperature-regulating purposes, however, as they demand too high a degree of supercooling and thus often fail to satisfy the mechanical properties. Furthermore, due to the significant melt transition at the low temperature, the fiber does not have good spinnability.

The image of the prepared fiber in Fig. 9.13 shows that the fiber surface is smooth and uniform. The 100 dtex fiber has a tenacity of about 0.7 cN/dtex. The fiber also shows an obvious neck-type phenomena, which is frequently observed in the static stress–strain curves of crystalline polymers.

The high molecular weight polyol provides the fiber with high breaking elongation ratios of more than 400% and the hydrogen bonded hard segments contribute to the mechanical strength, which is initially as low as 0.7 cN/dtex. The fiber mechanical strength can also be improved by decreasing the soft segment content, though this does reduce fiber heat storage capacity.

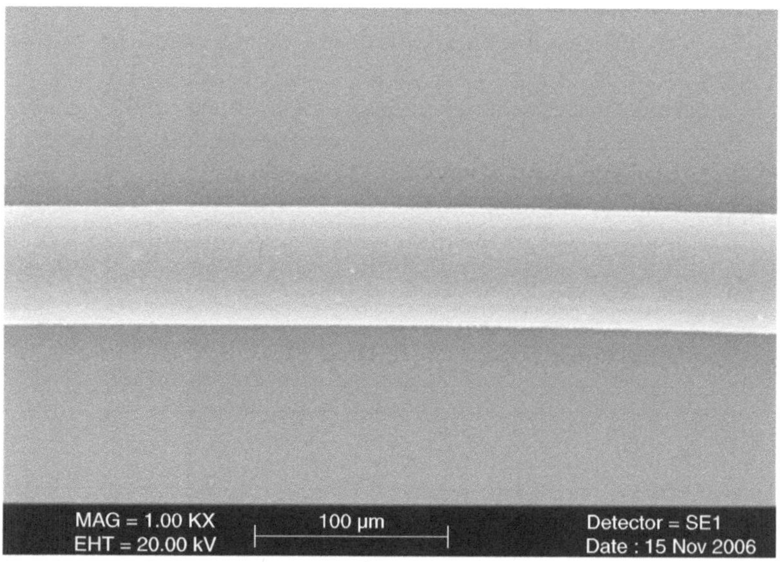

9.13 SEM surface image of fiber.

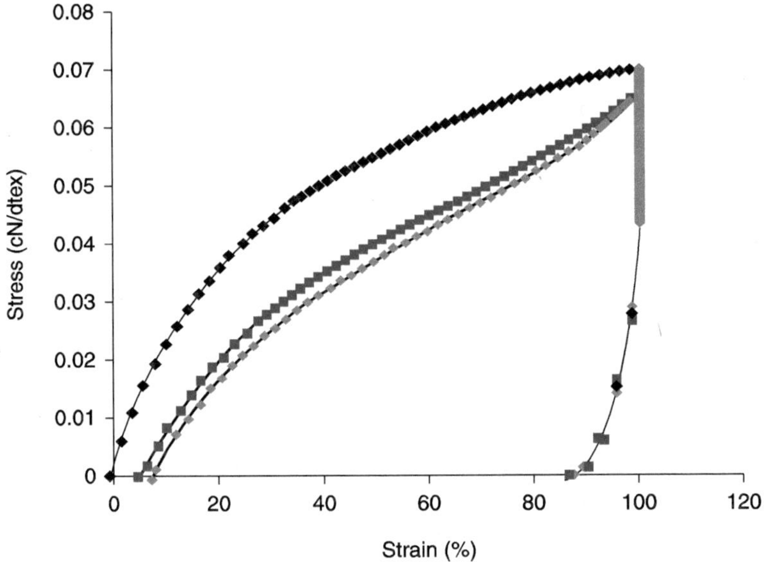

9.14 The cyclic strain–stress curves of the shape memory fiber.

The cyclic strain–stress curves of fiber obtained by thermo-mechanical cyclic tensile testing are shown in Fig. 9.14. The fiber has a fixity ratio of more than 85.8% and a recovery ratio of more than 95.4%. The detailed shape fixity ratios and recovery ratios are recorded in Table 9.3. At 60°C, in the first cycle, the maximum stress at full elongation is 0.07 cN/dtex.

The fiber SME accompanied with phase change effect can be illustrated as follows. In the unstretched state, the PEG phase in the fiber has high crystallinity (X_c), as demonstrated by differential scanning calorimetry (DSC) analysis. When the fiber is heated from the ambient temperature to 60°C – which is above the PEG phase melting temperature (T_{trans}) – the fiber absorbs heat energy and the PEG phase melts to assume a random state. However, the fiber does not break because of the 'physical cross-links' within the hard segment, which restrict the movement of the soft segment phase. When the fiber is stretched, the soft segments

Table 9.3 Detailed shape fixity ratios and recovery ratios of fiber

Circle no.	$\varepsilon_p(N)$	$[R_f(N)]$	Stress at 100% strain(cN/dtex)	$R_{r,tot}(N)$	$R_r(N)$
1	0.0%	85.8%	0.070	100.00%	
2	4.6%	86.7%	0.065	95.40%	95.4%
3	6.6%	87.2%	0.064	93.40%	97.9%

in the fiber are extended. If the temperature is cooled to below T_{trans}, the latent heat is released and consequently the soft segments crystallize. As a result, the internal stress is stored in the fiber and associated deformation is fixed temporally. If the fiber is reheated to a temperature above T_{trans}, the soft segment phase absorbs heat energy and becomes flexible and the fiber resumes to the original length as a result of releasing internal stress stored among the hard segments.

9.11 Summary

To conclude, SMPs with indirect heating SME can be developed in various ways, and have many current and potential applications in various fields. Conventionally, the thermal-responsive shape recovery of SMPs is induced by directly heating the polymer to a temperature above the switching temperature. When electrical conductive materials are incorporated into thermally-active SMPs, certain levels of electric conductivity can be achieved. When a current passes through the network of conductive materials within the SMPs, the induced Joule heating may raise the internal temperature to above the switch transition temperature of the polymer, and trigger shape recovery.

When ferromagnetic fillers are employed in SMPs, magnetic-active SME can also be achieved by inducing heat using a magnetic field. Water molecules can be used to decrease the switch temperature of the SMPs to induce shape recovery of deformed SMPs, and SME can be accelerated further if the polymer contains water soluble ingredients. Light absorbed by deformed SMPs can also increase SMP temperature to the switch transition temperature, to trigger shape recovery.

Moving onto the practical application of these polymers, we have seen some of the dilemmas which researchers and producers face. On the one hand, the heat generated by the body at a high temperature or during strenuous activity has to be released into the environment through the clothing we wear. On the other hand, the heat has to be maintained by the body by the barrier of clothing at a low temperature. PCMs can interactively respond to an individual's unique physiological condition, and so absorb, store and release heat to help the body to remain comfortable. They store or release heat energy over a narrow temperature range, thus helping the body to maintain its natural temperature in fluctuating conditions.

To obtain an efficient thermal-regulating effect for comfort, a number of parameters have to be considered, such as the phase change temperature of PCMs, PCM quantity, the duration and application of the garment, and the temperature profile of the body. The phase change temperature range needs to match the application temperature. The polymer-based PCMs should be applicable even at high degrees of supercooling. The conductivity of the PCM and the garment also significantly affect the efficiency of temperature-regulating garments. Thinner garments may have high heat conductivity; while thicker garments may have low heat conductivity, which may promote or delay the temperature-regulating effect. The temperature at different parts of the body and during different activities should also be taken into account. The quantity of heat produced by the human body by different physical activities can vary from 100 W while resting, to over 1000 W during maximum physical performance. Sometimes, different parts of garments for various activities may use different quantities of PCMs.

9.12 References

Ahir, S. V., Terentjev, E. M. (2005), Photomechanical actuation in polymer-nanotube composites, *Natural Materials*, **4**, 491–5.

Andrews, R., Weisenberger, M. C. (2004), Carbon nanotube polymer composites, *Current Opinion in Solid State and Materials Science*, **8**, 31–7.

Behl, M. (2008), Shape-memory polymers, *Third International Conference on Nanotechnology and Smart Textiles for Industry, Healthcare and Fashion*, 19 March, The Royal Society, London.

Biercuk, M. J., Llaguno, M. C., Radosavljevic, M., Hyun, J. K., Johnson, A. T., Fischer, J. E. (2002), Carbon nanotube composites for thermal management, *Applied Physics Letters*, **80**, 2767–9.

Borca-Tasciuc, T., Mazumder, M., Son, Y., Pal, S. K., Schadler, L. S., Ajayan, P. M. (2007), Anisotropic thermal diffusivity characterization of aligned carbon nanotube-polymer composites, *Journal of Nanoscience and Nanotechnology*, **7**, 1581–8.

Bruno, J. S., Vigo, T. L. (1994), Thermal properties of insolubilized polyacetals derived from non-formaldehyde cross-linking agents, *Thermochimica Acta*, **243**, 155–9.

Bryant, Y. G. (1999), Melt spun fibers containing micro-encapsulated phase change material. *Advances in Heat and Mass Transfer in Biotechnology – The ASME International Mechanical Engineering Congress and Exposition*, Nashville, TN.

Bryant, Y. G., Colvin, D. P. (1988), *Fiber with reversible enhanced thermal storage properties and fabrics made therefrom*, US Patent 091550/4756958.

Buckley, C. P., Prisacariu, C., Caraculacu, A. (2007), Novel triol-crosslinked polyurethanes and their thermo-rheological characterization as shape-memory materials, *Polymer*, **48**, 1388–96.

Buckley, P. R., Mckinley, G. H., Wilson, T. S., Iv, W. S., Benett, W. J., *et al.* (2006), Inductively heated shape memory polymer for the magnetic actuation of medical devices, *IEEE Transactions on Biomedical Engineering*, **53**, 2075–83.

Cadek, M., Coleman, J., Ryan, K., Nicolosi, V., Bister, G., Fonseca, A. (2004), Reinforcement of polymers with carbon nanotubes: The role of nanotube surface area, *Nano Letters*, **4**, 353–6.

Chen, S., Hu, J. L., Yuen, C., Chan, L. (2009), Novel moisture-sensitive shape memory polyurethanes containing pyridine moieties, *Polymer*, **50**, 4424–8.

Chen, W., Tao, X. M. (2005), Self-organizing alignment of carbon nanotubes in thermoplastic polyurethane, *Macromolecular Rapid Communications*, **26**, 1763–7.

Cho, J. W., Kim, J. W., Jung, Y. C., Goo, N. S. (2005), Electroactive shape-memory polyurethane composites incorporating carbon nanotube, *Macromolecular Rapid Communications*, **26**, 412–16.

Colvin, D. P. (1998), Protective clothing containing encapsulated phase change materials, *Proceedings of the 1998 ASME International Mechanical Engineering Congress and Exposition*, Anaheim, CA.

Colvin, D. P. (1999), Enhanced thermal management using encapsulated phase change materials – An overview, *Advances in Heat and Mass Transfer in Biotechnology – The ASME International Mechanical Engineering Congress and Exposition*, Nashville, TN.

Cooper, C. A., Ravich, D., Lips, D., Mayer, J., Wagner, H. D. (2002), Distribution and alignment of carbon nanotubes and nanofibrils in a polymer matrix, *Composites Science and Technology*, **62**, 1105–12.

Cox, R. (1998), Synopsis of the new thermal-regulating fiber outlast, *Chemical Fibers International*, **48**, 475–9.

Cox, R. (2001), Repositioning acrylic fibers for the new millennium, *Chemical Fibers International*, **51**, 118–20.

Cuevas, J. M., Alonso, J., German, L., Iturrondobeitia, M., Laza, J. M., *et al.* (2009), Magneto-active shape memory composites by incorporating ferromagnetic microparticles in a thermo-responsive polyalkenamer, *Smart Materials and Structures*, **18**, 075003.

Demczyk, B. G., Wang, Y. M., Cumings, J., Hetman, M., Han, W., *et al.* (2002), Direct mechanical measurement of the tensile strength and elastic modulus of multiwalled carbon nanotubes, *Materials Science and Engineering A*, **334**, 173–8.

Dincer, I., Rosen, M. A. (2002), *Thermal Energy Storage, Systems and Applications*, John Wiley & Sons, Chichester, UK.

Erkan, G. (2004), Enhancing the thermal properties of textiles with phase change materials, *Research Journal of Textile and Apparel*, **8**, 57–64.

Gommans, H. A. J., Tashiro, H., Park, J., Magnuson, J., Rinzler, A. (2000), Fibers of aligned single-walled carbon nanotubes: Polarized Raman Spectroscopy, *Journal of Applied Physics*, **88**, 2509–14.

Gong, X. L. J., Baskaran, S., Voise, R. D., Young, J. S. (2000), Surfactant-assisted processing of carbon nanotube/polymer composites, *Chemistry of Materials*, **12**, 1049–52.

Goo, N. S., Paik, I. H., Yoon, K. J. (2007), The durability of a conducting shape memory polyurethane actuator, *Smart Materials and Structures*, **16**, N23–N26.

Haggenmueller R. G. H., Gommans, H. H., Rinzler, A., Fischer, J. E., Winey, K. L. (2000), Aligned single-wall carbon nanotubes composites by melt processing methods, *Chemical Physics Letters*, **330**, 219–25.

Han, D., Hubbell, J. (1997), Synthesis of polymer network scaffolds from lactide and poly(ethylene glycol) and their interaction with cells, *Macromolecules*, **30**, 6077–83.

He, Q., Zhang, W. (2001), A study on latent heat storage exchangers with the high-temperature phase-change material, *International Journal of Energy Research*, **25**, 331–41.

Hou, H. Q., Ge, J. J., Zeng, J., Li, Q., Reneker, D. H., *et al.* (2005), Electrospun polyacrylonitrile nanofibers containing a high concentration of well-aligned multiwall carbon nanotubes, *Chemistry of Materials*, **17**, 967–73.

Hu, J., Yu, H., Chen, Y. M., Zhu, M. (2006), Study on phase-change characteristics of PET-PEG copolymers, *Journal of Macromolecular Science, Part B: Physics*, **45**, 615–21.

Hu, J. L., Meng, Q. H., Zhu, Y., Lu, J., Zhuo, H. T. (2007), *Shape memory fibers prepared by wet, reaction, dry, melt, and electro spinning*, US Patent 11/907,012.

Huang, H. M., Liu, I. C., Chang, Y. U., Tsai, H. C. (2004), Preparing a polystyrene-functionalized multiple-walled carbon nanotube via covalently linking acyl chloride functionalities with living polystyryllithium, *Journal of Polymer Science Part A: Polymer Chemistry*, **42**, 5802–10.

Hwang, J., Gommans, H. H., Ugawa, A., Tashiro, H., Haggenmueller, R., *et al.* (2000), Polarized spectroscopy of aligned single-wall carbon nanotubes, *Physical Review B*, **62**, 310–13.

Jia, Z. J., Wang, Z. Y., Xu, C. L., Liang, J., Wei, B. Q., *et al.* (1999), Study on poly(methyl methacrylate)/carbon nanotube composites, *Materials Science & Engineering, Part A: Structural Materials: Properties, Microstructure and Processing*, **271**, 395–400.

Jiang, M., Song, X., Ye, G., Xu, J. (2008), Preparation of PVA/paraffin thermal regulating fiber by *in situ* micro-encapsulation, *Composites Science and Technology*, **68**, 2231–7.

Jiang, Y., Ding, E., Li, G. (2002), Study on transition characteristics of PEG/CDA solid–solid phase change materials, *Polymer*, **43**, 117–22.

Jung, Y. C., Sahoo, N. G., Cho, J. W. (2006), Polymeric nanocomposites of polyurethane block copolymers and functionalized muti-walled carbon nanotubes as cross-linkers, *Macromolecular Rapid Communications*, **27**, 126–31.

Koerner, H., Price, G., Pearce, N. A., Alexander, M., Vaia, R. A. (2004), Remotely actuated polymer nanocomposites-stress recovery of carbon-nanotube-filled thermal-plastic elastomers, *Natural Materia* **3**, 115–20.

Kwon, J., Kim, H. (2005), Comparison of the properties of waterborne polyurethane/multiwalled carbon nanotube and acid-treated multiwalled carbon nanotube composites prepared by *in situ* polymerization, *Journal of Polymer Science Part A: Polymer Chemistry*, **43**, 973–85.

Lan, X., Huang, W. M., Liu, N., Phee, S. Y., Leng, J. S., Du, S. Y. (2008), Improving the electrical conductivity by forming NI powder chains in a shape-memory polymer filled with carbon black, *Proceedings of SPIE – The International Society for Optical Engineering*, vol. 6927, *Electroactive Polymer Actuators and Devices (EAPAD) 2008*, 692717, San Diego, CA.

Langer, R. (1990), New methods of drug delivery, *Science*, **249**, 1527–33.

Lau Kin-Tak, H. D. (2002a), Effectiveness of using carbon nanotubes as nano-reinforcements for advanced composite structures, *Carbon*, **40**, 1605–6.

Lau Kin-Tak, H. D. (2002b), The revolutionary creation of new advanced material-carbon nanotube composite, *Composites Part B*, **33**, 263–77.

Leng, J., Lv, H., Liu, Y., Du, S. (2007), Electroactivate shape-memory polymer filled with nanocarbon particles and short carbon fibers, *Applied Physics Letters*, **91**, 144105.

Leng, J., Lu, H., Liu, Y., Du, S. (2008a), Conductive nanoparticles in electro activated shape memory polymer sensor and actuator, *Proceedings of SPIE – The International Society for Optical Engineering*, vol. 6931, *Nanosensors and Microsensors for Bio-Systems 2008*, 693109, San Diego, CA.

Leng, J., Lv, H., Liu, Y., Du, S. (2008b), Synergic effect of carbon black and short carbon fiber on shape memory polymer actuation by electricity, *Journal of Applied Physics*, **104**, 104917.

Leng, J. S., Huang, W. M., Lan, X., Liu, Y. J., Du, S. Y. (2008c), Significantly reducing electrical resistivity by forming conductive NI chains in a polyurethane shape-memory polymer/carbon-black composite, *Applied Physics Letters*, **92**, 204101–3.

Leng, J. S., Lan, X., Liu, Y. J., Du, S. Y., Huang, W. M., *et al.* (2008d), Electrical conductivity of thermo-responsive shape-memory polymer with embedded micron-sized NI powder chains, *Applied Physics Letters*, **92**, 014103–4.

Leskovsek, M., Jedrinovic, G., Stankovic-Elesin, U. (2004), Spinning polypropylene fibers with microcapsules, *Proceedings of the Second ITC & DC*, Faculty of Textile Technology, University of Zagreb, 84–8.

Li, W. D., Ding, E. Y. (2007), Preparation and characterization of cross-linking PEG/MDI/PE copolymer as solid–solid phase change heat storage material, *Solar Energy Materials & Solar Cells*, **91**, 764–8.

Liang, X., Guo, Y. Q. (1995), Crystalline-amorphous phase transition of a poly (ethylene glycol)/cellulose blend, *Macromolecules*, **28**, 6551–5.

Liu, Y., Lv, H., Lan, X., Leng, J., Du, S. (2008), Review of electro-active shape-memory polymer composite, Compscitech.2008.08.016. *Composites Science and Technology*, **69**, 2064–8.

Lourie, O. W. H. (1999), Evidence of stress transfer and formation of fracture clusters in carbon nanotube-based composites, *Composites Science and Technology*, **59**, 975–7.

Lv, H., Leng, J., Du, S. (2008a), Electro-induced shape-memory polymer nanocomposite containing conductive particles and short fibers, *Proceedings of SPIE – The International Society for Optical Engineering*, vol. 6929, *Behavior and Mechanics of Multifunctional and Composite Materials*, 69291L, 10–13 March 2008, San Diego, CA.

Lv, H., Leng, J., Liu, Y., Du, S. (2008b), Shape-memory polymer in response to solution, *Advanced Engineering Materials*, **10**, 592–5.

Magill, M. C., Hartmann, M. H., Haggard, J. S. (2006), *Multi-component fibers having enhanced reversible thermal properties and methods of manufacturing thereof*, PCT/US2005/041030.

Mattila, H. R. (2006), *Intelligent Textiles and Clothing*, vol. 3, Woodhead Publishing Limited.

Meng, Q. H., Hu, J. (2008), A poly(ethylene glycol)-based smart phase change material, *Solar Energy Materials and Solar Cells*, **9**, 1260–8.

Meng, Q. H., Hu, J. L., Yeung, L. Y. (2007a), An electro-active shape memory fiber by incorporating multi-walled carbon nanotubes, *Smart Materials and Structures*, **16**, 830–6.

Meng, Q. H., Hu, J. L., Zhu, Y., Lu, J., Liu, Y. (2007b), Morphology, phase separation, thermal and mechanical property differences of shape memory fibers prepared by different spinning methods, *Smart Materials and Structures*, **16**, 1192–7.

Miaudet, P., Badaire, S., Maugey, M., Derré, A., Pichot, V., *et al.* (2005), Hot-drawing of single and multiwall carbon nanotube fibers for high toughness and alignment, *Nano Letters*, **5**, 2212–15.

Mohr, R., Kratz, K., Weigel, T., Lucka-Gabor, M., Moneke, M., Lendlein, A. (2006), Initiation of shape-memory effect by inductive heating of magnetic nanoparticles in thermoplastic polymers, *Proceedings of the National Academy of Sciences*, **103**, 3540–5.

Mondal, S. (2008), Phase change materials for smart textiles – An overview, *Applied Thermal Engineering*, **28**, 1536–50.

Nelson, G. (2001), Microencapsulation in textile finishing, *Review of Progress in Coloration and Related Topics*, **32**, 57–64.

Ounaies, Z., Park, C., Wise, K. E., Siochi, E. J., Harrison, J. S. (2003), Electrical properties of single-wall carbon nanotube reinforced polyimide composites, *Composites Science and Technology*, **63**, 1637–46.

Paik, I. H., Goo, N. S., Yoon, K. J., Jung, Y. C., Cho, J. W. (2005), Electric resistance property of a conducting shape memory polyurethane actuator, *Key Engineering Materials*, **297–300**, 1539–44.

Paik, I. H., Goo, N. S., Jung, Y. C., Cho, J. W. (2006), Development and application of conducting shape memory polyurethane actuators, *Smart Materials and Structures*, **15**, 1476–82.

Pause, B. (1995), Development of heat and cold insulating membrane structures with phase change material, *Journal of Industrial Textiles*, **25**, 59–68.

Pause, B. (2000), Textiles with improved thermal capabilities through the application of phase change material (pcm) microcapsules, *Melliand Textilberichte*, **81**, 753–4.

Pause, B. (2001), *Building conditioning technique using phase change materials*, US patent 6230444.

Pause, B. (2003), Non-woven protective garments with thermo-regulating properties, *Journal of Industrial Textiles*, **33**, 93–9.

Rahul, S., Bin, Z., Daniel, P., Mikhail, I., Hui, H., James, L. (2004), Preparation of single-walled carbon nanotube reinforced polystyrene and polyurethane nanofibers and membranes by electrospinning, *Nano Letters*, **4**, 459–64.

Razzaq, M. Y., Anhalt, M., Frormann, L., Weidenfeller, B. (2007a), Mechanical spectroscopy of magnetite filled polyurethane shape memory polymers, *Materials Science and Engineering: A*, **471**, 57–62.

Razzaq, M. Y., Anhalt, M., Frormann, L., Weidenfeller, B. (2007b), Thermal, electrical and magnetic studies of magnetite filled polyurethane shape memory polymers, *Materials Science and Engineering: A*, **444**, 227–35.

Sahoo, N. G., Jung, Y. C., Goo, N. S., Cho, J. W. (2005), Conducting shape memory polyurethane-polypyrrole composites for an electroactive actuator, *Macromolecular Materials and Engineering*, **290**, 1049–55.

Sahoo, N. G., Jung, Y. C., Cho, J. W. (2007a), Electroactive shape memory effect of polyurethane composites filled with carbon nanotubes and conducting polymer, *Materials and Manufacturing Processes*, **22**, 419–23.

Sahoo, N. G., Jung, Y. C., Yoo, H. J., Cho, J. W. (2007b), Influence of carbon nanotubes and polypyrrole on the thermal, mechanical and electroactive shape-memory properties of polyurethane nanocomposites, *Composites Science and Technology*, **67**, 1920–9.

Schmidt, A. M. (2006), Electromagnetic activation of shape memory polymer networks containing magnetic nanoparticles, *Macromolecular Rapid Communications*, **27**, 1168–72.

Sholz, M. T. (2000), *Water soluble films used in synthetic casting tapes*, US Patent 6,077,240.

Small, I. W., Wilson, T. S., Benett, W. J., Loge, J. M., Maitland, D. J. (2005), Laser-activated shape memory polymer intravascular thrombectomy device, *Optics Express*, **13**, 8204–13.

Son, C., Moreshouse, H. L., Jeffrey, H. (1991), An experimental investigation of solid-state phase-change materials for solar thermal storage, *Journal of Solar Energy Engineering*, **113**, 244–9.

Tao, X. M. (2001), *Smart Fibers, Fabrics and Clothing*, vol. 3: Woodhead Publishing Limited.

Thostenson, E. R. Z., Chou, T. W. (2001), Advances in the science and technology of carbon nanotubes and their composites: A review, *Composites Science Technology*, **61**, 899–912.

Thostenson, E. R. Z., Chou, T. W. (2002), Aligned multi-walled carbon nanotube-reinforced composites: Processing and mechanical characterization, *Journal of Physics D: Applied Physics*, **35**, L77–L80.

Vaia, R. (2005), Nanocomposites: Remote-controlled actuators, *Nature Materials*, **4**, 429–30.

Varga, Z., Filipcsei, G., Zrínyi, M. (2006), Magnetic field sensitive functional elastomers with tuneable elastic modulus, *Polymer*, **47**, 227–33.

Veronese, M., Monfardini, C., Caliceti, P., Schiavon, O., Scrawen, M. D., Beer, D. (1996), Improvement of pharmacokinetic, immunological and stability properties of asparaginase by conjugation to linear and branched monomethoxy poly(ethylene glycol), *Journal of Controlled Release*, **40**, 199–209.

Vigo, T. L., Frost, C. M. (1983), Temperature-adaptable hollow fibers containing polyethylene glycols, *Journal of Coated Fabrics*, **12**, 243–54.

Vigo, T. L., Frost, C. M. (1985), Temperature-adaptable fabrics, *Textile Research Journal*, **55**, 737–43.

Wang, X. W., Lu, E. R., Lin, W. X., Wang, C. Z. (2000), Micromechanism of heat storage in a binary system of two kinds of polyalcohols as a solid–solid phase change material, *Energy Conversion and Management*, **41**, 135–44.

Wei, Z. G., Sandstrom, R., Miyazaki, S. (1998), Shape-memory materials and hybrid composites for smart systems, Part I: Shape-memory materials, *Journal of Materials Science*, **33**, 3743–62.

Weigel, T., Mohr, R., Lendlein, A. (2009), Investigation of parameters to achieve temperatures required to initiate the shape-memory effect of magnetic nanocomposites by inductive heating, *Smart Materials and Structures*, **18**, 025011.

Xie, T., Rousseau, I. A. (2009), Facile tailoring of thermal transition temperatures of epoxy shape memory polymers, *Polymer*, **50**, 1852–6.

Yakacki, C. M., Satarkar, N. S., Gall, K., Likos, R., Hilt, J. Z. (2009), Shape-memory polymer networks with FE_3O_4 nanoparticles for remote activation, *Journal of Applied Polymer Science*, **112**, 3166–76.

Yang, B., Huang, W. M., Li, C., Li, L. (2006), Effects of moisture on the thermo-mechanical properties of a polyurethane shape memory polymer, *Polymer*, **47**, 1348–56.

Ye, H., Ge, X. (2000), Preparation of polyethylene-paraffin compound as a form-stable solid–liquid phase change material, *Solar Energy Materials & Solar Cells*, **64**, 37–44.

Zalba, B., Marín, J. M., Cabeza, L. F., Mehling, H. (2003), Review on thermal energy storage with phase change: Materials, heat transfer analysis and applications, *Applied Thermal Engineering*, **23**, 251–83.

Zamora-Ledezma, C., Blanc, C., Anglaret, E. (2009), Orientational order of single-wall carbon nanotubes in stretch-aligned photoluminescent composite films, *Physical Review B: Condensed Matter and Materials Physics*, **80**, 113407.

Zhang, S., Yu, Z., Govender, T., Luo, H., Li, B. (2008), A novel supramolecular shape memory material based on partial A-CD-PEG inclusion complex, *Polymer*, **49**, 3205–10.

Zhang, H., Wang, H., Zhong, W., Du, Q. (2009a), A novel type of shape memory polymer blend and the shape memory mechanism, *Polymer*, **50**, 1596–601.

Zhang, W., Chen, L., Zhang, Y. (2009b), Surprising shape-memory effect of polylactide resulted from toughening by polyamide elastomer, *Polymer*, **50**, 1311–15.

Zhang, W. D., Shen, L., Phang, I. Y., Liu, T., Yeem, L. T. (2004), Carbon nanotubes reinforced nylon-6 composite prepared by simple melt-compounding, *Macromolecules*, **37**, 256–9.

Zhang, X. X., Wang, X. C., Hu, L. (1999), Spinning and properties of PP/PEG composite fibers for heat storaging and thermo-regulated, *Journal of Tianjin Institute Textiles Science and Technology*, **18**, 1–3.

Zhang, X. X., Fan, Y. F., Tao, X. M., Yick, K. L. (2005a), Crystallization and prevention of supercooling of micro-encapsulated N-alkanes, *Journal of Colloid Interface Science*, **281**, 299–306.

Zhang, X. X., Wang, X. C., Tao, X. M., and Yick, K. L. (2005b), Energy storage polymer/micropcms blended chips and thermo-regulated fibers, *Journal of Materials Science*, **40**, 3729–34.

Zhang, X. X., Wang, X. C., Tao, X. M., and Yick, K. L. (2006), Structures and properties of wet spun thermo-regulated polyacrylonitrile-vinylidene chloride, *Textile Research Journal*, **76**, 351–9.

Zhu, Y., Hu, J. L., Yeung, L. Y., Lu, J., Meng, Q. H., *et al.* (2007), Effect of steaming on shape memory polyurethane fibers with various hard segment contents, *Smart Materials and Structures*, **16**, 969–81.

Zuckerman, J. L., Pushaw, R. J., Perry, B. T., Wyner, D. M. (2003), *Fabric coating containing energy absorbing phase change material and method of manufacturing same*, US patent 6514362.

10 Shape memory finishing treatments for smart textiles

DOI: 10.1533/9780857098542.259

Abstract: Shape memory polyurethane (SMPU) solution is used as an easy-care finishing agent on cotton fabric to provide a recovery force during heating which, upon thermal stimulation, facilitates a flatter appearance and crease retention. Compared to traditional easy-care finishing methods, SMPU finishing has the advantages of high strength retention and zero formaldehyde content. Liquid ammonia (LA) is a newly developed material for modification of textile materials and dimethyloldihydroxylethyleneurea (DMDHEU) is widely used for easy-care finishing. When SMPU was applied to fabrics after LA and/or DMDHEU treatment, the resulting mechanical properties and surface appearances proved that SMPU can be used with existing treatments.

Key words: smart textiles, shape memory finishing, DMDHEU, wrinkle-free, surface morphology.

10.1 Introduction

Whilst polyurethane (PU) coating is commonly used in the textile industry, there is a growing interest in the use of its shape memory properties as a value-added finishing technique for textile fabrics. After the discovery of shape memory properties of polymers, some researchers turned their interest to transferring these into textile products. Some, such as Mitsubishi Heavy Industries, Ltd. (Kobayashi *et al.*, 1992), even studied the possibility of using a T_g type of shape memory polyurethane (SMPU) as the textile material. However, the influence of SMPU treatments on textile materials in terms of physical and structure properties remains unclear.

Traditional easy-care finishing agents, such as dimethyloldihydroxylethylene-urea (DMDHEU), can improve the surface appearance of textile materials, but as most are formaldehyde-related products, they will also weaken the fabric strength and produce harmful vapors. However, a liquid ammonia (LA) pre-treatment provides a soft hand feel, improves dimensional stability and increases the breaking strength of the treated fabric. However, it will not contribute to the surface appearance of fabrics.

In this chapter, the morphology, structure and physical properties of SMPU treated cotton fabrics are examined to provide a clear picture of the potential application of SMPU treatments on textile materials. To enhance fabric performance, an SMPU solution can also be mixed with traditional finishing agents, DMDHEU and LA. The effects of this process on cotton fabrics will also be discussed.

10.2 Reaction of shape memory polyurethane (SMPU), dimethyloldihydroxylethyleneurea (DMDHEU) and liquid ammonia (LA) as finishing agents with cellulosic materials

According to reports, SMPU treatment contributes to physical surface modifications of textile materials, while DMDHEU and LA treatments cause chemical modifications to the cellulose structure (Greenwood, 1987; Han and Chen, 1998; Jang *et al.*, 1993; Liu *et al.*, 2005; Trask-Morrell *et al.*, 1996; Yanai, 2001; Yanai *et al.*, 2005, 2006; Yanai and Shimizu, 2006; Zeronian *et al.*, 1990). As will be described below, there are significant differences to the working principles of the SMPU, DMDHEU and LA treatments respectively, when they are applied to cellulose materials.

10.2.1 Shape memory polyurethane

SMPU is a long polymer chain, and with the exception of some special cases, its molecular weight is generally larger than 10 000. It is impossible for such a large chain to penetrate into the cellulose structure and react with the inner part of the cellulose fibers. Therefore, SMPU must be coated or grafted onto textile materials.

Due to the flexibility of molecular design and the possibility of attaching special functional groups into the polymer side chains, the SMPU used in textile can be designed to either be reactive or not. In this chapter, only non-reactive SMPU is used, and the role of SMPU in the SMPU-cotton composition is simply that of a coating on the fiber/fabric surface. Even though the SMPUs described are not expected to react with the cellulose structure, a pad-dry-cure finishing process is still needed for SMPU treatment. Considering the basic principle that an SMPU can be designed with a permanent shape, this shape is set in the curing process.

10.2.2 Dimethyloldihydroxylethyleneurea

Dimethyloldihydroxylethyleneurea (DMDHEU) molecules penetrate into the cellulose structure, cross-link with the hydroxyl groups in the amorphous region and thus reduce the slipping of cellulose chains. Macroscopically, fabric appearance, crease retention and wrinkle recovery performance can be increased by DMDHEU treatment, while the fabric strength can be significantly reduced.

10.2.3 Liquid ammonia

Liquid ammonia (LA) treatment is a newly developed treatment for cellulose fabrics. Ammonia, a colorless and corrosive solution, is liquefied at 240 K (−33°C). LA has very low viscosity and surface tension. As a result, it is able to penetrate into the cellulose structure, destroy the boundary of crystals, increase the ratio of the amorphous region and even change some crystal structures from cellulose I to cellulose IIII. Such a conversion of the crystal structure is reversible in hot and moist conditions. Moreover, cotton fabrics will swell after LA treatment and fiber convolution will be reduced, increasing the fabric tensile strength, dimension stability and hand feel of the samples (Bertoniere and King, 1989; Greenwood, 1987; Lee *et al.*, 2005; Patel and Varghese, 1983; Tao, 2005; Yanai, 2001; Yanai *et al.*, 2005, 2006; Yanai and Shimizu, 2006; Zeronian *et al.*, 1990).

As a result of the reduction of fiber twists and destruction of the crystal region, the weak-link effect of cellulose is also reduced, increasing the breaking stress of the fabric. In summary, the LA treatment can increase fabric strength and dimensional stability. However, the fabric appearance will not be improved simply as a result of LA treatment.

10.3 Shape memory finishing treatments for cotton

The microscopic structure of SMPU treated cotton was studied by SEM. The aim of this investigation was to ensure the presence of SMPU on the fabric. The macroscopic properties of the treated fabrics, such as flat appearance, crease retention and breaking strength, were then measured. The flat appearance of the fabric was measured after five cycles of washing in hot water and tumble-drying at 66 ± 5°C. Hot air drying can activate the shape memory effects of SMPUs and therefore improve the appearance of the fabric (Li *et al.*, 2005). Finally, the internal structure of the fabrics treated was studied by X-ray diffraction and Raman spectroscopy. Based on the results, a model of the reaction mechanism of these treatments on cotton fabric was proposed.

A 100% cotton plain weave (F_1) and twill weave (F_2) are used as shirting materials which are provided by a local shirt manufacturer for this study. The specification of the two fabrics is shown in Table 10.1.

PCL4000 was purchased from Daicel Chemical Industries, Ltd.; methylene diphenyl diisocyanate (MDI) was purchased from the Aldrich Chemical Company; and butandiol (BDO) from Acro Organics. They were used to synthesize the T_m type of PCL-based SMPUs. In this PCL-MDI-BDO composition,

Table 10.1 Specifications of fabrics used for SMPU treatment

Fabric code	Fabric sett (threads per inch)		Yarn count (tex)		Weave structure	Thickness (mm)	Fabric weight (g/m^2)
	Warp	Weft	Warp	Weft			
F_1	133	68	40	40	Plain	0.205	150
F_2	108	58	20	20	Twill	0.328	200

polycaprolactone (PCL) acts as the soft segment, whereas MDI and BDO act as the hard segments.

PCL was dried in a vacuum oven at 70°C for 12 hours. MDI was taken from the refrigerator and kept in a fume hood until it reached room temperature, before being further dried in a vacuum oven at 70°C for one hour. BDO was also dried in a vacuum oven for one hour before synthesis. The synthesis process followed a pre-polymerization method (Hu *et al.*, 2005; Ji *et al.*, 2006; Lin and Chen, 1998), whereby the PCL first reacts with excess MDI in a dimethyl formamide (DMF) solution at 90°C for two hours to produce the 'pre-polymer', before the chain is extended with BDO for another two hours at 85°C. The solid content is about 25 wt.%, and the hard segment content is about 20%.

As well as the PCL-based SMPU, a poly(propylene glyol) (PPG, purchased from Elf Atochem) based SMPU with a smaller molecular weight (2000, denoted as PPG2000) was also synthesized and used as a finishing agent. In using these two kinds of SMPUs and fabrics, the aim was to investigate the effects of the molecular weight of an SMPU on the surface morphology of SMPU treated cotton fabrics, and to ensure the surface grafting of the SMPU layers.

The low-formaldehyde cross-linking agent used in this project is a modified DMDHEU, which is marketed under the brand name of 'Fixapret™ CL' by BASF. To provide an acidic environment for the reaction between DMDHEU and cotton, magnesium chloride ($MgCl_2$, purchased from Sigma) was used as the catalyst.

The SMPU solution prepared was used directly in solution form, and was prepared by mixing 80 g of SMPU solution with 1 L of water. The process employed was the traditional pad-dry-cure process, whereby the cotton fabric is immersed into the finishing solution on a padding machine (Rapid 354) and padded twice between two rubber rollers at a pressure of 3 MPa. After the padding process, the wet pick-up of the fabric was about 75% (based on a completely dry sample). It was subsequently dried at 80°C for two minutes, before being cured at 120°C for three minutes.

Following the instructions of BASF, the DMDHEU was diluted with water with a concentration of 80 g/L, and 26 g/L of $MgCl_2$ was added to the finishing solution as the catalyst. Theoretically, an SMPU treatment can be applied before, after or together with the DMDHEU finishing. The effects of the two treatments on the overall performance of the 100% cotton fabrics will also be examined in this chapter.

The LA treatment was prepared by the Printing and Dyeing and Chemistry Department at the Chengdu Textile College (China). An intermittent type of LA treatment machine (Zheng, 2002) (Fig. 10.1) was used. The fabric specimens were treated with LA for three minutes, washed with diluted sulfuric acid (1–2%) and water, and dried for subsequent treatment. The specimens that were treated with the above process are denoted as 'LA-treated' in the following section.

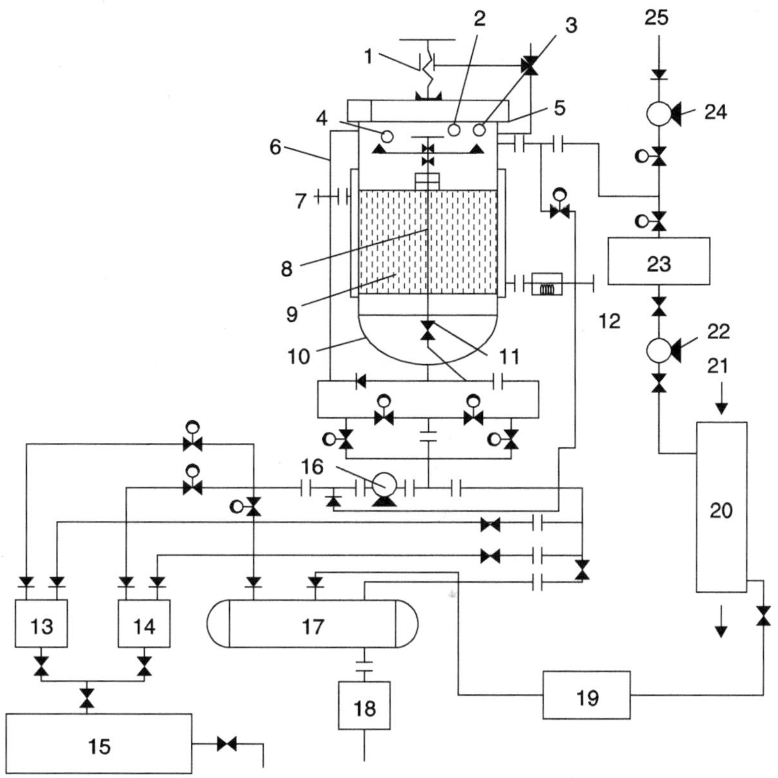

10.1 Equipment for intermittent liquid ammonia treatment: 1 = pressured equipment; 2 = pressure gauge; 3 = thermometer; 4 = safety valve; 5 = seal-up set-up; 6 = liquid level indicator; 7 = gas inlet; 8 = warp beam; 9 = fabric; 10 = liquid ammonia reactor; 11 = warp beam connector; 12 = gas outlet; 13 = water; 14 = acid; 15 waste water; 16 = recycle pump; 17 = liquid ammonia storage; 18 = emergency valve; 19 = degreaser; 20 = condenser; 21 = water; 22 = nitrogen compressor; 23 = degreaser; 24 = vacuum pump; 25 = air.

10.4 Surface morphology of cotton fabrics treated with shape memory polyurethane (SMPU)

We now examine the morphologies and properties of SMPU, DMDHEU and LA, and their effect on cotton fabric. The surface morphologies of two cotton fabrics with different structures, treated with two different kinds of SMPUs, are also investigated. Furthermore, the effects of mixing the DMDHEU and LA with SMPU treatment are studied, and the tensile properties, surface appearance and crystal structures of cotton fabrics after different treatments are discussed, while in the last section, the micro-mechanics of the SMPU treatment are studied.

10.4.1 Surface morphology of SMPU treated cotton fabrics

The surface morphologies of untreated and mercerized cotton are presented in Figs 10.2 and 10.3(a), respectively. It is clear that a rough surface can still be observed, even though the cotton samples were finished with a sodium hydroxide (NaOH) mercerization. Figure 10.2(b) presents the surface morphology of a PCL4000-based SMPU treated sample, which has very smooth fiber surfaces, a property which could be explained by the grafting of the SMPU layers onto the

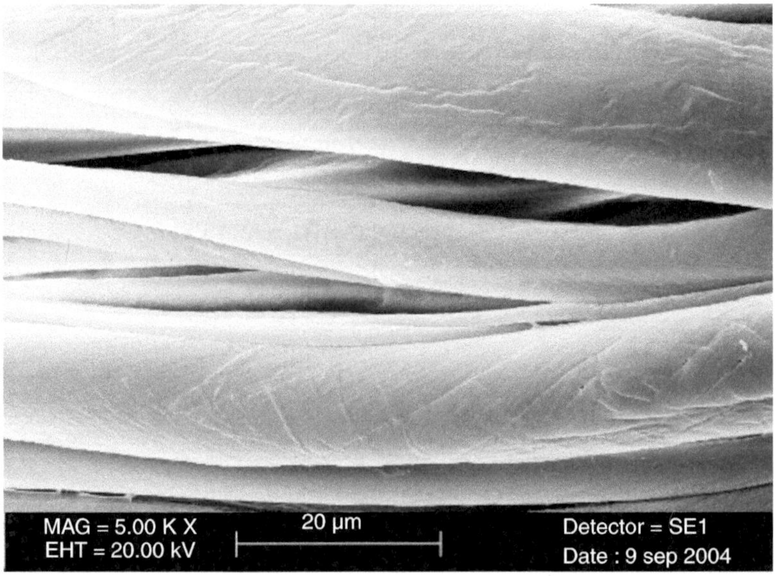

10.2 Surface morphology of untreated cotton.

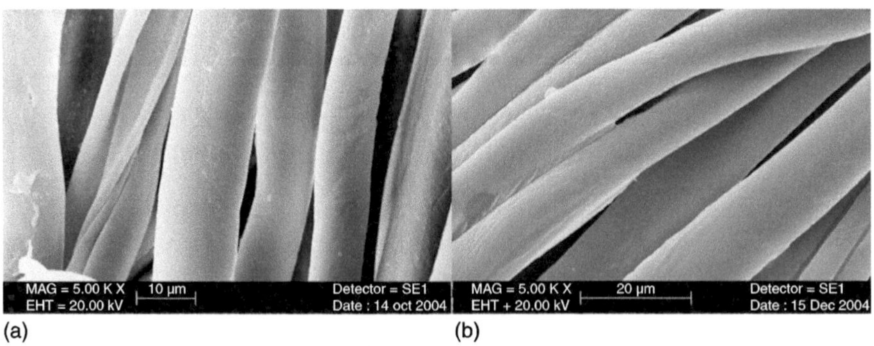

10.3 Surface morphology of: (a) mercerized cotton; and (b) SMPU treated cotton.

fiber surfaces. Since the average molecular weight of SMPUs is 50 000 to 200 000, such a macromolecule is not able to penetrate into the cellulose structure of the fiber. Instead, SMPU layers would be grafted onto the cotton fibers and even 'bridged' between them (Fig. 10.3(b)).

The surface morphology of the two shape memory fabrics are shown in Fig. 10.3(a) and (b). A similar morphology was observed in both fabrics F_1 and F_2; the shape memory polymer is grafted onto individual cotton fiber or bridged within two or more cotton fibers.

Figure 10.4(a,b) shows the PPG2000-based SMPUs, while Fig. 10.5(a,b) illustrates the PCL4000-based SMPUs. It is worth noting that fabrics treated with an SMPU that has a lower molecular weight tend to resemble Fig. 10.4(a), where the shape memory polymer is grafted onto individual cotton fibers. Fabrics that are treated with an SMPU with a high molecular weight, meanwhile, would emulate Fig. 10.4(b), where the shape memory polymer is bridged among the cotton fibers.

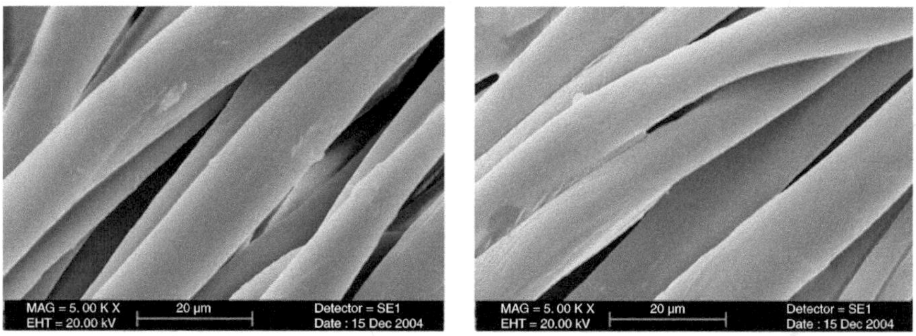

(a) (b)

10.4 SMPU: (a) on the fibers; and (b) between the fibers (F_1).

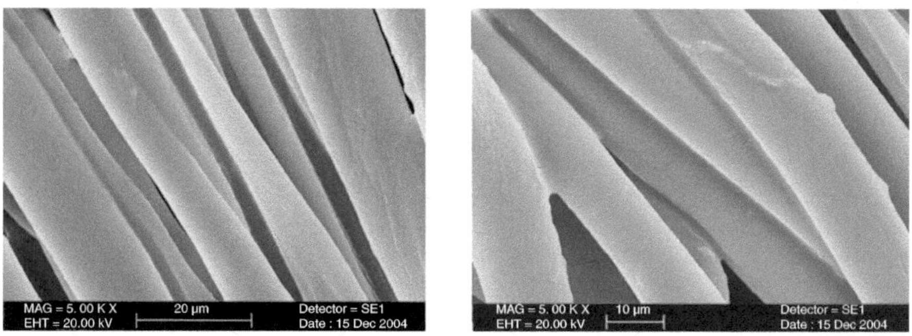

(a) (b)

10.5 SMPU: (a) on the fibers; and (b) between the fibers (F_2).

The following section examines the effects of combining an SMPU treatment with DMDHEU and LA. First, the macroscopic properties, such as tensile strength and surface appearance, are measured; second, the cross-section morphology is studied by optical and electronic microscopes, and finally, the crystal structures of these treated cotton fabrics are investigated.

10.5 Fabric tensile properties and surface appearance

Due to the limitations of the LA treatment on cotton fabric, only fabric F_1 is described in this section. The tensile properties of cotton fabrics were measured by ASTM standard 5305, while the flat appearance and crease retention of the fabric samples were studied by AATCC standards 124 and 88C, respectively. The specimens were first washed and then rated against standard replicas, on a scale of 1 to 5.

Table 10.2 presents the breaking load and surface appearances of a plain woven cotton fabric treated by different methods. Note that both the flat appearance and crease retention of LA treated cotton has the same rating as the untreated one, which means that the LA treatment provides dimensional stability and soft hand feeling, but cannot improve the surface appearance. Furthermore, either DMDHEU or SMPU causes an improvement on the surface appearance. When DMDHEU and PU treatments are combined with the LA treatment, the cotton fabrics develop good surface appearance, hand feel and dimensional stability.

The tensile properties of the samples are studied by standard ASTM D5035 (stripe). Table 10.2 also shows the breaking load for the samples. The SMPU finishing has no significant effect on the tensile properties. The result suggests that

Table 10.2 Breaking load and surface appearance of various treatments on fabric F_1

Treatments	Breaking load (N)		Surface appearance	
	Warp	Weft	Flat appearance 1	Crease retention 2
Untreated	474 (100%)	359 (100%)	1	1
SMPU	455 (96%)	341 (95%)	3	3
DMDHEU	324 (72%)	220 (65%)	3	3
LA	536 (119%)	428 (126%)	1	1
LA+SMPU	520 (109%)	411 (114%)	3	3
LA+DMDHEU	366 (82%)	252 (74%)	3	3
SMPU+DMDHEU	320 (71%)	206 (61%)	3.5	3.5
LA+SMPU+ DMDHEU	370 (85%)	250 (74%)	3	3.5

Source: Yeung *et al.* (2007)

the reduction of the breaking stress by DMDHEU can be improved via the LA treatment. This result corresponds to the previous result of the LA treatment on cotton fabric strength (Wakida *et al.*, 2000).

10.5.1 Fabric cross-sections

The cross-section of the above specimens was further investigated to determine the changes in the fiber morphology as well as the fabric properties. In the cross-sectional study, both warp and weft yarns of the specimens were carefully removed from the fabric, and then investigated by a traditional optical microscopy and a SEM. In the optical microscope, the sample was prepared by procedures described in standard AATCC20: a bundle of yarns were used to pack the hole of a stainless steel plate, then a sharp razor blade was used to produce a smooth cut on both sides of the plate (AATCC, 2004). The steel plate was then examined by a transmitted light microscope with magnification of 500×.

In the SEM, the samples were first embedded in the epoxy resin Epon-812 (SPI Supplies). The embedded resin was prepared with the formula shown in Table 10.3. A single yarn was first placed in an embedding mold made with silicone gel, and then the epoxy resin was added. The mold was initially kept at 30°C for 6 hours, then 60°C for 10 hours and finally 90°C for 8 hours. The oven was slowly cooled down to room temperature before the embedding mold was removed.

Table 10.3 Recipe of embedding material

Materials	DMP-30a	DDSAb	NMAc	Epon-812
Volume (mL)	0.6	10	8.9	16.2

10.5.2 Optical microscopy

The cross-sections of cotton fibers are first investigated by a Nikon optical microscope in this section. The effects of the various treatments can be observed in Figs 10.6 and 10.7. According to reports, the original untreated cotton fibers show lumens and bean-shaped cross-sections, while the LA treated cottons all show well-swelled shapes.

10.5.3 Scanning electron microscopy

The cross-section of the cotton fibers was prepared by an RMC PT-XL ultra-microtome system, and then investigated by SEM (Leica Stereoscan 440).

When the images from the optical microscope and SEM were compared, it became clear that the cotton fibers were fully swelled after the LA treatment, but

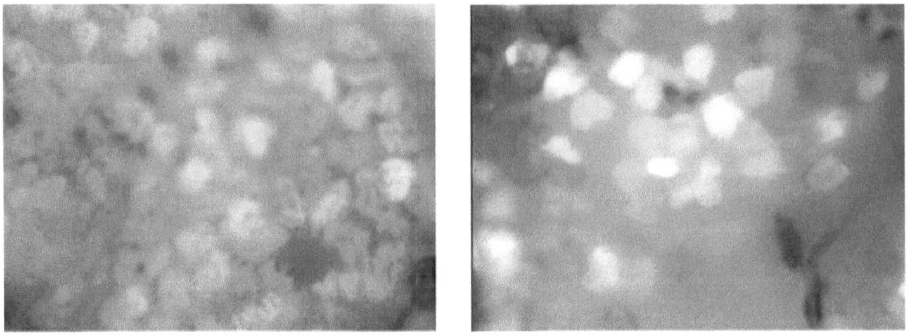

(a) (b)

10.6 Cross-sections of: (a) untreated cotton; and (b) LA treated cotton.

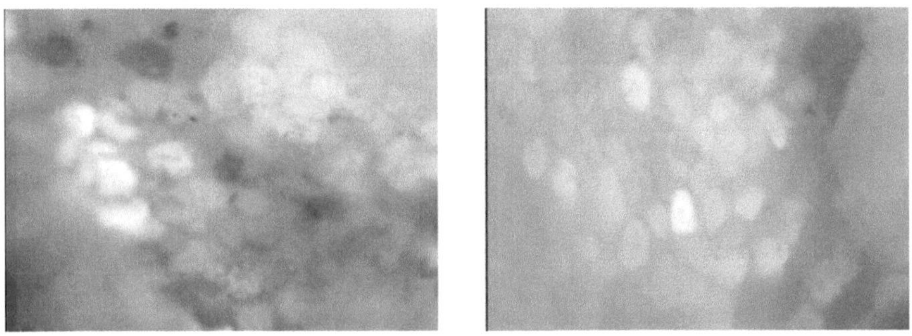

(a) (b)

10.7 Cross-sections of: (a) LA + DMDHEU cotton; and (b) LA + SMPU treated cotton.

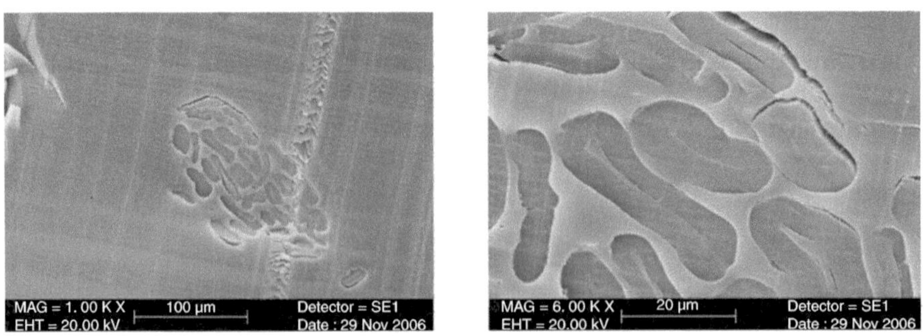

(a) (b)

10.8 Cross-section of untreated cotton.

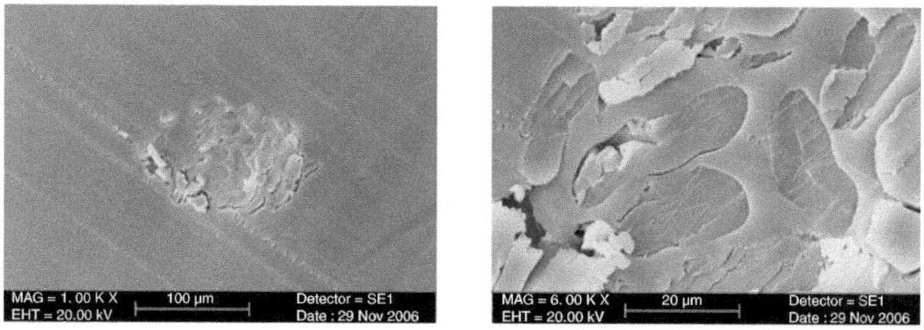

10.9 Cross-section of DMDHEU treated cotton.

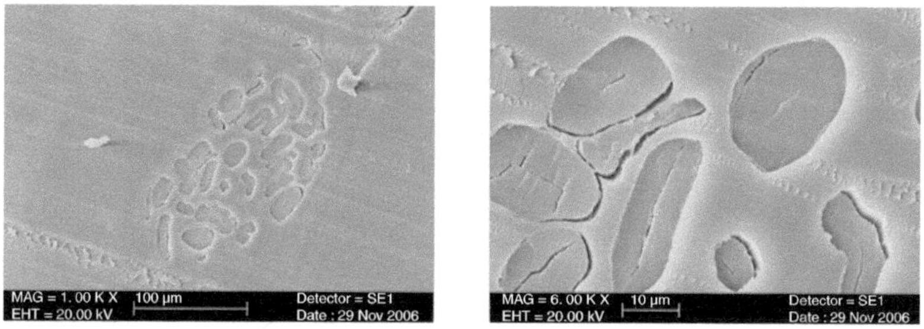

10.10 Cross-section of LA treated cotton.

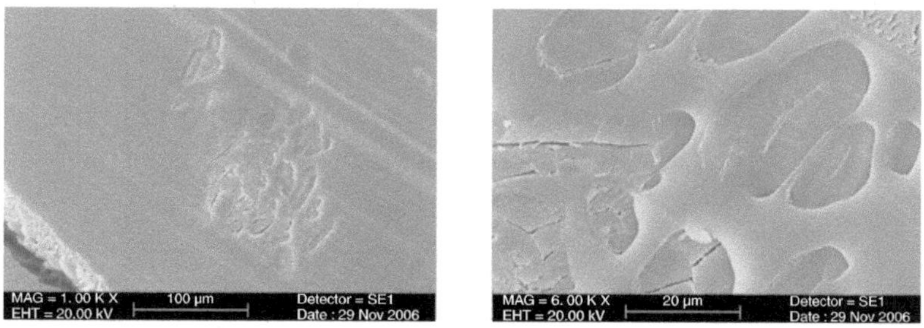

10.11 Cross-section of LA and PU treated cotton.

shrunk slightly after the DMDHEU treatment, as shown in Figs 10.8(a,b) and 10.9(a,b). Since the aim of the LA treatments is to fully swell cotton fibers and improve the evenness of stress distribution and dimensional stability, the emergence of fiber shrinkage reduced their advantages on cotton fabrics. Figures 10.10 to 10.12 show the cross-section image of LA-SMPU treated cotton. The fibers swelled equally as well as the LA treated sample when an MPU finishing was applied.

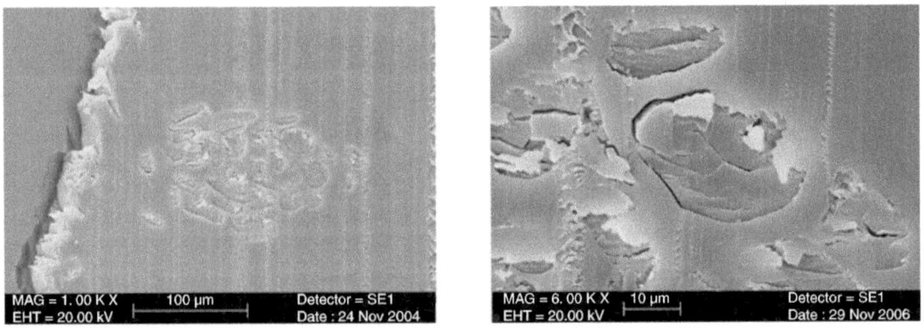

10.12 Cross-section of LA and DMDHEU treated cotton.

10.6 Fabric structure

The crystal structure of the SMPU, LA and DMDHEU treated cotton fabrics was investigated by a Philips Xpert XRD system in θ to 2θ scanning mode within 10° to 40°. 'Xpert', software developed by Philips, is used for profile fitting. The crystallinity (X_c) of the specimens is calculated, as previously discussed, by the ratios of area of the crystal peaks to an amorphous region in the fitting profiles.

Figure 10.13 presents the XRD results of untreated cotton, and LA and DMDHEU treated cotton. Peaks were observed at 11.2, 20.7 and 28.1°, respectively, which supports the existence of a cellulose III structure. No major differences were

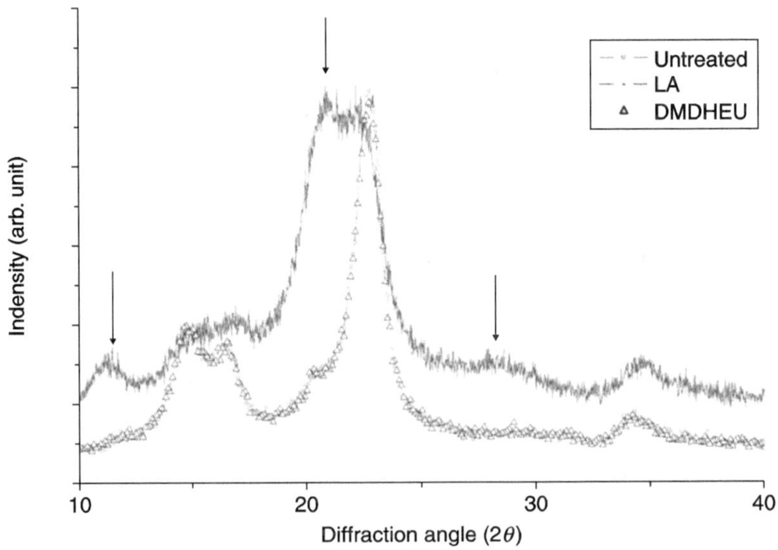

10.13 XRD results for untreated cotton, and LA and DMDHEU treated cotton.

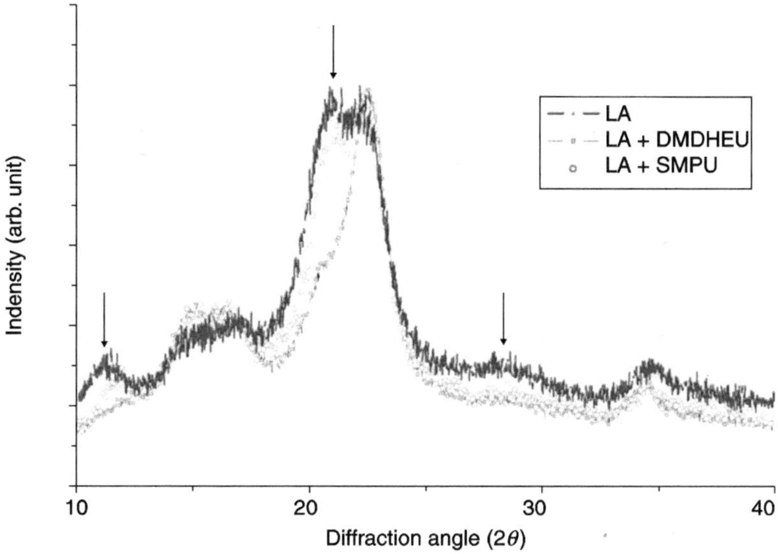

10.14 XRD results of LA, LA and PU, and LA and DMDHEU treated samples.

observed between the untreated specimen and the one treated by DMDHEU. The results show that the DMDHEU treatment has no effect on the crystal structure of cotton. However, Fig. 10.14 shows that the peak observed at 20.7°C was slightly reduced after the DMDHEU treatment. This indicates that some of the cellulose IIII structure changes back to cellulose I after the DMDHEU treatment, or under hot and moist conditions, as previously suggested (Yanai and Shimizu, 2006). It is clear that the crystal structures of LA and LA+PU treated cotton are almost the same, and that SMPU plays a role in obstructing a reverse conversion from cellulose IIII to I.

The X_c of LA, LA+PU, LA+DMDHEU and DMDHEU, together with the untreated cotton samples, are listed in Table 10.4. The fabric X_c drops from 66 to 49% when the LA treatment is applied. It is further reduced to 35% with the LA+DMDHEU treatment, although there is no effect under SMPU treatment. It is interesting to discover that only the LA treatment reduces the X_c of the 100% cotton fabric, with the X_c dropping from 49 to 45%.

Table 10.4 Crystallinity of cotton fabrics after different treatments

	Untreated	LA	LA+ DMDHEU	LA+ water	LA+ SMPU	DMDHEU
Crystallinity (%)	65.6	49.4	34.5	52.4	44.7	64.7

10.7 Effects of shape memory polyurethane (SMPU) on LA/DMDHEU treated cotton

As shown above, it is clear that SMPU can be used together with either the DMDHEU or LA treatment on cellulose cotton fabrics. Fabric breaking loads are significantly improved when subjected to LA+DMDHEU+SMPU treatment. The surface appearance of the fabric, measured by flat appearance and crease retention, improved after the SMPU treatment. As SMPU is suggested as a polymer layer grafted onto the fabric surface, the SMPU finishing does not conflict with the chemical modification. According to the result reported above, the SMPU finishing provides an additional effect on the cotton fabric based on the previous treatment (LA or DMDHEU).

In the crystal structure of the LA treated cotton sample, the LA broke and transferred part of the original crystal structure from cellulose I into cellulose III. The X_c of these fabrics further decreased after the DMDHEU treatment, but returned to the original cellulose I structure in an environment containing heat and moisture.

The reduction of X_c can be explained by the reaction mechanisms of DMDHEU (Chang and Chen, 1996; Han and Chen, 1998; Jang *et al.*, 1993) and LA (Greenwood, 1987; Yanai, 2001). LA treatment had destroyed some of the cotton crystal structure, which turned into a pseudo-stable state in the hot wet environment. The DMDHEU, meanwhile, cross-linked with the cellulose structure within the destroyed crystal structure and prevented the transformation of cellulose chains from the pseudo-stable state to a crystal structure. As a result, the crystal structure of the fabric was fixed at the lowest X_c state. Nevertheless, the SMPU can form a polymer layer on the cotton fibers. It provides good protection to the LA treated cotton, and reduces the effect of the hot-moisture transformation of the crystal structure from cellulose III to cellulose I.

10.8 Internal stresses in cotton fabrics

In this section, we examine the internal stress within the cotton fibers treated with SMPU and DMDHEU. Eichhorn and Young (Eichhorn *et al.*, 2001; Eichhorn and Young, 2001, 2003; Eichhorn *et al.*, 2003) showed that the internal stress of cellulose fibers can be measured by shifts of C-O-C vibrations in the Raman spectrum: an increase in the wave intensity indicates a reduction of the internal stress, and vice versa. Using the strain-induced shift in the special Raman peak, incorporated with micro-Raman spectroscopy, is a technique unique to micro-mechanics. A Raman spectrum of untreated cotton is shown in Fig. 10.15. The C-O-C peaks are 1096 and 960 cm^{-1}, depending on the internal stress of the cellulose chain.

Change of internal stress of cotton after SMPU/DMDHEU treatment (Figs 10.16 to 10.19) show the Raman spectrums of the C-O-C peak at nearly 1095 cm^{-1} for PU, DMDHEU and PU mixed DMDHEU treated cotton.

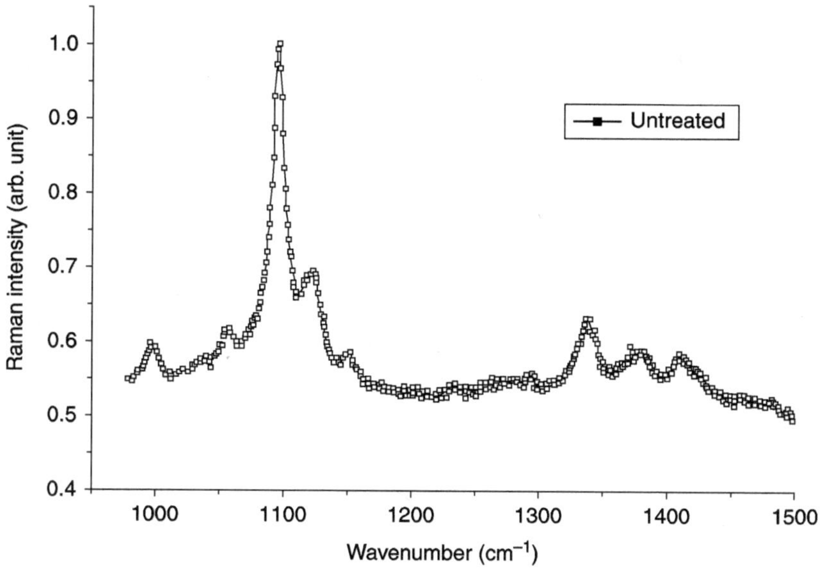

10.15 Raman spectrum of untreated cotton.

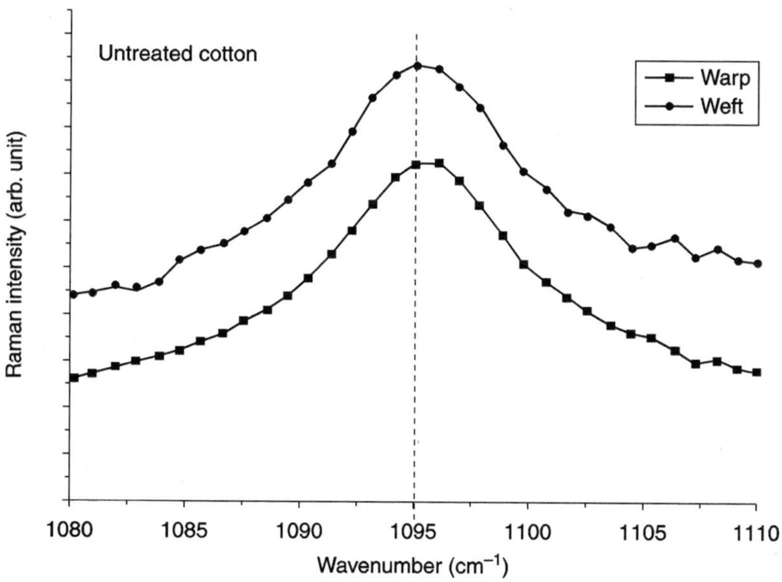

10.16 C-O-C peak of untreated cotton.

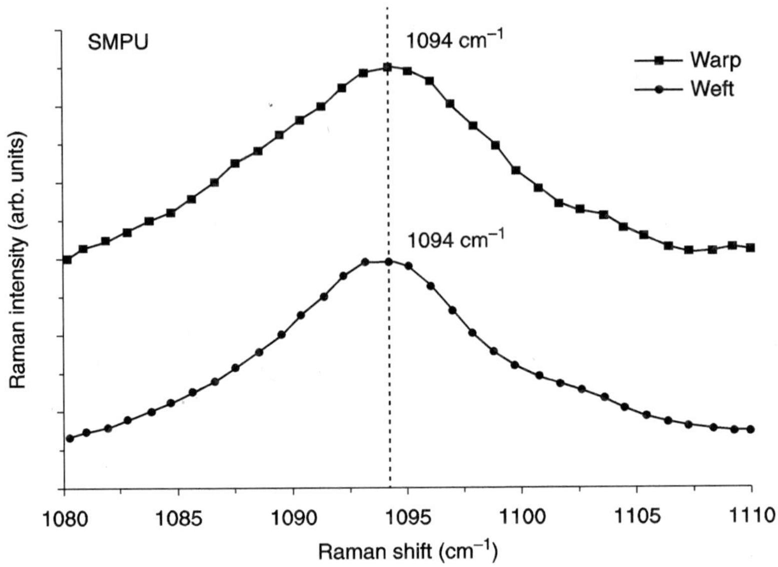

10.17 C-O-C peak of SMPU treated cotton.

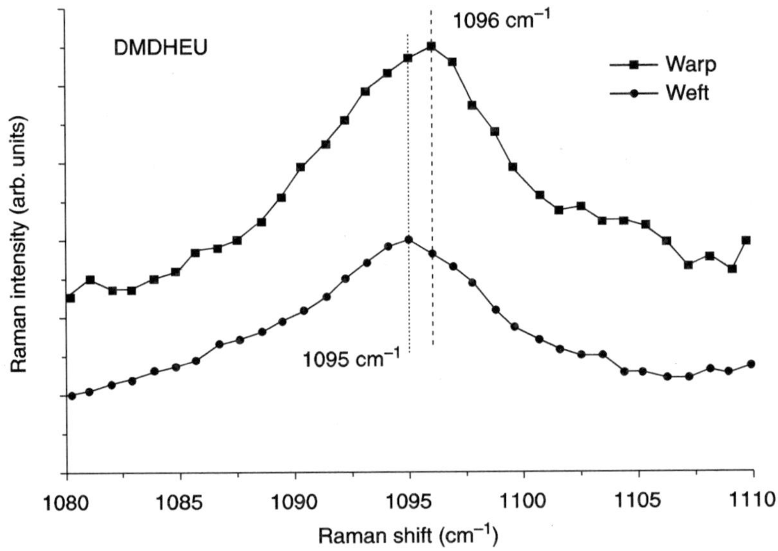

10.18 C-O-C peak of DMDHEU treated cotton.

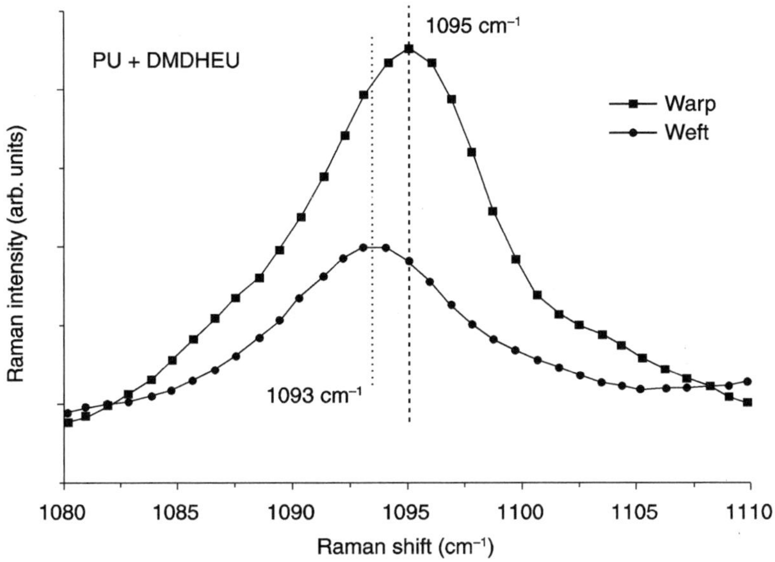

10.19 C-O-C peak of SMPU+ DMDHEU treated cotton.

A summary of this process is also presented in Fig. 10.20. The 1096 cm^{-1} peaks in the DMDHEU treated cotton are unchanged. However, the peaks appear at a lower wave frequency in PU and PU+DMDHEU treated cotton. Also observed from the results is the return of the Raman shift in the PU+DMDHEU treated cotton. The warp yarns of the PU+DMDHEU treated cotton sample show a Raman peak at 1095 cm^{-1}, whilst such a phenomenon cannot be observed in the weft yarns.

The shifts of internal stress obtained above could be explained by the differences in the finishing process and the reaction between the SMPU and DMDHEU on the cotton fabrics.

The results indicate that DMDHEU reacts with the hydroxyl groups of the celluloses and act as a cross-linker. It is worth noting that, due to this cross-linking, the 1096 cm^{-1} peak appears at a lower wave frequency (1 cm^{-1}) in the weft fiber (this is less apparent in the warp component). This phenomenon can be explained by the shrinkage mechanism proposed by Hamburger and Fox (1956), in which shrinkage is determined by the difference in swelling and change of radius of curvature.

Figure 10.21 presents the schematic of the yarns subjected to a drying and curing process. When the cross-sectional area of the warp yarns (depicted by circles) increases due to swelling, the radius of the curvature of the weft yarns (depicted by stripes) also increases. This results in extra extension, and causes shrinkage in fabrics. Most of this extension can be recovered when the fabric is dried. However, the results in this study indicate that the extension cannot be recovered if DMDHEU (shown by the dots in Fig. 10.21(c,d)) is used as the

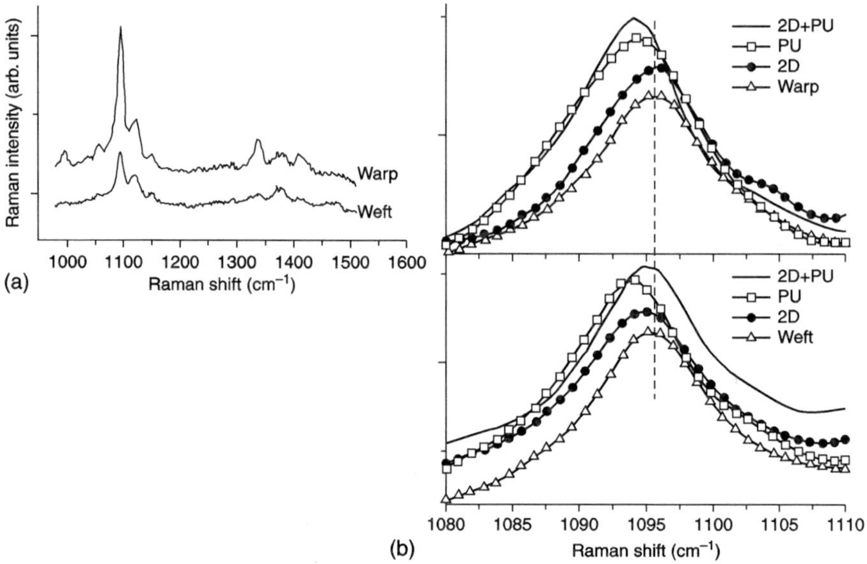

10.20 Raman spectrum of: (a) untreated cotton; (b) cotton treated with SMPU, DMDHEU (2-D); and SMPU+DMDHEU mixture in warp (top) and weft (bottom) directions.

finishing agent. As shown Fig. 10.21(e), the formation of cross-linkages between cotton and DMDHEU results in induced-residual strain inside the fibers.

Induced strain occurs in both weft and warp yarns, though it is more pronounced in the former. As warp yarns are highly packed, the swelling effect is diminished

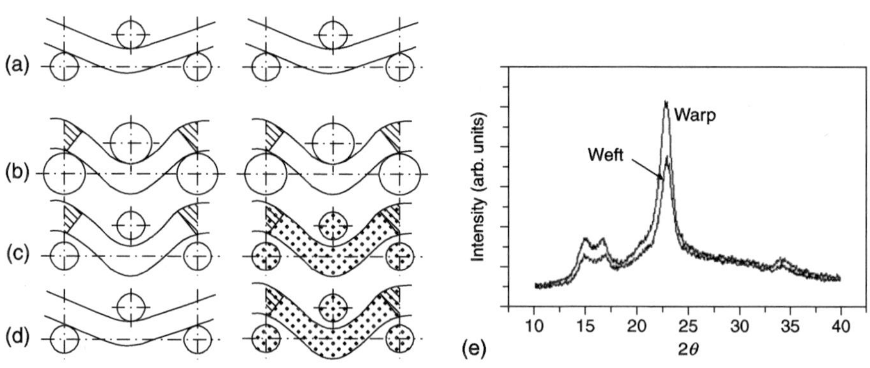

10.21 Schematic of: (a) untreated; (b) swollen; (c) dried; and (d) cured warp (depicted by circles) and weft (depicted by stripes) fibers. The dots in (c) and (d) represent the cross-link sites inside a fiber. (e) XRD result of both the fibers after processes (a)–(d).

by the inherent physical constraint. Figure 10.21(e) presents the X-ray diffraction data of untreated warp and weft yarns. It is observed that the X_c of warp yarns is higher than that of the weft yarns, which indicates a relatively large amorphous region in the weft yarns in which the chemical reaction can occur. Deformation due to swelling is an elastic type of deformation in fibers. The cross-linking effect produced by DMDHEU therefore freezes the elastic deformation and traps the induced stress inside the fibers. In contrast, less amorphous areas exist in the warp yarns for the chemical reaction to occur, thereby limiting deformation and resulting in less induced stress in the fibers.

Figure 10.22 shows the shift of the same Raman peak of the cotton cellulose when pure PU is used as the finishing agent for the cotton fabrics. Both the warp and weft fibers undergo a reduction in wave frequency by approximately 2 cm^{-1}, which suggests that some physical interactions exist between the agent and the cellulose fibers. The PU is unable to penetrate the cellulose due to its large molecular size. The macromolecular size of SMPU can only be grafted onto or between the cotton fibers. It is believed that the SMPU coating on fiber surfaces only affects the swelling of fibers in a completely dry state during the curing process and also induces residual stress in the fibers.

The Raman spectrum of PU+DMDHEU treated cotton is presented in Fig. 10.22. In the warp direction, a similar trend of reduced wave frequency is

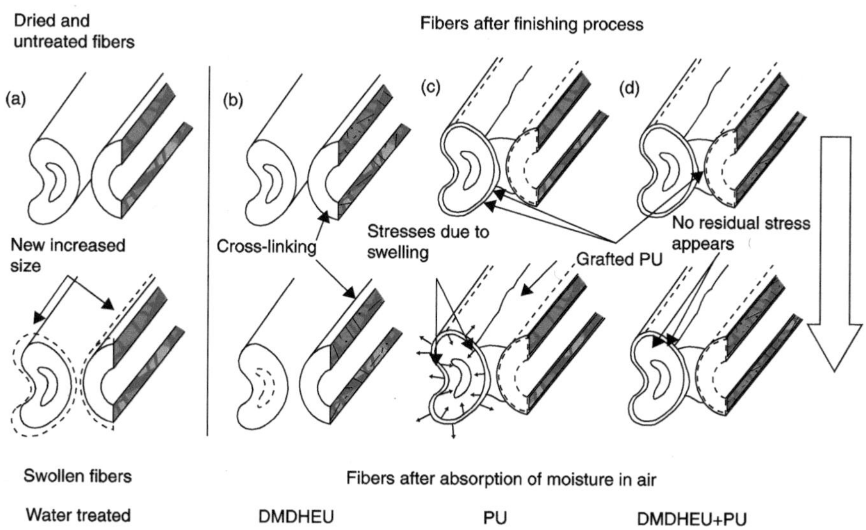

10.22 Schematic of two adjacent weft fibers subjected to various finishing treatments. In (a), swelling of the fibers by absorbing the moisture in air; in (b)–(d), various treated fibers initially dried and increasing their size by swelling.

observed between PU treated and PU+DMDHEU (2-D) treated cotton samples. The effects of PU+DMDHEU in the warp direction are not as significant as those in the PU treated cotton sample, but it is worth noting that the expected reduction in wave frequency in the weft direction is reduced and returned to nearly its original untreated value. A returned Raman shift in the weft fibers such as this usually indicates the relaxation of internal stress within the fibers.

10.9 The role of shape memory polyurethane as a finishing agent for cotton fabrics

Based on the experiment data, it is possible to picture the role of the SMPUs during the finishing process. At this stage, the SMPU has provided necessary physical contact with fibers, but altered the overall chemical reaction between DMDHEU and the fibers in particular. The originality of the finding resides in the fact that the return of the reduction in wave frequency (~2 cm^{-1}) is observed only in weft yarns, not warp yarns. In a completely dry state, the PU acts as a surface coating on the fibers; however, the PU+DMDHEU treated fabric (weft fibers) gradually absorbs moisture from the air. Accordingly, the swelling is hindered due to an increase in induced stress by the presence of DMDHEU, which reacts with amorphous regions of the fibers and cross-links the fibrils together.

Figure 10.22 illustrates the chemical and physical interactions during the finishing process. The cellulose fibers swell up and increase their cross-sectional area, as shown in Fig. 10.22(a). It is worth emphasizing that the treated fabrics were initially kept in a dry state when taken from the oven. They immediately absorbed the moisture in the air, however, and increased their size. Figure 10.22(b,c) illustrates the fibers treated by DMDHEU, PU and FMDHEU+P, respectively.

The finishing agent DMDHEU reacts with the cellulose fibrils by forming cross-links between them. The links are represented by short dash lines inside the fibers, as shown in Fig. 10.22(b). DMDHEU is largely useless in hindering the longitudinal deformation, but does reduce the degree of radial swelling due to the cross-linking between fibrils. There are a considerable amount of induced stresses (as shown by the arrows in Fig. 10.22(c)) at the perimeter of the fiber after being treated by PU as a result of swelling. However, the DMDHEU+PU treated fibers have less induced stress, due to the swelling force of the fibers which are balanced by the cross-links. The induced stress exists in the weft fibers due to the smaller amount of cross-linking bonds inside the fibrils. It is then comparable to the case presented in Fig. 10.22(c). The fabrics are, therefore, composed of stressed warp yarns, but stress-free weft fibers.

10.10 Summary

In this chapter, the performance of SMPU finishing as well as its mixture with traditional treatments on cotton fabrics has been investigated in terms of fabric surface appearance, mechanical properties and crystal structure. The role of SMPU as a finishing agent of cotton fabrics has also been discussed, according to the results obtained from Raman spectroscopy.

We have also examined the interactions of SMPU treatment with a traditional DMDHEU easy-care finishing method and a LA pre-treatment. The results indicate a surface grafting of SMPU on cotton fibers, and a model of interaction between SMPU and DMDHEU/LA treatments has been proposed accordingly. During the pad-dry-cure process, an SMPU layer forms on cotton fibers. The SMPU layer acts as a gas-permeable layer, which can remember the original shape upon the first deformation at a high temperature (which occurs at the curing process). However, the DMDHEU and LA do react with cellulose at a microscopic level.

The shape memory effects of SMPU as a finishing agent of cotton have been reviewed at macroscopic and microscopic levels. At the macroscopic level, the shape memory effects enhance the surface appearance of cotton fabrics. The fabrics form a permanent shape after the curing process, which are deformed in normal use and then returned to a permanent shape after thermal stimulation. At the microscopic level, the SMPU layer deforms and recovers in the dry-cure process, resulting in changes of internal stress within the cellulose chains.

10.11 References

AATCC (2004), Test method 20-2004, fiber analysis, *Qualitative, Technical Manual of the American Association of Textile Chemists and Colorists*.

Bertoniere, N. R., King, W. D. (1989), Effect of scouring/bleaching, caustic mercerization, and liquid ammonia treatment on the pore structure of cotton textile fibers, *Textile Res. J.*, **59**(2), 114–21.

Chang, H.-L., Chen, C.-C. (1996), Cross-linking of cotton with DMDHEU in the presence of sodium chloride, *Textile Res. J.*, **66**(12), 803–9.

Eichhorn, S. J., Young, R. J. (2001), The Young's modulus of a microcrystalline cellulose, *Cellulose*, **8**, 197–207.

Eichhorn, S. J., Young, R. J. (2003), Deformation micromechanics of natural cellulose fibre networks and composites, *Comp. Sci. Technol.*, **63**, 1225–30.

Eichhorn, S. J., Sirichaisit, J., Young, R. J. (2001), Deformation mechanisms in cellulose fibres, paper and wood, *J. Mater. Sci.*, **36**, 3129–35.

Eichhorn, S. J., Young, R. J., Davies, R. J., Riekel, C. (2003), Characterisation of the microstructure and deformation of high modulus cellulose fibres, *Polymer*, **44**, 5901–8.

Greenwood, P. F. (1987), Mercerisation and liquid ammonia treatment of cotton, *J. Soc. Dyer. & Colour.*, **103**(10), 342–9.

Hamburger, W. J., Fox, K. R. (1956), A new process for compacting textile materials, Part I: Research and development, *Textile Res. J.*, **26**, 441–7.

Han, T.-Y., Chen, C.-C. (1998), Cross-linking of sulfonated cotton cellulose, Part I: Cross-linking and physical properties of DMDHEU-treated fabrics, *Textile Res. J.*, **68**(2), 115–20.

Hu, J. L., Yang, Z. H., Yeung, L. Y., Ji, F. L., Liu, Y. Q. (2005) Cross-linked polyurethanes with shape memory properties, *Polym. Int.*, **54**, 854–9.

Jang, T.-R., Sheu, T.-C., Sheu, J.-J., Chen, C.-C. (1993), Cross-linking of cotton fabrics premercerized with different alkalis, Part III: Cross-linking and physical properties of DMDHEU-treated fabrics, *Textile Res. J.*, **63**(11), 679–86.

Ji, F. L., Zhu, Y., Hu, J. L., Liu, Y., Yeung, L. Y., Ye, G. L. (2006), Smart polymer fibers with shape memory effect, *Smart. Mater. Struct.*, **15**(6), 1547–54.

Kobayashi, H., Kazuyuki, Hayashi, N. (1992), *Woven Fabric Made of Shape Memory Polymer.* Available from http://www.google.com/patents?id=5r8iAAAAEBAJ&printsec=abstract&zoom=4#v=onepage&q&f=false

Lee, M., Wakida, T., Tokuyama, T., Doi, C., Lim, Y. J., Jeon, S. K. (2005), Liquid ammonia treatment of regenerated cellulosic fabrics, *Textile Res. J.*, **75**(1), 13–18.

Li, Y. K., Yeung, L. Y., Chung, S. P., Hu, J. L. (2005), *Evaluation Method of Innovative Shape Memory Fabrics*, NTCTIAWC 2005, Carolina.

Lin, J. R., Chen, L. W. (1998), Study on shape-memory behavior of polyether-based polyurethanes, Part I: Influence of the hard-segment content, *J. Appl. Polym. Sci.*, **69**, 156–74.

Liu, Y. Q., Hu, J. L., Zhu, Y., Yang, Z. H. (2005), Surface modification of cotton fabric by grafting of polyurethane, *Carbonhyd. Polym.*, **61**, 276–80.

Patel, S. K., Varghese, J. (1983), Reparatory treatments and their influence on dyeing of cotton – a review, Part II: Mercerisation and liquid ammonia treatment, *Colourage*, **30**(5), 3–12.

Tao, Q. X. (2005), Latest development of liquid ammonia treatment, *Dyeing & Finishing*, **31**(15), 48–9.

Trask-Morrell, B. J., Andrews, B. K., Vinyard, B. T. (1996), Thermo-analytical study of durable press reactant levels on cotton fabrics, Part III: Finishes based on DMDHEU, *Textile Res. J.*, **66**(3), 172–83.

Wakida, T., Moriya, T., Lee, M., Yoshioka, H., Yanai, Y. (2000), Effect of liquid ammonia, sodium hydroxide/liquid ammonia, and subsequent cellulase treatments on mechanical properties of cotton fabrics, *Textile Res. J.*, **70**(2), 161–5.

Yanai, Y. (2001), The mechanism of liquid ammonia treatment (in Japanese), *Cellulose Commum.*, **8**(3), 135–9.

Yanai, Y., Shimizu, Y. (2006), The liquid ammonia treatment of cotton fibers – structural changes of cotton fibers after liquid ammonia and hot water treatment (in Japanese), *SEN-I GAKKAISHI*, **62**(5), 101–5.

Yanai, Y., Hamada, K., Shimizu, Y. (2005), The liquid ammonia treatment of cotton fibers – the structural changes of cotton fibers under various treatment conditions in practical units (in Japanese), *SEN-I GAKKAISHI*, **61**(11), 287–93.

Yanai, Y., Hamada, K., Shimizu, Y. (2006), The liquid ammonia treatment of cotton fibers – comparison and combination with mercerization using a practical unit (in Japanese), *SEN-I GAKKAISHI*, **62**(4), 81–8.

Yeung, L. Y., Hu, J. L., Zheng, G. H. (2007), Investigation on liquid ammonia treatment of shape memory cellulose fabrics, *Dyeing & Finishing*, **33**(6), 1–5.

Zeronian, S. H., Bertoniere, N. R., Alger, K. W., Xin, Q. (1990), The effect of cross-linking with DMDHEU on the mechanical properties of liquid-ammonia-treated cotton fibres, *Journal of Textile Institute*, **81**(3), 310–18.

Zheng, G. H. (2002), *Intermittent Liquid Ammonia Treatment Equipment*, China Patent ZL02289238.9.

11
Manufacture of T_g and T_m shape memory polyurethane (SMPU) polymer fibers

DOI: 10.1533/9780857098542.281

Abstract: In this chapter, T_g-type and T_m-type shape memory polyurethane (SMPU) fibers are discussed. The preparation of T_g-type SMPU fibers by melt spinning, and the preparation of T_m-type SMPU fibers by wet spinning are described. The chapter reviews the shape memory effects (SMEs) achieved with these fibers.

Key words: shape memory fiber, shape memory polyurethane, shape memory hollow fiber, shape memory effect.

11.1 Introduction

Though there has been much research on shape memory polymers (SMPs) as films, foams and finishing coatings in the last two decades (Hyashi, 1993; Kim and Lee, 1998; Kim *et al.*, 1996; Lendlein and Kelch, 2002; Lin and Chen, 1998; Liu *et al.*, 2002, 2007; Tang and Stylios, 2006; Tobushi *et al.*, 1998), studies on shape memory fibers (SMFs) remain rare. In principle, shape memory polyurethane (SMPU) fibers can be prepared by using dry, wet, chemical or melt spinning technologies. Of these different methods, melt spinning is the most effective in terms of health, safety, and environmental and economy concerns, because it does not involve the use of harmful solvents such as N,N-dimethylformamide (DMF) or N,N-dimethylacetamide) (DMAc) and requires no coagulation bath. Thus, a SMF was prepared by melt spinning.

A T_m-type SMPU was synthesized and corresponding SMFs were fabricated by melt spinning. Polyester polycaprolactone (PCL) was chosen as the soft segment instead of polyether, because it is reported that PCL-based polyurethanes have good properties. Small-sized diol butandiol (BDO) was selected as a molecular extender instead of diamine, because the diamines forming urea-urethane groups are known to increase the processing temperature of the polyurethane to a level too high for processing in a molten state (Fourne, 1999). The switching temperature of the SMF obtained was the soft segment phase melting temperature, which was higher than ambient temperature (Li *et al.*, 1996; Lin and Chen, 1998; Yang *et al.*, 2003). Further to this, the stress–strain behavior and thermal-mechanical properties were investigated, with particular focus on the shape memory effect (SME) of the SMF.

Wet spinning is a more traditional method of producing synthesized fibers. Continuous development over the past few years has resulted in considerable development of mechanical design and process of wet spinning, and resulted in higher productivity than that of melt spinning. Nevertheless, a major problem with the wet-spinning method is the use of organic solvents in the spinning solution. Even though equipment is available that adopts photocatalysts for waste water treatment, the treatment of waste water in the wet-spinning process is still expensive and time-consuming. The use of an organic solvent in the spinning process results in water pollution and thus limits further development of wet spinning, due to public emphasis and attention on green issues.

Even though wet spinning may cause water pollution, it has so far been widely used for the production of synthesis fibers. Thus, the properties of T_g-SMPU as well as the effects of thermal-humidity conditions on the structure, the thermal-mechanical properties and shape memory properties of wet-spun SMPU fibers have been closely studied. Since the concentration of an organic solvent in the water bath may influence both the recycling process of waste water, and the behavior and properties of wet-spun SMPU fibers, its effects on SMPU fibers will also be described.

11.2 T_m-type shape memory fibers prepared by melt spinning

11.2.1 Synthesis strategies

The SMPUs for melt spinning should have high thermal stability and relatively high molecular weight to obtain good mechanical properties. As mentioned before, PCL (Daicel Chemical Industrial) was chosen as the soft segment during the polyurethane synthesis, because it was reported that the PCL-based polyurethanes have comprehensively good properties (Fourne, 1999), while a small-sized diol was used as a molecular extender instead of diamine, because diamines forming urea-urethane groups increase the temperature of the polymer, which is too high for processing in a molten state.

The polyurethane was synthesized using methylene diphenyl diisocyanate (MDI) (Aldrich Chemical Company) and BDO (1,4-butanediol) (Acros Organics) as the hard segment. PCL diol was dried and degassed at 80°C under 1 to 2 mm Hg for 12 hours prior to use. Extra pure grade of MDI was melted at 50°C without further treatment, while the BDO was dried by molecular sieving beforehand. The polyurethane was prepared by the pre-polymerization technique – pre-polymers were prepared by terminating PCL with excessive MDI at both ends (Cherubini, 1995; Zhang, 2004), before the remaining MDI was added. BDO was added during the last stage. The polyurethane was further cured for 24 hours. The molecular weight of the prepared SMPU was 127 128 Daltons, with a polydispersity index of 1.34.

Polyurethanes show poor thermal stability due to the labile urethane linkages and ester bonds. It is reported that the initial decomposition point of urethane formed by MDI and poly(ethylene adipate) is 227°C (Krol, 2008). The decomposition of the SMPU yields diisocyanate and polyols. The MDI hard segment undergoes secondary reactions and produces more stable urea and isocyanurate structures (Krol, 2008). SMPU is also prone to hydrolysis in the presence of small amounts of moisture, yielding an amine and carbonic acid. The carbonic acid formed is unstable and must be decarboxylated immediately. As a result, the hydrolytic degradation products of a SMPU include amines, alcohols and carbon dioxide. At high temperatures, other types of degradation reactions are also possible. To protect against hydrolysis, a carbodiimide (acid acceptor)-type antihydrosis agent was added to SMPUs. To protect against thermal degradation, a phenolic phosphite-type stabilizer (reducer of oxidized free radical) was used.

The PCL-based polyurethane chips were prepared using a single-screw extruder. The chips were dried in a vacuum oven for 12 hours to decrease the relative humidity to below 100 ppm (as recorded by a Micro-water Analyzer CHY-8 (Donghua University, China)) before spinning. The SMFs were fabricated using a 20 cm single-screw extruder spinning machine. The spinning temperature was 210°C.

11.2.2 Molecular structure analysis

According to International Chemical Safety Cards (ICSC International Programme on Chemical Safety, 0298), inhalation of MDI will cause headache, nausea, shortness of breath and a sore throat, while contact with the skin will cause redness. The –NCO group of MDI has four characteristic IR peaks: 2260 to 2280 cm^{-1} as a result of asymmetry stretching vibration; 1375 to 1395 cm^{-1} as a result of symmetry stretching vibration; 600 to 650 cm^{-1} from out-plane bending vibration; and 590 to 600 cm^{-1} from in-plane bending vibration. The peak at 2260 to 2280 cm^{-1} is prominent, so is frequently used to determine the existence of –NCO. The Fourier transform infrared (FT-IR) spectra of MDI and SMPUs at 2000 to 2400 cm^{-1}, with increasing curing time at 110°C, are presented in Fig. 11.1. The main IR-band

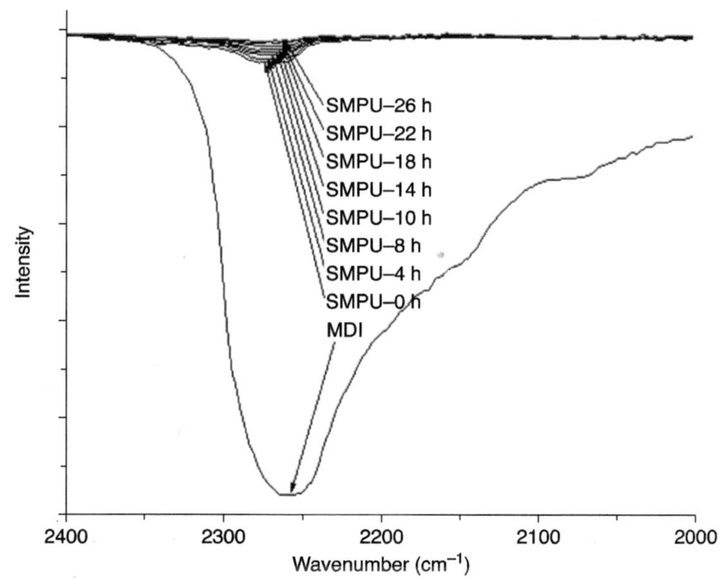

11.1 The change of –NCO FT-IR intensity with increasing curing time.

Table 11.1 Band assignment and domain origin

Energy (cm^{-1})	Assignment	Domain origin
3322	ν(N–H)	Hard
2952	ν(C–H)	Hard/soft
2939	ν(C–H)	Hard/soft
2850	ν(C–H)	Hard/soft
1726, 1702	ν(C=O)	Hard/soft
1612	phenyl ring mode 8a (Schoonover *et al.*, 2001)	Hard
1596	phenyl ring mode 8b (Schoonover *et al.*, 2001)	Hard
1524	ν(C–N)+ δ(N–H)	Hard
1412	phenyl ring mode 19b	Hard
1358	CH_2 wag	Soft
1308, 1222	ν(C–N)+ δ(N–H)	Hard
1204	Phenyl ring mode ag	Hard
1064	ν(C–O–C)	Soft
1018	Phenyl ring mode 18a	Hard

N = stretching, δ = in-plane bending

assignment of the SMPU spectra and the domain origin are tabulated in Table 11.1. It can be seen that, with increased curing time, the area of the peak at 2260 to 2280 cm^{-1} decreases. After 22 hours cure, the peak almost disappears. In the following experiments, SMPUs have usually been cured for 24 hours at 110°C.

11.2.3 Morphology of shape memory fibers

From the SMP chips, PCL-based SMFs were fabricated by melt spinning. The prepared SMFs are shown in Fig. 11.2.

11.2 PCL-based SMFs prepared by melt spinning.

The SMF cross-section image taken using an optical microscope is shown in Fig. 11.3, while Fig. 11.4 shows the SEM surface image of the SMF. The cross-section of the SMF is round and the surface is smooth.

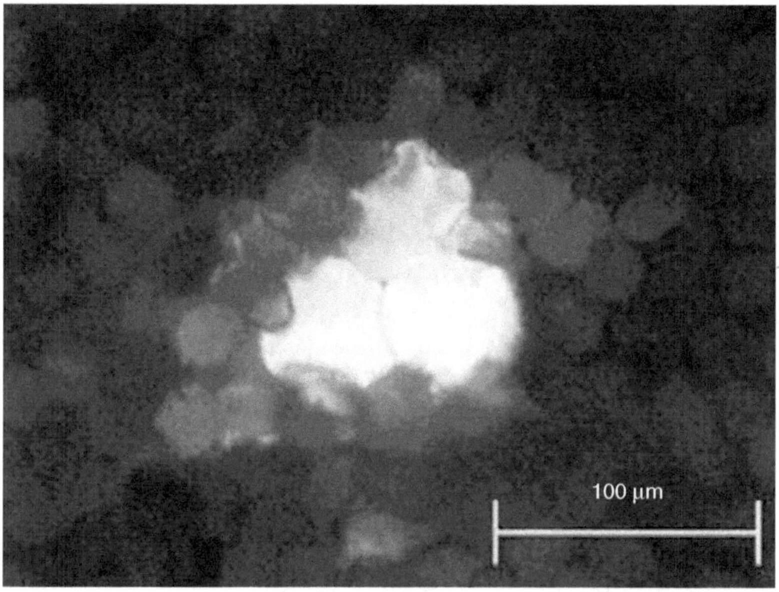

11.3 Cross-section of the PCL-based SMFs prepared by melt spinning.

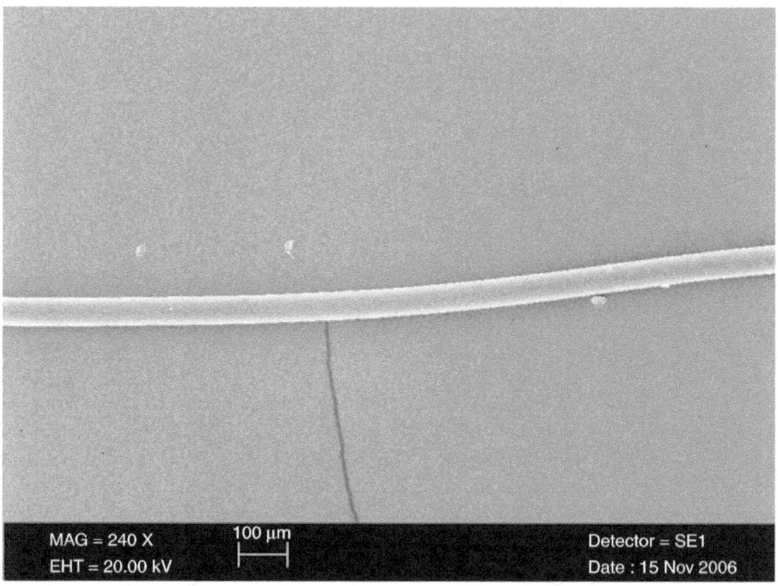

11.4 The surface image of the PCL-based SMFs prepared by melt spinning.

11.2.4 Mechanical properties

The prepared SMF has a tenacity of about 1.0 cN/dtex, and a strain at break of 562~600%. This tenacity is generally suitable for textile applications. In comparison with most other man-made fibers, such as Polyester and Nylon, the mechanical strength of SMF is lower. Polyester and Nylon usually have high tenacity – above 3.0 cN/dtex – and low breaking elongation ratios. The mechanical properties of Polyester and Nylon fibers are attributed to their higher overall orientation, strong intermolecular bonding in polyamide and the high crystallinity (X_c) of the molecular chain in Polyester (Yao *et al.*, 1990). However, for the SMFs which show high shape fixity ratios and shape recovery ratios, the elongation at break is much higher compared to that of Polyester and Nylon. The mechanical properties of SMF can be further improved using the pre-polymer cross-linking method or pre-end-capping method, which have already been applied in melt spinning of commercialized polyurethane elastic fibers (Yoshihaya *et al.*, 2003).

11.2.5 Shape memory properties

This section looks at the following properties:

1. drawing at high temperature and thermal recovery cyclic tensile properties;
2. drawing at low temperature and thermal recovery cyclic tensile properties;

Figure 11.5 illustrates the stress–strain behaviors of the SMF by drawing at high temperature and thermal recovery cyclic tensile testing, and the data of the fixity ratio, recovery ratio and stress at 100% strain are tabulated in Table 11.2.

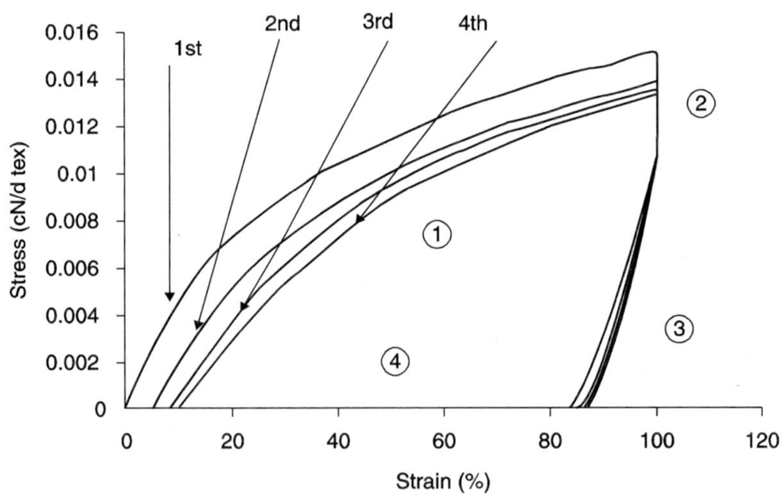

11.5 Thermo-mechanical cyclic tensile testing of the fiber.

Table 11.2 The data of the fixity ratio, recovery ratio and stress at 100% strain at high temperature

Cycle no.	ε_p (N)	ε_u (R_f(N))	Stress at 100% strain (cN/dtex)	$R_{r,tot}$(N)	R_r(N)
1	0.0%	84.0%	0.0152	100.00%	–
2	5.0%	85.5%	0.0139	95.00%	95.0%
3	8.6%	86.0%	0.0136	91.40%	96.2%
4	10.0%	86.5%	0.0134	90.00%	98.5%

The SMF has a fixity ratio of more than 84% and recovery ratio of up to 95%. The recovery stress at 100% elongation is about twice that of a SMPU without molecular orientation (Hu *et al.*, 2005a,b; Ping *et al.*, 2005). The significant difference between the first thermal cycle and the remaining cycles is partially due to the reorganization of fiber molecules which occurs in molecular orientation, crystallization, or the breaking of a weak point during elongation. After one cycle, the stress–strain behaviors become similar.

Figure 11.5 also suggests that the SMF cannot fix the temporary elongation completely while cooled from a high to an ambient temperature. The instantaneous elastic deformation recovers once the external stress is released. It can be deduced that the fixity ratio can be improved by increasing the X_c of the soft segments. Furthermore, the SMF cannot recover its original length completely, because of the molecular slippage and breakage during the fiber cyclic elongation.

The shape memory mechanism of the SMF can be illustrated as follows: during melt spinning, at a temperature which is higher than T_{perm} (222°C in the given experiment), the fiber is extruded from a spinneret. Upon cooling to an ambient temperature, which is below T_{trans}, the fiber is wound up and the permanent fiber shape is cast.

The model in Fig. 11.6 can be employed to illustrate the SME of the SMF. In Fig. 11.6, the zig-zags represent coiled or folded chains of polyols, while circles represent isocyanate groups. In the unstretched state, the fiber's molecules are slightly orientated with some crystallized soft segments. The hard segments remain, but have a tendency to adhere to each other through strong hydrogen bonding. When the fiber is heated to 70°C, which is above the soft segment phase melting temperature (T_{trans}), the soft segments are in a random state. When it is stretched, the soft segments are extended. If the temperature is cooled to below T_{trans}, the soft segments crystallize. As a result, the internal stress is stored in the fiber and associated deformation is fixed temporally. If the SMF is reheated to above T_{trans}, the soft segments become flexible. Consequently, the soft segments resume to the folded configuration with the release of the internal stress stored among the hard segments, and the SMF recovers its original length.

The drawing at low temperature and thermal recovery testing was conducted to study the SME in SMFs. When characterizing the SME of SMPs, the deformation

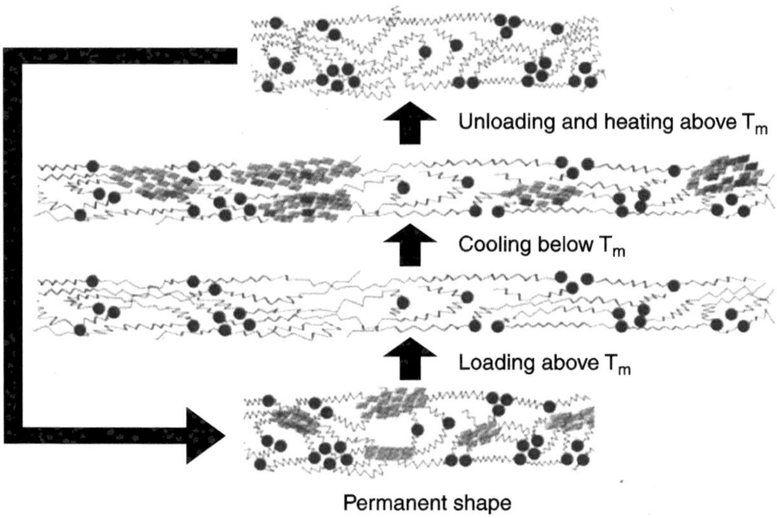

11.6 The molecular mechanism of the shape memory effect of the SMF.

is usually conducted at a temperature higher than the switching transition temperature, to allow the deformation to be more easily developed. In practice, however, the deformation in SMFs, such as the creation of a crease in the cloth, usually happens at the ambient temperature.

The drawing at low temperature and thermal recovery cyclic tensile curves of the SMF are shown in Fig. 11.7. The data of fixity ratio, recovery ratio and stress at

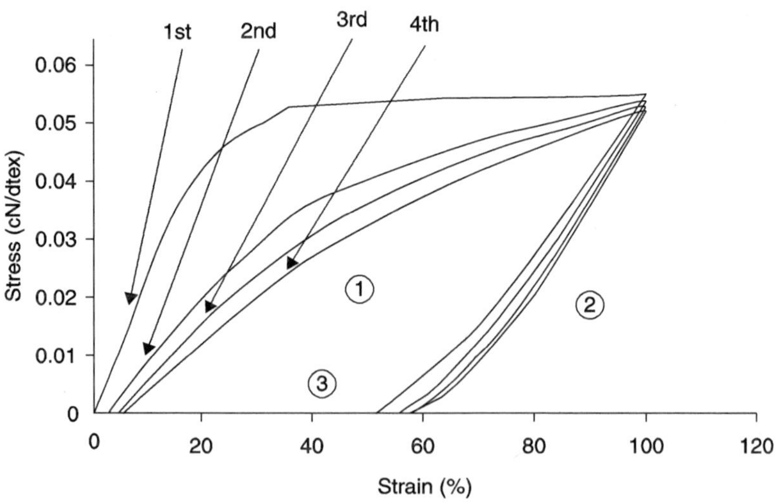

11.7 The drawing at low temperature and thermal recovery cyclic tensile curves of the SMF.

Table 11.3 The data of the fixity ratio, recovery ratio and stress at 100% strain at low temperatures

Cycle no.	ε_P (N)	ε_u (R_f(N))	Stress at 100% strain (cN/dtex)	$R_{r,tot}$(N)	R_r(N)
1	0.00%	51.00%	0.0550	100.00%	–
2	2.50%	55.00%	0.0540	97.50%	97.5%
3	4.00%	57.00%	0.0533	96.00%	98.5%
4	5.00%	57.50%	0.0527	95.00%	99.0%

100% strain are summarized in Table 11.3. The SMF also has high shape recovery ratios of about 95%. The fixity ratios are about 55%, which is much lower than those of SMF tested by drawing at high temperature and thermal recovery.

11.2.6 Thermal properties

The DSC results of the SMF are shown in Fig. 11.8. The melting temperature and crystalline enthalpy are tabulated in Table 11.4. The soft segments of the SMF show a melting transition at 47°C; the hard segments display a moderate endothermic peak at about 215°C. The X_c of the soft segments in the SMF is 14.42%, which is calculated from the enthalpy data ΔH of the crystallization peak, using the 140 J/g enthalpy value for fusion of 100% crystalline PCL (Crescenzi *et al.*, 1972; Luo *et al.*, 1996).

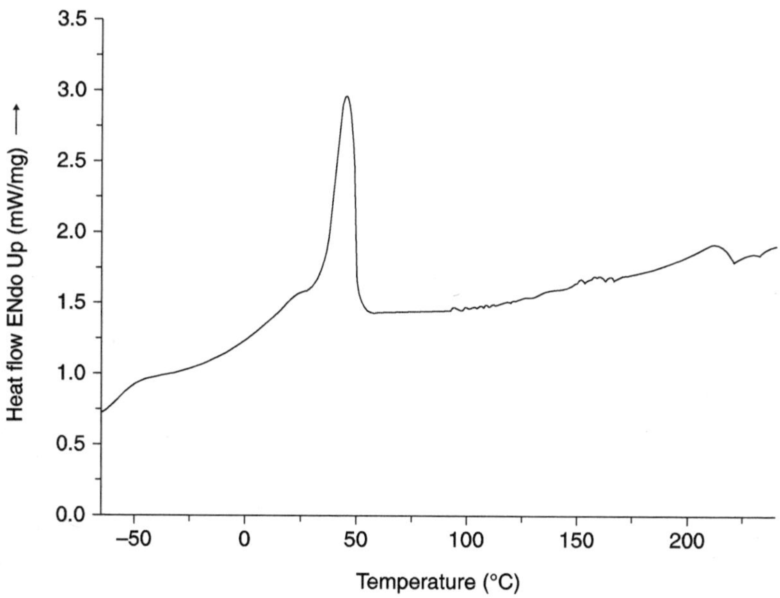

11.8 The DSC results of the SMF.

Table 11.4 The melting temperature and crystalline enthalpy

	Value
Soft segment	
T_m (°C)	47.02
ΔH (J/g)	20.19
Hard segments	
T_m (°C)	214.54
ΔH (J/g)	1.286

At the ambient temperature, the soft segments are partially crystallized. As a result, the fiber is not completely elastic. It can thus be deduced that the SME of the SMF can be improved by increasing the soft segment X_c; however, this is at the expense of the elasticity of the SMF.

11.3 T_g-type shape memory fibers prepared by wet spinning

11.3.1 Synthesis strategies

The soft segment of the SMPU used for wet spinning was a polybutylene adipate (PBA) with a molecular weight of 600 (PBA-600). The SMPU solution was prepared by a pre-polymer method, using MDI and BDO, which has a hard segment content of 70%. In the pre-polymerization process, PBA was reacted with excess MDI, while dimethylacetamide (DMAc) was used as the solvent. The pre-polymerization process was conducted at 80°C for two hours in a nitrogen environment. The pre-polymer was then chain extended with BDO. The NCO:OH ratio was kept at 1:1 with a hard segment content of 55%, as controlled by the ratio of PBA to MDI.

In wet spinning, a polymer solution with a suitable viscosity was prepared by dissolving the polymer into DMAc. In this project, the solid concentration of the final SMPU solution in the DMAc was adjusted to a range of 20 to 30 wt.% to meet the required viscosity in the wet-spinning process. During the spinning process, the PU solution was extruded horizontally through 30 spinneret capillary holes in a coagulation bath, with a spinning speed of 15 m/min^{-1}, until a multi-filament structure was formed by coagulation. The filaments were then passed through rollers in a wash-draw machine to further remove the residual organic solvent. Finally, the fibers were dried with hot air at a temperature of 60°C and wound onto a paper cone, as shown in Fig. 11.9.

During the fiber formation process, a surface layer was formed on the outermost surface of the solution when the solution was passed through the spinneret holes into the water bath. The solvent diffused from the inside of the fiber and the surface layer further coagulated toward the same point. The rate of diffusion was dependent

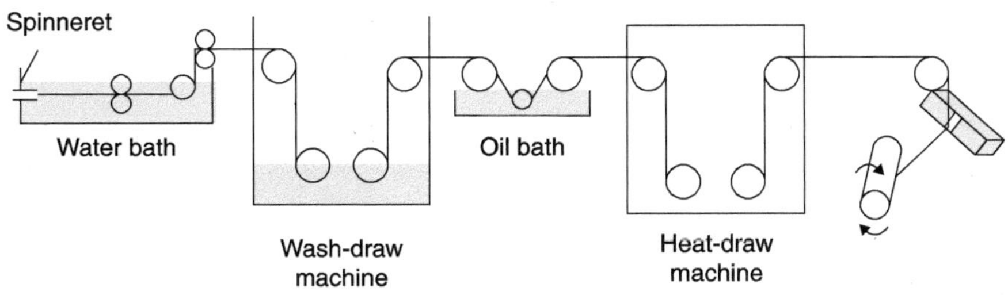

11.9 Sketch of wet-spinning equipment.

on the solvent concentration between the polymer solution and water bath. A more detailed explanation of the filament formation process is shown in Fig. 11.10.

The PBA-based, wet-spun SMPU fibers were fixed onto paper cones and placed in a humidity cabinet at various temperatures and humidity conditions. The samples were stored in a specifically conditioned room (21°C, 65 RH%) for at least one week before examination. The effects of the thermal-humidity conditionings on the thermal, mechanical and shape memory properties of the wet-spun fibers were then examined. Table 11.5 shows the conditions and coding of the temperature-humidity conditioning.

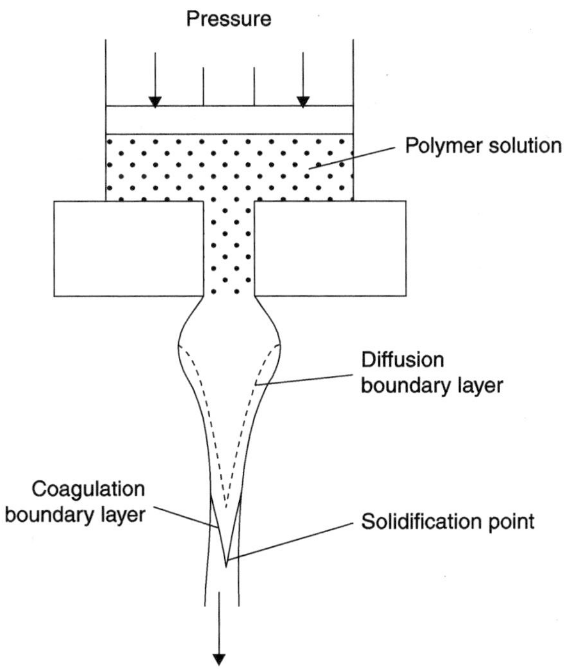

11.10 Formation of synthesis fiber in wet spinning.

Table 11.5 Codes and conditions of thermal-humidity treatments on SMPU fibers

Time (hours)	Temperature of 50°C		Temperature of 80°C	
	RH 30%	RH 80%	RH 30%	RH 80%
40	e-40	f-40	g-40	h-40
70	e-70	f-70	g-70	h-70
100	e-100	f-100	g-100	h-100
190	e-190	f-190	g-190	h-190

Note: Condition e: 50°C, 30RH%; f: 50°C, 80RH%; g: 80°C, 30RH%; h: 80°C, 80RH%

11.3.2 Thermal-humidity conditioning on T_g-type SMPU fibers

Huang and colleagues (Huang *et al.*, 2005; Pan *et al.*, 2008; Yang *et al.*, 2006) studied the moisture sensitivity of T_g types of SMPU fibers that were prepared by MHI, which was synthesized from MDI, BDO, adipic acid, ethylene glycol, ethylene oxide, polypropylene oxide (PPO) and bisphenol A. MHI did not disclose the exact SMPU formula, but it is assumed to be of PPO/MDI/BDO composition. This chapter therefore aims to reveal the effects of heat and moisture conditions on the mechanical and thermal properties of wet-spun SMPU fibers.

11.3.3 Tensile properties of wet-spun SMPU fibers

The tensile properties of a T_g-type SMPU fiber were tested by an Instron 4411 universal tensile tester with a specimen length of 5 cm, and strain rate of 10 cm/min. Ten specimens were tested for each sample, and the mean values of the breaking tenacities and stresses of the specimens under tensile testing are shown in Table 11.6.

The stress–strain curves of the SMPU fibers are presented in Fig. 11.11. The fibers were treated at various thermal-humidity conditions for 190 hours. The changes in the breaking tenacities and strains of the PBA-based SMPU fibers are less significant than that of the PCL-based T_m-type fibers. The increase of temperature and/or humidity results in small augmentations in the breaking strains, and a decrease in the tenacity of the SMPU fibers. When the treatment temperature is 80°C, humidity is 80% and the time is 190 hours, the fiber breaking strain increases by 10% and the tenacity decreases by 14% for the PBA-based SMPU fibers. The changes in the breaking strain and tenacity for PCL-based SMPU fibers under the same conditions are 119 and −75%, respectively. A statistical analysis was conducted to evaluate the effects of thermal-humidity treatments on PBA-based SMPU fibers.

The initial modulus of the specimens was also examined. The mean values of the initial modulus, calculated with a slope at 10% strain in the stress–strain curves, are shown in Table 11.7. The initial moduli of all specimens after a thermal-humidity treatment were lower than those of the untreated specimens, with this becoming even more noticeable when the treatment temperature was

Table 11.6 Tenacities and breaking strains of PCL-based SMPU fibers after various treatments

Sample	Mean		Standard derivation	
	Tenacity (cN/dtex)	Breaking strain (%)	Tenacity (cN/dtex)	Breaking strain (%)
Untreated	0.532	159.25	0.025	11.92
e-40	0.477	152.54	0.012	6.58
e-70	0.484	151.65	0.018	11.94
e-100	0.503	159.24	0.022	12.49
e-190	0.520	149.84	0.011	6.40
f-40	0.498	162.05	0.019	8.69
f-70	0.520	151.40	0.012	8.13
f-100	0.497	142.58	0.014	11.40
f-190	0.519	147.22	0.022	8.47
g-40	0.487	165.90	0.013	5.04
g-70	0.489	162.81	0.011	10.51
g-100	0.456	159.92	0.024	15.53
g-190	0.435	162.56	0.023	11.02
h-40	0.492	180.44	0.013	14.00
h-70	0.506	188.93	0.008	7.05
h-100	0.476	176.58	0.013	8.39
h-190	0.456	175.42	0.019	15.57

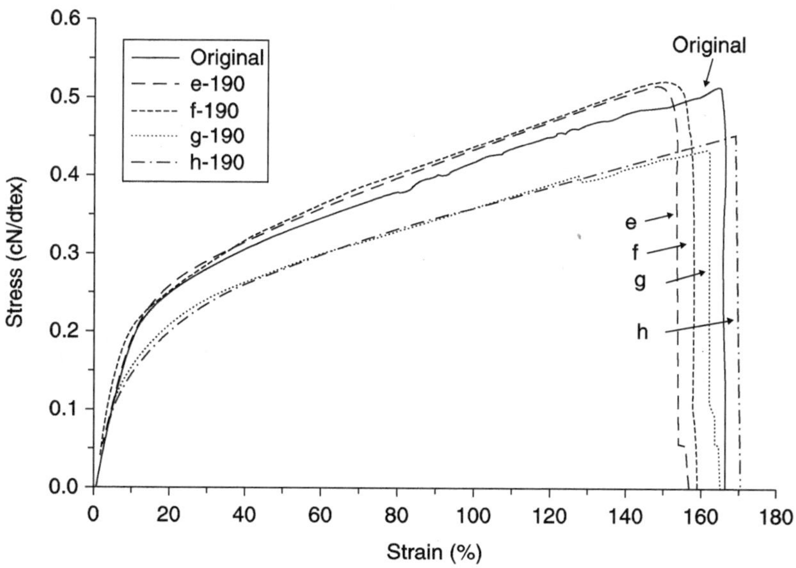

11.11 Typical stress–strain curves for PBA-based SMPU fibers after 190 hours of conditioning.

Table 11.7 Initial modulus of PBA-based SMPU fibers

Sample	Modulus (cN/dtex)	Standard derivation
Untreated	1.63	0.09
e-40	1.13	0.07
e-70	1.12	0.17
e-100	1.26	0.11
e-190	1.50	0.04
f-40	1.26	0.04
f-70	1.53	0.09
f-100	1.53	0.04
f-190	1.59	0.12
g-40	1.12	0.03
g-70	1.12	0.04
g-100	1.01	0.03
g-190	1.04	0.03
h-40	1.06	0.03
h-70	0.98	0.03
h-100	0.94	0.04
h-190	0.89	0.03

increased from 50 to 80°C. However, the moduli increased with changes in humidity conditions when the temperature was set at 50°C, but decreased at 80°C.

A statistical analysis based on t-tests was carried out, taking into account the changes of the tensile property results of PBA-based SMPU fibers (Tables 11.8 and 11.9). The significant level of the statistical analysis is 0.05, while H_0 and H_1 are respectively defined as 'the tensile property results of the treatment specimen which are the same as the untreated sample' and 'the tensile property results of the treatment specimen which are not the same as the untreated sample'. The hypotheses and t-value are defined as follows.

$$t = \frac{(\bar{x}_1 - \bar{x}_2) - (\bar{\mu}_1 - \bar{\mu}_2)}{\sqrt{\frac{s_1^2}{n_1} + \frac{s_2^2}{n_2}}}$$

The statistics of the tenacity and breaking strain for the conditioned specimen vs untreated specimen are shown in Tables 11.8 and 11.9, comparing the initial modulus of the SMPU fibers obtained in dry (30 RH%) and wet (80 RH%) conditions. The results show that the initial modulus of SMPU fibers is different in wet conditions to that in dry conditions.

Tables 11.8 and 11.9 show the statistics of the breaking strain and tenacity of thermal-humidity treated PBA-based SMPU fibers, when compared with an untreated specimen. Condition e has a temperature of 50°C and relative humidity of 30%. When condition e occurs, the breaking tenacity drops from 0.532 to 0.477 cN/dtex, and then increases with treatment time. Finally, the breaking tenacity of specimen e-190 returns to the original value of 0.52 cN/dtex. However, the differences in the

Table 11.8 T-test for tenacity (comparisons with untreated specimen)

	Mean	Standard derivation	t	P (T<t, two-tail)	Accept or reject of null hypothesis (H_0)
Untreated	0.532	0.025			
e-40	0.477	0.012	−5.95	1.25E-05	Reject
e-70	0.484	0.018	−4.67	1.89E-04	Reject
e-100	0.503	0.022	−2.61	1.76E-02	Reject
e-190	0.52	0.011	−1.32	2.04E-01	Accept
f-40	0.498	0.019	−3.25	4.46E-03	Reject
f-70	0.52	0.012	−1.30	2.11E-01	Accept
f-100	0.497	0.014	−3.66	1.77E-03	Reject
f-190	0.519	0.022	−1.17	2.57E-01	Accept
g-40	0.487	0.013	−4.79	1.46E-04	Reject
g-70	0.489	0.011	−4.72	1.70E-04	Reject
g-100	0.456	0.024	−6.58	3.52E-06	Reject
g-190	0.435	0.023	−8.57	9.13E-08	Reject
h-40	0.492	0.013	−4.26	4.73E-04	Reject
h-70	0.506	0.008	−2.97	8.18E-03	Reject
h-100	0.476	0.013	−5.96	1.22E-05	Reject
h-190	0.456	0.019	−7.26	9.47E-07	Reject

Table 11.9 T-test for breaking strain (compared with untreated specimen)

	Mean	Standard derivation	t	P (T<t, two-tail)	Accept or reject of null hypothesis (H_0)
Untreated	159.25	11.9			
e-40	152.54	6.6	−1.48	0.078	Accept
e-70	151.65	11.9	−1.35	0.097	Accept
e-100	159.24	12.5	0.00	0.499	Accept
e-190	149.84	6.4	−2.09	0.026	Reject
f-40	162.05	8.7	0.57	0.288	Accept
f-70	151.40	8.1	−1.63	0.060	Accept
f-100	142.58	11.4	−3.03	0.004	Reject
f-190	147.22	8.5	−2.47	0.012	Reject
g-40	165.90	5.0	1.54	0.070	Accept
g-70	162.81	10.5	0.67	0.255	Accept
g-100	159.92	15.5	0.10	0.460	Accept
g-190	162.56	11.0	0.61	0.274	Accept
h-40	180.44	14.0	3.46	0.001	Reject
h-70	188.93	7.1	6.43	0.000	Reject
h-100	176.58	8.4	3.57	0.001	Reject
h-190	175.42	15.6	2.47	0.012	Reject

breaking strains between the untreated specimen and the specimens treated under condition e are largely insignificant, with the notable exception of specimen e-190.

When humidity conditions increase from 30 to 80%, the situation becomes more complex. A general trend cannot be obtained in the relationship between the breaking properties and treatment time. In combination with the above observations for the SMPU fibers treated at 50°C, it is suggested that there has to be some other time-dependent reaction, which resulted in the opposite effects of plasticization. The data obtained above is the final result of the competition between different reaction mechanisms in the thermal-humidity treatments.

When the treatment temperature is increased to 80°C, the breaking tenacities of the treated SMPU fibers decrease in comparison with those of the untreated specimen. Treatment time can also alter breaking tenacity, which remains largely constant under a treatment time of 40 to 70 hours, but drops significantly when the treatment time is further increased.

Even though the changes in breaking tenacities are similar for conditions g and h, the changes in breaking strains differ. The breaking strains for the SMPU fibers after condition g (80°C, 30 RH%) are nearly the same as those of the untreated specimen, while in condition h, the breaking strains increase by 10 to 15% (80°C, 80 RH%).

Tables 11.10 and 11.11 compare the initial modulus of the specimens with the corresponding thermal and humidity conditions. The initial modulus of the SMPU fibers increases with humidity for treatments at 50°C, and decreases with humidity for treatments at 80°C. The initial modulus of the SMPU fibers decreases as the treatment temperature rises from 50 to 80°C, with the same drop even more rapid in higher humidity conditions.

Based on the preceding hypothesis, it is possible to draw the following conclusions of the PBA-based SMPU fibers studied above. The breaking tenacity decreases at the treatment temperature of 80°C, the breaking strains increase after being treated at 80°C in 80 RH%, and the breaking strains remain unchanged after being treated at 80°C in 30 RH%. The initial modulus of the specimens increases further, when the humidity rises from 30 to 80%, at treatment temperature of 50°C.

Table 11.10 T-test for initial modulus (compared with the corresponding 30 RH%)

	Mean	Standard derivation	t	P (T<t, two-tail)	Accept or reject of null hypothesis (H_0)
f-40	1.26	0.04	4.837	0.000	Reject
f-70	1.53	0.09	6.394	0.000	Reject
f-100	1.53	0.04	6.920	0.000	Reject
f-190	1.58	0.08	2.623	0.017	Reject
h-40	1.06	0.03	−4.243	0.000	Reject
h-70	0.98	0.03	−8.400	0.000	Reject
h-100	0.94	0.04	−4.200	0.001	Reject
h-190	0.89	0.03	−10.607	0.000	Reject

Table 11.11 T-test for initial modulus (compared with the corresponding temperature at 50°C)

	Mean	Standard derivation	t	P (T<t, two-tail)	Accept or reject of null hypothesis (H_0)
g-40	1.12	0.03	−0.39	6.98E−01	Accept
g-70	1.12	0.04	0.00	1.00E+00	Accept
g-100	1.01	0.03	−6.58	3.53E−06	Reject
g-190	1.04	0.03	−27.60	3.49E−16	Reject
h-40	1.06	0.03	−12.00	5.05E−10	Reject
h-70	0.98	0.03	−17.39	1.06E−12	Reject
h-100	0.94	0.04	−31.29	3.81E−17	Reject
h-190	0.89	0.03	−23.24	7.11E−15	Reject

However, at treatment temperature of 80°C, the initial modulus of the specimens decreases when the humidity changes from 30 to 80%. Finally, the initial modulus of the specimens decreases when treatment temperature increases from 50 to 80°C.

When the treatment temperature is 80°C, the breaking tenacity decreases when the humidity drops from 30 to 80 RH%. The reaction dynamics are complicated by low temperature treatments, and no conclusion can be drawn for this state. However, as the most interesting parameter in this study, the shape memory properties are nearly the same as the untreated specimen after conditions e and f. The reaction dynamics can therefore be left for further investigation without affecting the validity of the given study.

11.3.4 Thermal properties of shape memory fibers

Traditionally, T_g is measured by the step changes of heat flow on the DSC thermographs. However, these may sometimes be unmeasurable using the DSC. The glass transition, unlike melting behavior, is a second-order phase, and requires less energy than crystal melting. As a result, thermal analyses of T_g types of SMPU fibers are conducted by a dynamic mechanical analyzer (DMA).

The changes of $\tan\delta$, measured by DMA, and the changes of heat flow, measured by DSC upon glass transition and melting, oppose one another. In examination of the melting behavior, there is a step change in the DMA, while a melting peak is observed for DSC. In the case of glass transition behavior, a peak appears in the DMA and a step change is found in DSC. Figure 11.12 shows a DMA test result for untreated fiber, in which the T_gs are defined by peaks of $\tan\delta$ in the DMA.

Table 11.12 shows the T_gs of PBA-based SMPU fibers under different temperature and humidity treatments. In comparison with the untreated sample, the values of T_g are reduced for all specimens, with the exception of specimen h-190. This is similar to the result obtained in the change of the initial modulus, and such a decrement is more significant when the treatment temperature increases from 50 to 80°C.

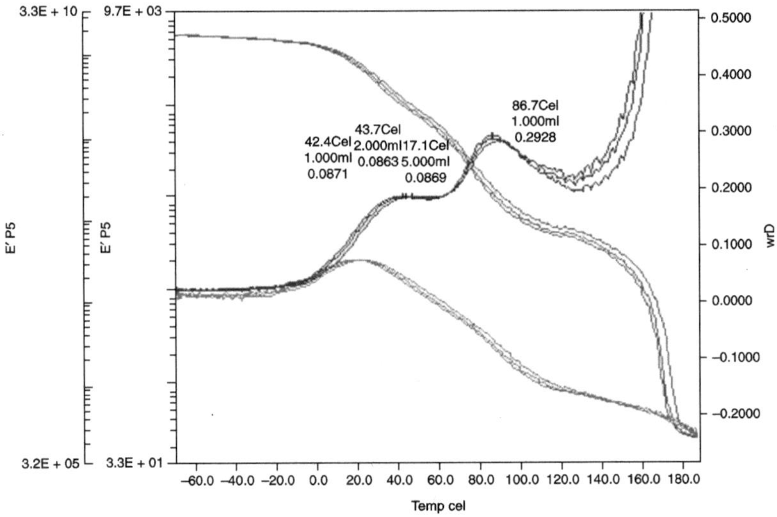

11.12 DMA result of the untreated PBA-based SMPU fibers.

Table 11.12 T_g of various treated PBA-based SMPU fibers tested by DMA

Time of treatment	Glass transition temperature (°C)			
	Untreated enote	F	g	h
40 hours	39.2	40.1	33.4	31.5
70 hours	39.6	35.1	31.5	36.1
100 hours	41.3	40.5	32.8	34.0
190 hours	39.2	40.0	31.8	54.7

Note: e: 50°C, 30 Rh%; f: 50°C, 80 Rh%; g: 80°C, 30 Rh%; h: 80°C, 80 Rh%

In thermal gravimetric analysis (TGA), the experiments were performed in air, and the temperature range was set as 30 to 550°C with a heating rate of 10°C/min. An empty sample holder was used as the baseline correction before testing. Air was used as the purge gas and the sample weights were in the range of 5 to 10 mg. The TGA results, together with the first derivative of TG, are shown in Figs 11.13 to 11.15.

No weight loss was observed for PBA-based SMPU fibers when the temperature was lower than 250°C (Fig. 11.14), which indicates the absence of free water within the fibers. However, a small percentage of weight loss was observed in specimens e-190 and f-190 at 400°C, whilst a more substantial percentage of weight loss was visible in specimen h-190. Of the four specimens studied, we can therefore conclude that specimen h is the more sensitive to air.

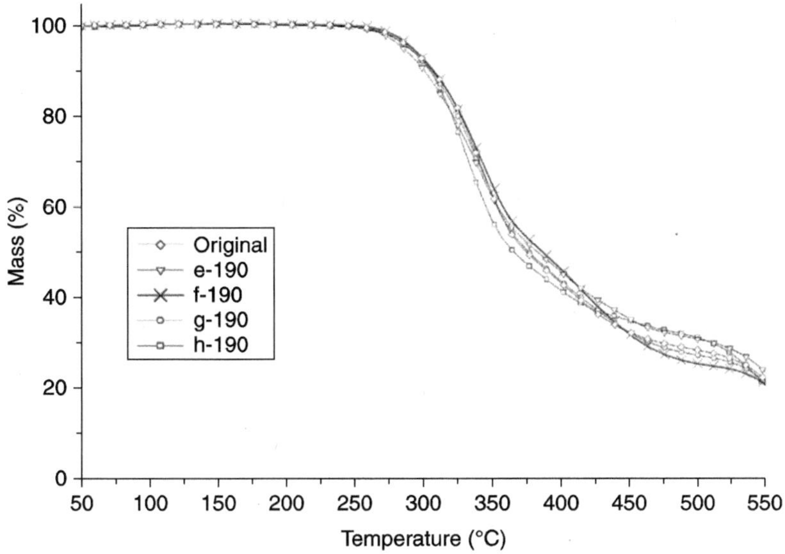

11.13 TGA results for PBA-based wet-spun SMPU fibers.

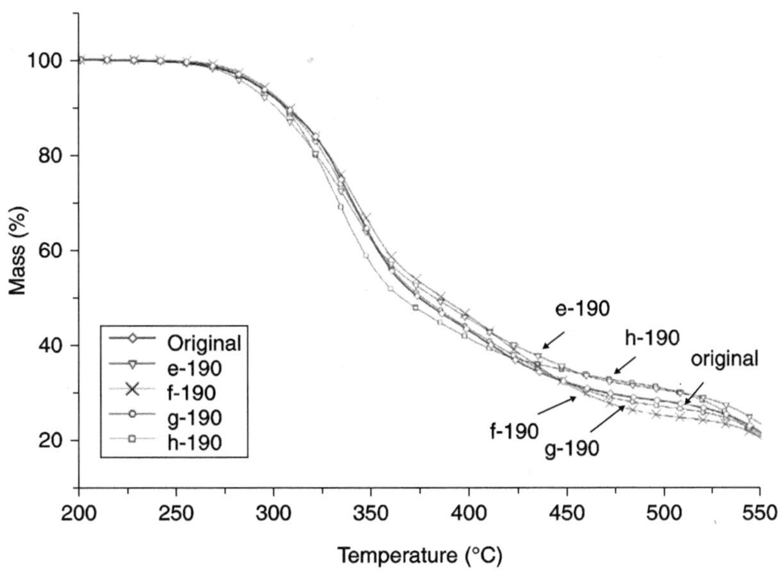

11.14 TGA for PBA-based T_g-type SMPU fiber at 200 to 550°C.

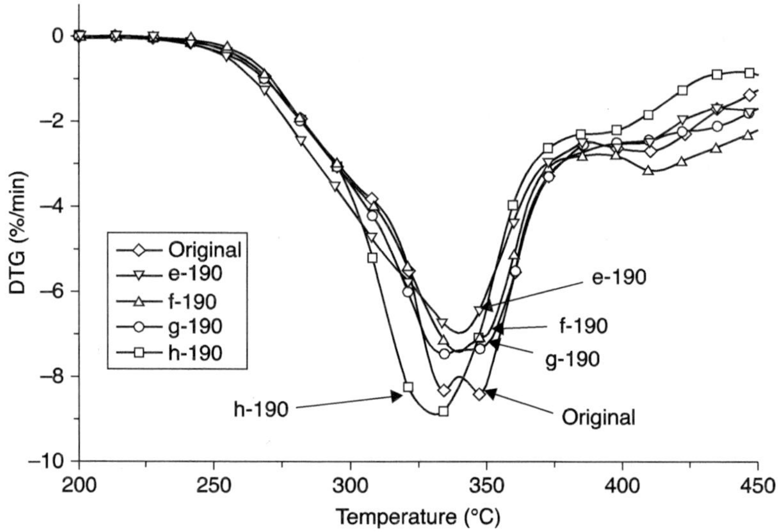

11.15 First derivatives of TGA for PBA-based SMPU fibers.

Previous studies on the degradation of segmented PU (Ferguson and Petrovic, 1976; Mondal and Hu, 2007; Petrovic *et al.*, 1994) indicated that the first degradation step corresponds to the hard segment, while the second step is associated with the soft segment. The increased weight loss of specimen h-190 is, therefore, considered to be the result of the weakening of chemical bonding between the hard and soft segments. The weakening of PBA-based SMPU fibers after a high-temperature/humidity treatment is therefore the result of the chemical changes of the SMPU structure.

11.3.5 Thermal-mechanical properties

The thermal-mechanical properties of these PBA-based, T_g-type SMPU fibers were studied with the DMA. A bundle of ten fibers was attached to the sample holder with a gauge length of 1 cm. The tensile gauge was then driven by a sinusoidal signal with a minimum force amplitude of 100 mN. Liquid nitrogen was used as a coolant and the temperature range was −70 to 200°C. The experiments stopped either when the equipment reached the target temperature or the fibers broke.

Table 11.13 shows the T_g as measured by the DMA and the ratio of |E| at 30°C to |E| at 75°C. In this study, the most significant difference between the T_g-type and T_m-type SMPU fibers is the ratio of modulus below and above the transition temperature. In a review of the values of |E(30°C)|/|E(75°C)|, the T_m-type SMPU fiber has a value of 55 to 168, which is about ten times the T_g-type SMPU fiber shown above. This is a result of a difference in polymer structure, as the bonding is much stronger within a crystal structure than within a glassy state.

Table 11.13 DMA results of T_g and change of |E| for PCL-based SMPU fibers

	T_g (peak of tanδ)	|E(30°C)|/|E(75°C)|
U	42.4	10.2
e-40	39.2	10.4
e-70	39.6	12.8
e-100	41.3	8.5
e-190	39.2	7.1
f-40	40.1	7.6
f-70	35.1	6.7
f-100	40.5	7.3
f-190	40	7.6
g-40	33.4	7.3
g-70	31.5	7.7
g-100	32.8	6.2
g-190	31.8	5.9
h-40	31.5	4.9
h-70	36.1	6.7
h-100	34	7.2
h-190	54.7	7.7

The change of tanδ against temperature for samples that are conditioned for 190 hours is shown in Fig. 11.16. As the only difference between the PBA-based and PCL-based fibers is the composition of the soft segment, the two peaks observed in Fig. 11.16 could be associated with the glass transition of the soft segment. The untreated specimens e-190, f-190 and g-190 all have two T_gs at about 35 to 40°C and 85 to 90°C, but these two transition peaks combine into one broad peak with the maximum at 54.7°C for specimen h-190. This is the result of phase mixing after elongated treatment at a high temperature and humidity level, which will be discussed in the FT-IR spectra later.

11.3.6 Shape memory properties of T_g-type shape memory fibers

The shape memory properties of a T_g-type of wet-spun SMPU fiber were also examined by an Instron 5566 universal tensile tester, equipped with a temperature cabinet. A bundle of eight fibers was first clamped onto the gauges, then heated to 75°C and stretched to 100% strain at a rate of 10 mm/min. The fibers were then cooled to room temperature under 100% strain, and held for 15 minutes. Finally, the clamps were returned to ground level, and the cabinet was heated to 75°C before the next tensile cycle started.

A load–strain curve of untreated PBA-based SMPU fibers under a cyclic tensile test is shown in Fig. 11.17. As the shape fixity and recovery ratios could change

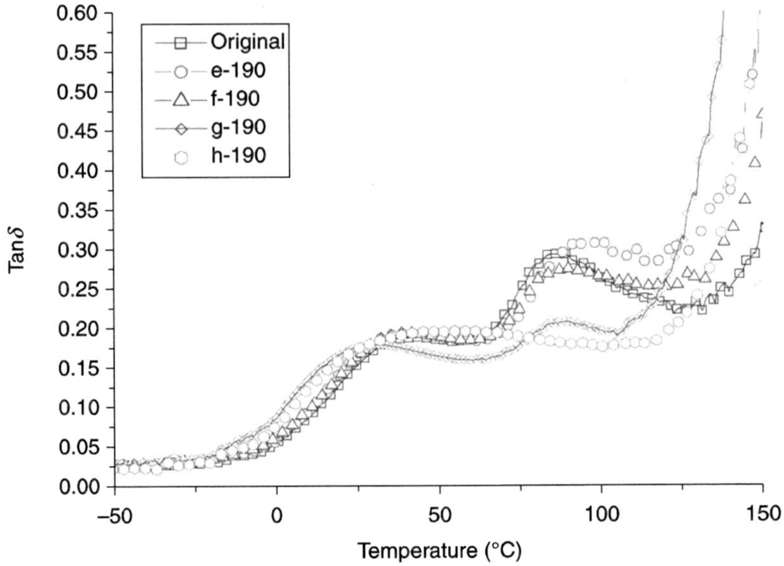

11.16 Tanδ of as-spun and conditioned PBA fibers.

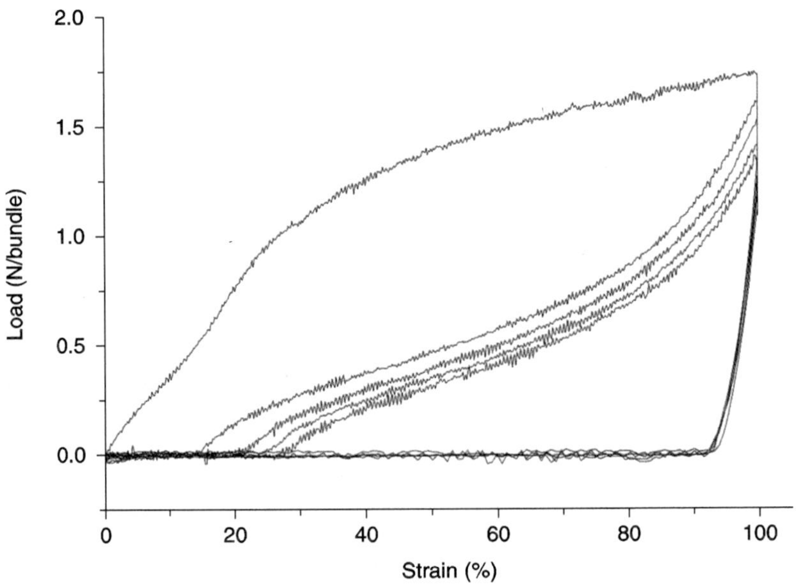

11.17 Cyclic tensile test of untreated PBA-based SMPU fibers.

Table 11.14 Shape fixity and shape recovery ratio of PBA-based SMPU fibers after thermal-humidity treatment for 100 and 190 hours

Sample	R_f (%)	R_r (%)	Sample	R_f (%)	R_r (%)
Untreated	92.4	79.2			
e-100	91.4	81.9	e-190	92.2	83.3
f-100	92.5	80.7	f-190	92.4	82.5
g-100	89.4	84.0	g-190	90.0	83.0
h-100	89.5	83.4	h-190	87.8	82.2

per the tensile cycles, the first tensile cycle is ignored, and the fixity and recovery ratios are calculated as the mean of the second to the fourth cycles. The average shape recovery and fixity ratios are shown in Table 11.14.

Table 11.14 shows the shape fixity and shape recovery ratios of the specimens after thermal-humidity treatments for 100 and 190 hours, respectively. The shape fixity ratios for specimens that are treated below 50°C are 91.4 to 92.5%, which are close to the 92.4% of the untreated specimen. After the treatments at 80°C, the shape fixity ratios reduce to 89 to 90% (except for specimen h-190). Furthermore, the shape recovery ratios increase from 79 to 81–84% after the thermal-humidity treatments.

11.3.7 Molecular structure

The effects of thermal-humidity conditioning on the chemical structure of PBA-based SMPU fibers were measured by an FT-IR and the results are shown in Figs 11.18 and 11.19. The scanning range was 450 to 4000 cm^{-1} and, on average, 16 scans were taken for each specimen.

Figure 11.18 shows that the intensity of specimen h-190 at 1700 cm^{-1} is stronger than the others, while Fig. 11.19 shows that intensity of specimen h-190 at 3325 cm^{-1} is weaker than the others. The literature indicates that the 3325 cm^{-1} peak corresponds to the hydrogen bonded on the N–H stretch to carbonyl groups, and the 1700 cm^{-1} peak corresponds to the hydrogen bonded to the urethane C=O stretch. Unlike the PCL samples, no absorbance peak is observed at 3650 to 3670 cm^{-1}, which implies the absence of a hydrogen-bonded O–H stretch in the specimens. In other words, no bonded water has formed in the samples.

It is interesting to note that the specimen treated at high temperature and humidity (sample h) has more hydrogen/non-hydrogen bonded urethane C=O groups, but fewer hydrogen groups bonded between the N–H and C=O. If we consider the chemical structure of the basic urethane group (R-NHCO), the above phenomenon can be concluded as the result of hydrogen bond breakage between PU chains under high temperature and humidity. Furthermore, the breakage of hydrogen bonds also reduces the stability of the SMPU, and therefore results in a high percentage of greater weight loss, as shown in the TGA results.

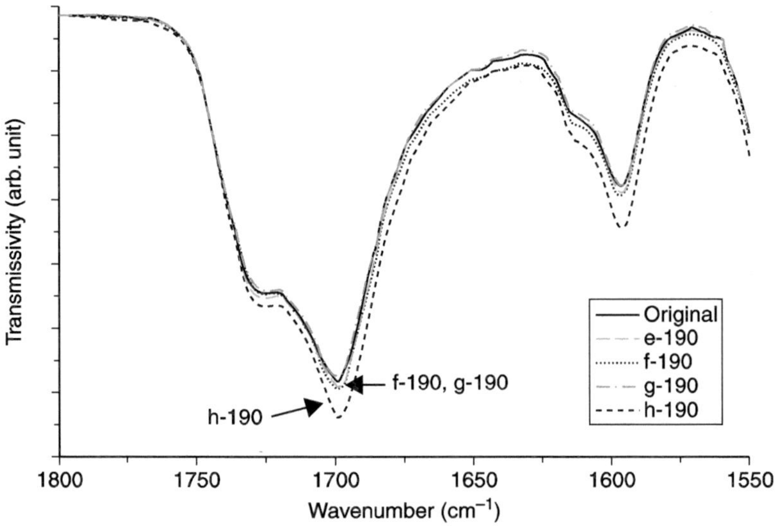

11.18 FT-IR spectra at lower wavenumber for PBA-based SMPU fibers after various conditioning.

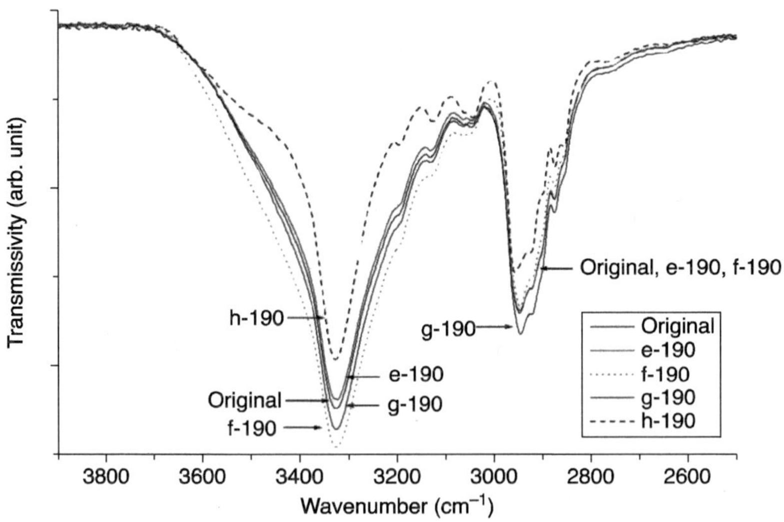

11.19 FT-IR spectra for PBA-based SMPU fibers after various conditioning (2550–3850 cm^{-1}).

11.3.8 Surface morphology

The surface morphology of the SMPU fibers was studied by a Leica Stereoscan 440 SEM. A single fiber of the produced multi-filament SMPU fibers was attached to the SEM sample holder. The accelerated voltage of the electron beam was 20 kV and the secondary electron images were recorded by the CCD system in the SEM.

Some holes appear on the surface of fibers that are treated at 80°C (Figs 11.20 to 11.24), and this distortion is one of the reasons for the reduction of shape fixity ratios in specimens after a high temperature treatment. When this is combined with the FT-IR results, it supports the argument that high temperature and humidity conditions will increase the decomposition rate of PBA-based fibers, and reduce shape fixity ratios.

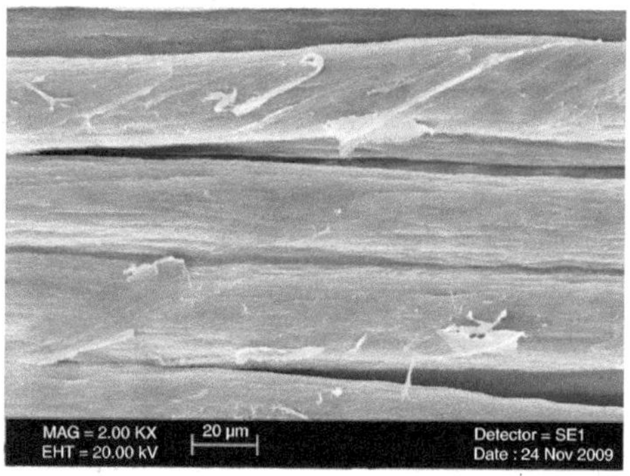

11.20 SEM images of untreated PBA fibers: (left) 500x; and (right) 2000x.

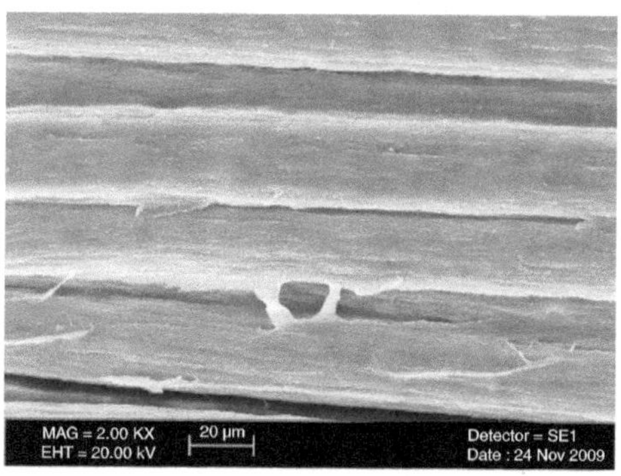

11.21 SEM images after treatment of PBA fibers at 50°C, 30 RH% for 190 hours: (left) 500x; and (right) 2000x.

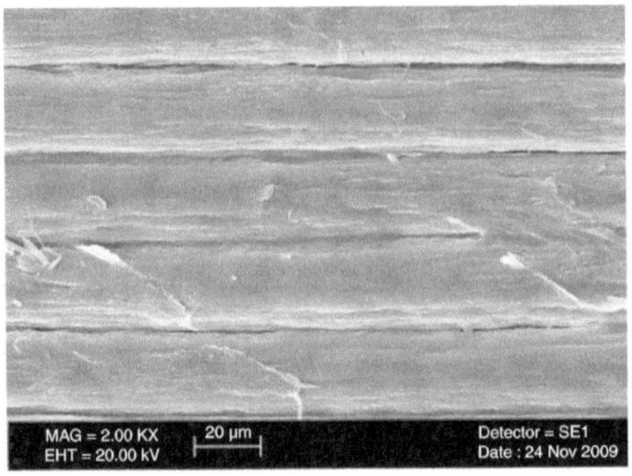

11.22 SEM images of PBA fibers after a treatment at 50°C, 80 RH% for 190 hours: (left) 500x; and (right) 2000x.

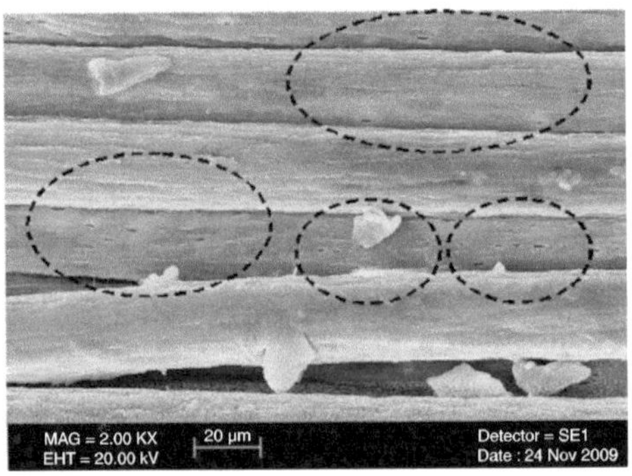

11.23 SEM images of treated PBA fibers at 80°C, 30 RH% for 190 hours: (left) 500x; and (right) 2000x.

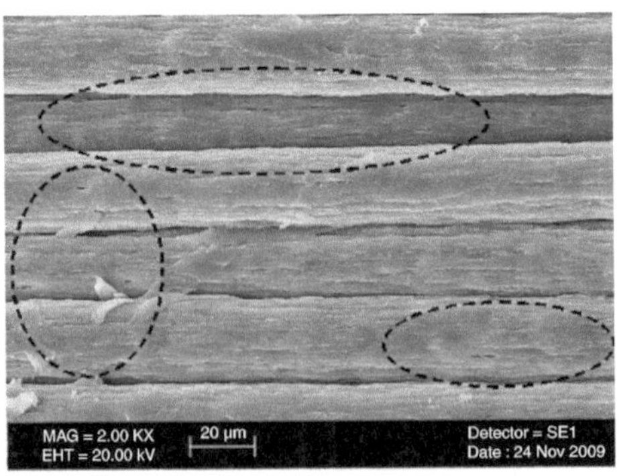

11.24 SEM images of treated PBA fibers at 80°C, 80 RH% for 190 hours: (left) 500x; and (right) 2000x.

11.3.9 Effects of solvent concentration in the water bath

As discussed in the Introduction, the solvent used in the spinning solution diffuses from the solution into the water bath. Therefore, the solvent will accumulate in the water bath. The concentration of an organic solvent such as this plays an important role in both the fiber properties and the environment, with the recycling process of the solvent being particularly significant.

DMF solvent has been in common use over the last decade. However, organizations such as the International Agency for Research on Cancer (IARC) reported that DMF is a carcinogen and results in the potential risk of cancer. Even though this is not proven by the Environment Protection Agency (EPA) in the USA, DMF is banned in some countries for environmental reasons. Therefore, a less toxic solvent, DMAc, is proposed.

When DMAc is used, a simple distillation system will generally suffice for the recycling process. The recycled solvent can be reused to prepare another batch of spinning solution and the cost of the solvent can thus be considerably reduced. The economic benefits of the recycling process primarily depend on the concentration of DMAc within the waste water. When the concentration is higher than a critical value (depending on the scale and efficiency of the recycling system), running such a recycling process is profitable. However, the solvent concentration will also affect the coagulation process and, by extension, the fiber quality (Gupta, 1997; Ziabicki, 1976). As a result, a study on the effect of the solvent concentration in the water bath on fiber properties was conducted.

In this study, a controlled volume of DMAc was added to adjust its concentration in the water bath. Four specimens of PCL-based SMPU fibers were spun through water baths with a DMAc concentration (V/V %) of 1, 2, 4 and 8%, respectively, and were labeled PCL-1%, PCL-2%, PCL-4% and PCL-8%, accordingly.

The breaking tenacity and strain of each SMPU fiber spun in different water baths was tested by an Instron universal tensile tester and the results are shown in Table 11.15. Some typical tensile test results of the above fibers are provided in Fig. 11.25. A significant change in fiber breaking tenacity and strain is observed

Table 11.15 Tenacity and breaking strain of PCL-based SMPU fibers spun in water baths with different concentrations of DMAc

Sample	Mean		Standard derivation	
	Tenacity (cN/dtex)	Breaking strain (%)	Tenacity (cN/dtex)	Breaking strain (%)
PCL-1%	1.426	51	0.154	5.61
PCL-2%	0.663	154	0.047	15.18
PCL-4%	0.550	170	0.027	12.64
PCL-8%	0.621	205	0.019	11.32

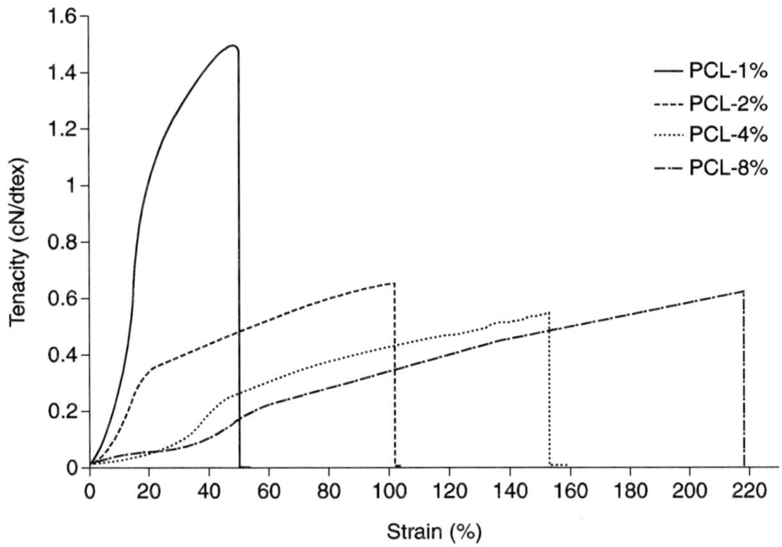

11.25 Typical stress-strain curves for SMPU fibers spun in water baths with various solvent concentrations.

when the concentration of DMAc increases from 1 to 2% (breaking tenacity was reduced from 1.426 to 0.663 cN/dtex, whilst the strain was increased from 51 to 154%). The breaking tenacities of the PCL-based fibers further decreased (0.55 cN/dtex) when the concentration of DMAc in the water bath was increased to 4%, but slightly increased as the concentration of DMAc was increased to 8%. However, the fiber breaking strain further increased with an increase of DMAc (205% for PCL-8%).

The initial modulus of the SMPU fibers spun in different water baths are shown in Table 11.16. The initial modulus of the SMPU fibers decreased from 0.027 to 0.0025 cN/dtex when the concentration of DMAc was increased from 1 to 4%, but remained at 0.0026 cN/dtex for the PCL-8% specimen. With regard to the typical stress–strain curves for the specimens, it was noted that the yield point shifted to the right as the solvent concentration was increased. The above phenomena can be

Table 11.16 Initial modulus of SMPU fibers spun in different water baths

Sample	Modulus (cN/dtex)	Standard derivation (cN/dtex)
PCL-1%	0.0270	0.0048
PCL-2%	0.0120	0.0009
PCL-4%	0.0025	0.0001
PCL-8%	0.0026	0.0003

explained by a reduction in fiber X_c and orientation, as well as the increment of molecular weight.

The melting temperatures and thermal-mechanical properties of the above SMPU fibers were studied by a DMA. The thermal degradation properties were studied by a TGA, as it is also important to test if any solvents had remained within the solidified polymer fibers. The thermal-mechanical properties of the PCL-based, T_m-type of SMPU fiber – which was spun by wet spinning – were also studied by a DMA. The entire set-up was maintained as previously reported, except for the number of fibers that would undergo testing. Bundles of 20 fibers instead of 10 were used due to the low modulus of the fibers.

Table 11.17 illustrates the T_m which is measured by a DMA and the ratio of the modulus |E| at 30°C to |E| at 75°C. The change of tanδ against temperature for the four specimens is shown in Fig. 11.26. The value of tanδ is similar before and

Table 11.17 DMA results of T_m and change of |E| for SMPU fibers spun in different water baths

	T_m (onset of tanδ)	|E(30°C)|/|E(75°C)|
PCL-1%	41.0	213
PCL-2%	43.8	161
PCL-4%	37.3	186
PCL-8%	37.3	75

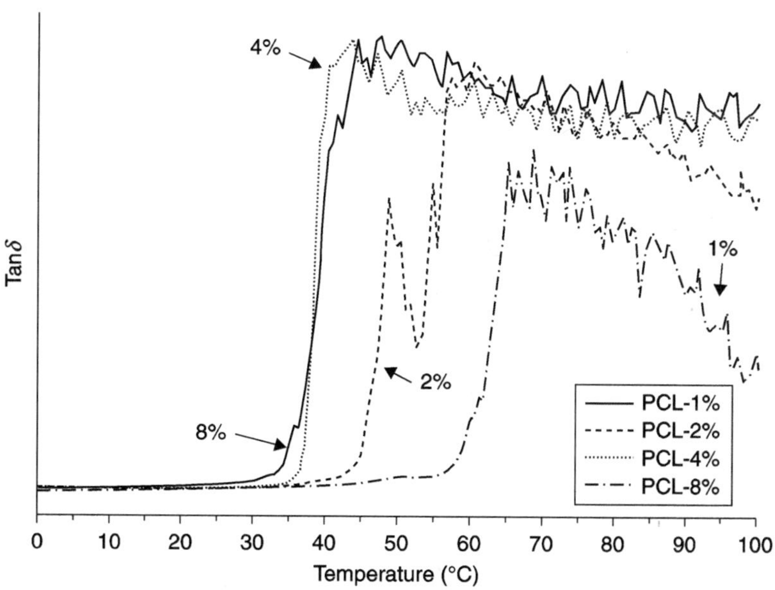

11.26 Tanδ of SMPU fibers spun in various water baths.

after the transition for all specimens, except for PCL-1%. Tanδ is related to the energy disperse in the loading-releasing process where a lower value represents more energy dispersed during a tensile cycle. This results in poor shape memory properties as the energy is most likely to disperse in polymer chain alignments, which produce an irrevocable deformation to the material.

In the TGA analysis, the temperature range was set from 30 to 550°C with a heating rate of 10°C/min (Fig. 11.27). Air was used as the purge gas and the sample weight was in the range of 5 to 10 mg.

The first derivative for the sample weights clearly shows changes in the decomposition process (Fig. 11.28). The first peak positions are 309.2, 319.6, 326.9 and 335.8°C for the four samples, which indicate a higher starting/rate in the decomposition temperature when the solvent concentration is increased from 1 to 8%. The second peaks correspond to the major decomposition process of the SMPU. These peaks are located at similar temperatures (~390°C).

According to the TGA curves (Fig. 11.28), no weight loss occurs when the temperature is lower than 230°C. This is because most of the remaining solvent is washed out in the water bath and, therefore, no solvent evaporation is observed. The onset temperature of the decomposition process (the first drop in the TGA curves) is related to two factors: first, the number of small molecules within the specimen, and second, the strong bonding of some major polymer chains. An increase in the onset temperature of the first weight loss thus indicates either a stronger bond formed, or

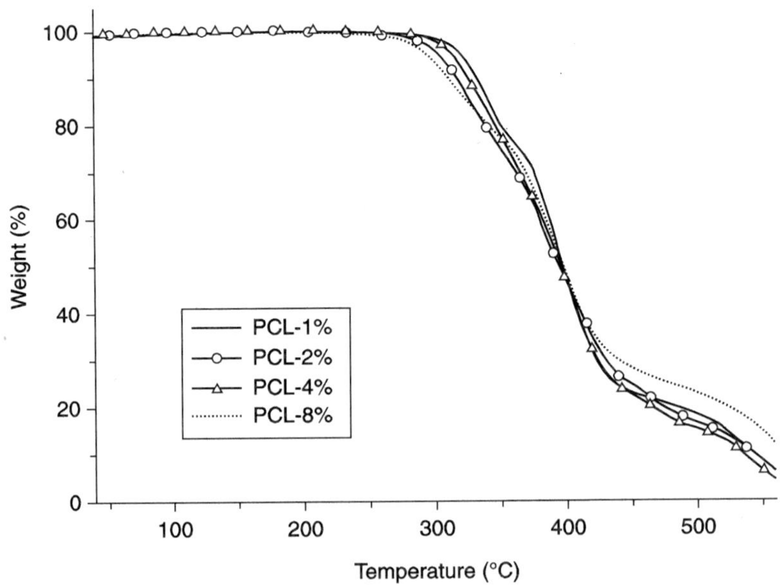

11.27 TGA results of PCL-based SMPU fibers spun in various water baths.

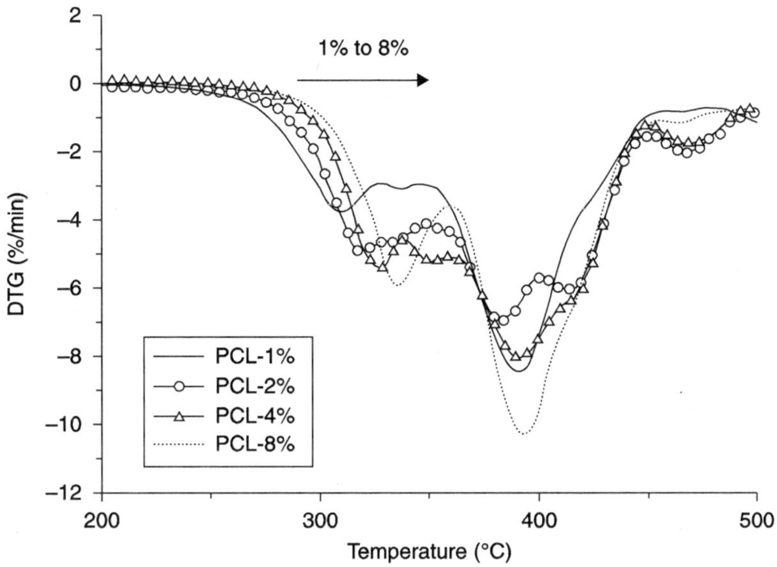

11.28 First derivative of TGA for PCL-based SMPU fibers spun in various water baths.

a number of small polymer fragments, which will decrease when the concentration of DMAc increases. The TGA result will be further discussed with the FT-IR spectra.

The crystal structure and X_c of the SMPU fibers spun in water baths with different concentrations of organic solvent were studied by XRD (Fig. 11.29). The X_c of the four specimens spun in 1, 2, 4 and 8% DMAc are 25.5, 8.2, 10.8 and 10.6, respectively. The huge drop in X_c results in decrements of breaking tenacity and increments of breaking extension (Table 11.18). Considering the changes in X_c of the wet-spun PCL-based SMPU fibers against the melt-spun samples studied, however, a reduction in X_c is not enough to explain the changes in tensile and shape memory properties. Some differences in the chemical structure need to be explained more precisely, and will be discussed later.

Table 11.18 R_f and R_r of PCL-based SMPU fibers spun in different water baths

Sample	R_f (%)	R_r (%)
PCL-1%	72.0	54.3
PCL-2%	90.9	53.6
PCL-4%	86.3	53.9
PCL-8%	89.9	60.4

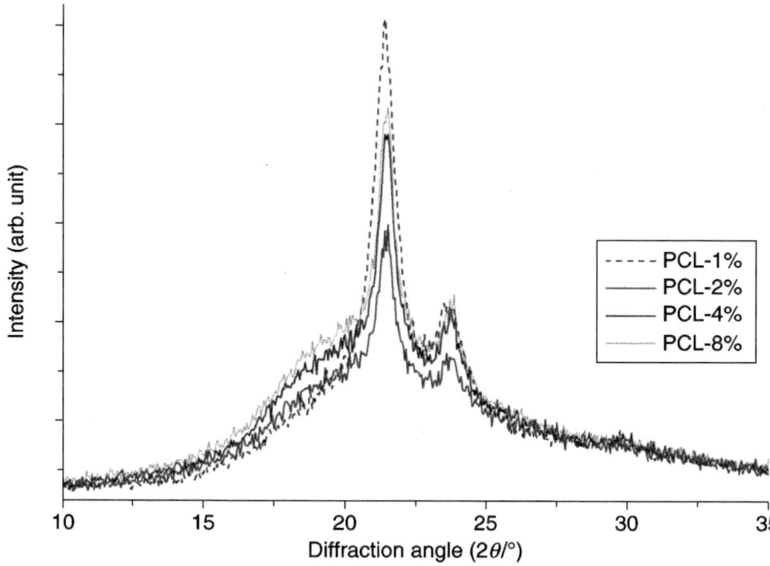

11.29 XRD pattern for SMPU fibers spun in various water baths.

The shape fixity and shape recovery ratios of SMPU fibers spun in water baths with different concentrations of organic solvent are shown in Table 11.18. The testing method and calculations are the same.

As indicated by the shape memory test results, when the concentration of solvent increases, the shape fixity ratio increases rapidly from about 70 to 90%, while the fixity ratio remains unchanged until the concentration is increased to 8%.

Taking the results of the XRD study and the shape memory test together, we observe that the shape memory properties of the PCL-based SMPU fibers increase, and X_c decreases. This result opposes the general understanding of SMEs due to the crystallization process. Therefore, some other major differences between the four fibers spun in water baths with different solvent concentrations are expected, and are shown in the following sections.

The FT-IR spectra of the concerned SMPU fibers are shown in Figs 11.30 and 11.31. The spectra are similar to each other, and only some FT-IR bands of PCL-1% are weaker than others. Band assignments of the FT-IR spectra are shown in Table 11.19. It is expected that the bands at 1410 to 1415 cm^{-1} and 1590 to 1620 cm^{-1} are identical for all specimens, as these two bands correspond to the benzene ring in the MDI structure, which does not take part in any chemical reaction during the wet-spinning process.

Some of the vibration bands increase for fibers spun in a higher solvent concentration. It is interesting to note that most of these bands with increased intensity correspond to the PU structure, while the bands which remain unchanged correspond to raw materials (such as PCL, MDI or BDO).

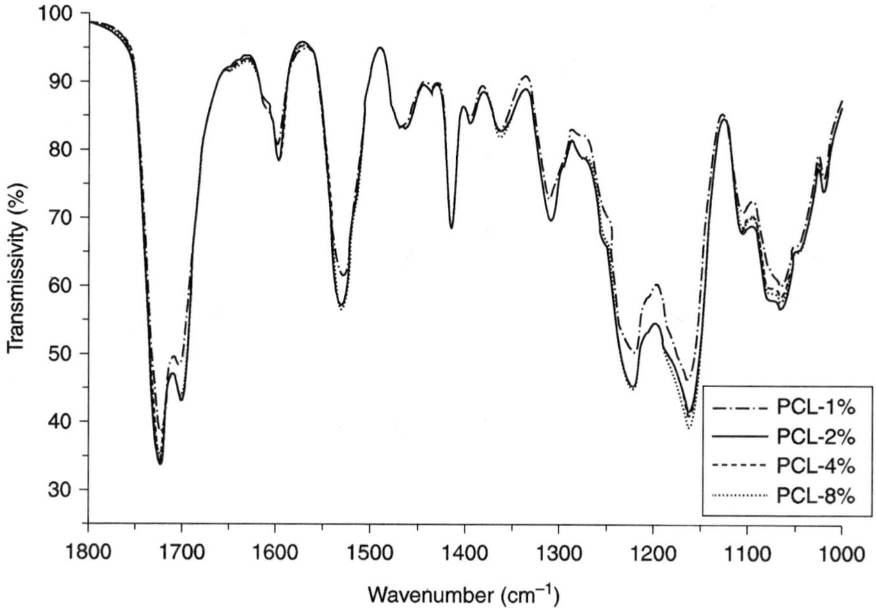

11.30 FT-IR spectrum for fibers spun in various water baths (1000–1800 cm^{-1}).

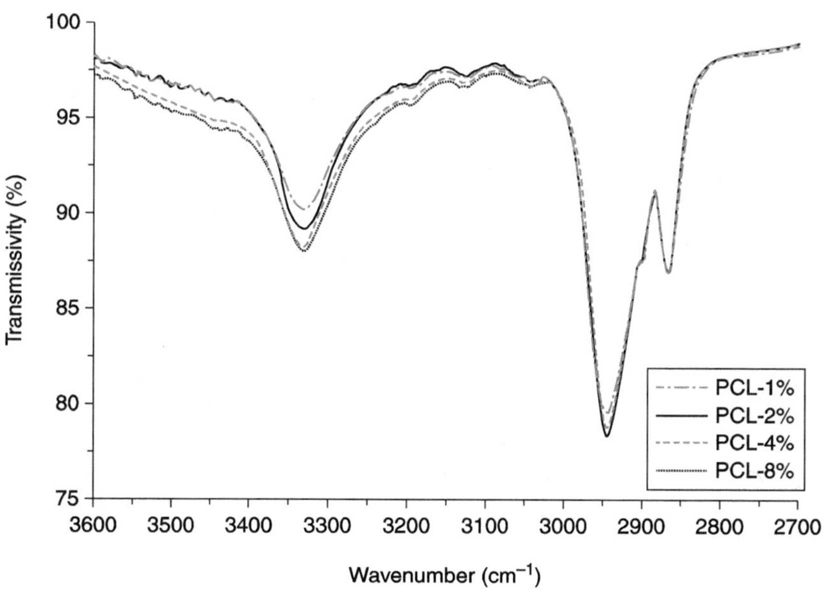

11.31 FT-IR spectrum for fibers spun in various water baths (2700–3600 cm^{-1}).

Table 11.19 Band assignments of FT-IR spectrum

Peak position (cm^{-1})	Assignment	Reference
3320–3335	N–H stretch (hydrogen bonded to carbonyl)	Brunette *et al.* (1982); Christenson *et al.* (1986)
2935–2945	Aliphatic antisymmetric CH stretch	Imada *et al.* (1965)
2790–2800	Aliphatic symmetric CH stretch	Imada *et al.* (1965)
1725–1735	Non-hydrogen bonded urethane C=O stretch	Christenson *et al.* (1986)
1700–1710	Hydrogen bonded C=O stretch	Christenson *et al.* (1986)
1610–1620	C=C stretch benzene ring	Marchant *et al.* (1987)
1595–1600	C–C stretch benzene ring	Christenson *et al.* (1986)
1530–1535	Urethane N–H bending + C–N stretch	Imada *et al.* (1974)
1465–1470	Aliphatic CH2 bending	Imada *et al.* (1965)
1410–1415	C–C stretch, benzene ring	Christenson *et al.* (1986)
1222–1230	C–N stretching	Dillon and Hughes (1987)
1080–1083	Urethane C–O–C stretch and aliphatic symmetric C–O–C stretch	Imada *et al.* (1965); Srichatrapimuk and Cooper (1978)

When the solvent concentration in the water bath increases, diffusion of the organic solvent from the spinning solution into the water bath will become slower. This results in an increased distance from the spinneret to the solidification point. In other words, this will increase the path and time for the solvent diffusion/solidification process from the polymer solution to the solid fiber. The FT-IR results indicate that such an increment of the solidification process enhances the formation of PU, which results in a heightened band strength around the urethane group.

As discussed above, the diffusion process of the organic solvent in the fiber formation process depends on the osmotic pressure between the spinning solution and the water bath. When the water bath has a high solvent concentration, diffusion will become slower, resulting in the fiber becoming smoother. In contrast, when the water bath has a low solvent concentration, a violent diffusion will occur and the fiber surface will become rougher. However, the concentration of the solvent also affects the formation of the droplet shape when the solution is first pulled out from the spinneret. When the concentration is low, the solvent diffuses out quickly and a smaller droplet shape is formed. Conversely, when the concentration of solvent is high, the solvent diffuses out slower and a larger droplet shape is formed. The former results in a straighter fiber, while the latter results in a fiber with many beads. In order to evaluate the effects of the solvent concentration in the water bath on fiber surface morphology, some SEM observations were conducted on the SMPU fibers.

The surface morphology of the four SMPU fibers is shown in Figs 11.32 to 11.35. First, when the concentration of solvent in the water bath is increased, the fiber morphology changes from chains to beads. This unevenness of the fiber cross-section is one of the reasons for it becoming weaker. Second, the surface of individual micro-fibers becomes smoother when the concentration of solvent is increased. The smoothness may be unsuitable for textile applications, as it reduces the surface friction and affects other processes between the spun-fiber and yarn states, such as twisting or fiber bending.

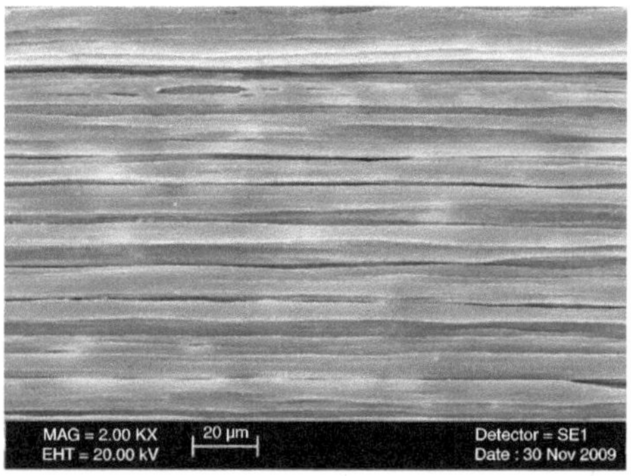

11.32 SEM image of PCL = 1%.

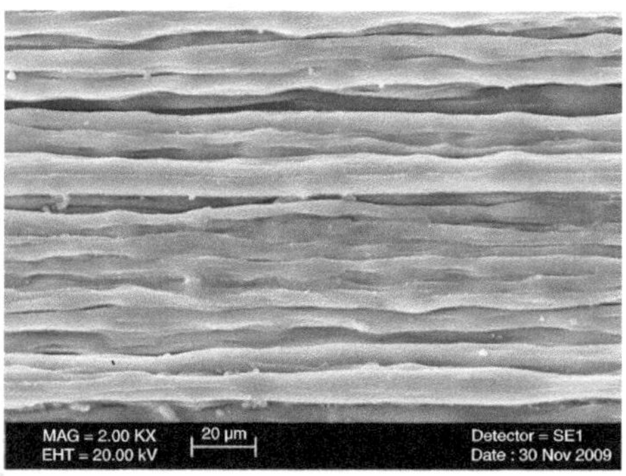

11.33 SEM image of PCL = 2%.

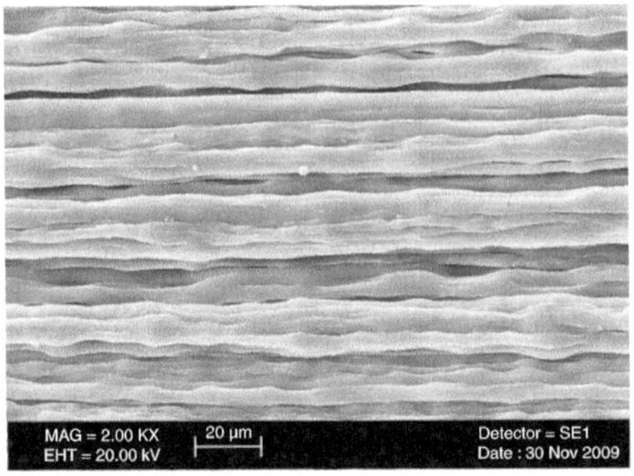

11.34 SEM image of PCL = 4%.

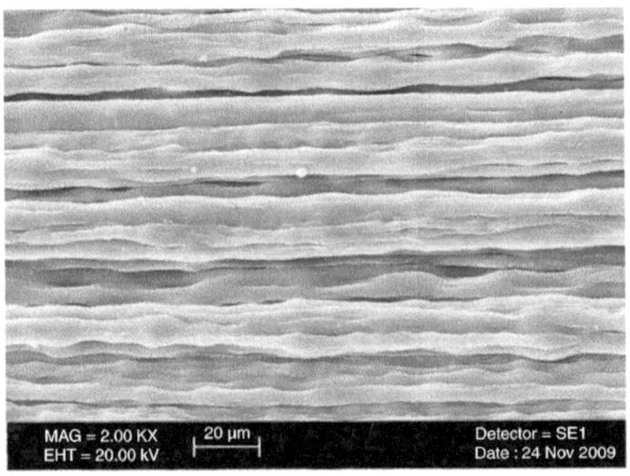

11.35 SEM image of PCL = 8%.

11.4 Summary

In the study summarized at the beginning of this chapter, a T_m-type SMF was fabricated by using the SMPU. The switching transition temperature of the SMF was the PCL segment phase melting temperature at 47°C. The partially crystallized PCL segment phase provided the SMF with partial elasticity at ambient temperature, and the ability to fix the temporary length once the fiber was cooled to ambient temperature. The SMF could recover its original length by reheating the fiber above 47°C, at which temperature the crystallized soft segments melt.

Also studied were PBA- and PCL-based SMPU fibers, spun by the wet-spinning process. The effects and mechanisms of thermal-humidity treatments on the shape memory properties of PBA-based SMPU fibers were examined based on mechanical, thermal and chemical analyses. The experimental results show that PBA-based SMPU fibers are not as sensitive to water as PCL-based fibers. The reduction of shape memory properties after high-temperature conditioning is probably a result of heat shrinkage in the radial direction of the multi-filament fiber, and will not occur in PCL-based melt-spun fiber as it does a single fiber.

Even though melt spinning has many advantages over wet spinning, the control of the viscosity of the melt flow within the spinning system is a major difficulty which researchers have yet to overcome. Unlike traditional PET, PU has a complex reaction in high temperature melting states, which results in a temperature–time dependent melt-flow index. This complexity is a major challenge for the industry, as it highlights the need to fine tune the chemical structure and spinning conditions. Therefore, it is still necessary to consider the possibility of using wet spinning to produce T_m-type SMPU fibers.

11.5 References

Brunette, C. M., Hsu, S. L., Macknight, W. J. (1982), Hydrogen bonded properties of hard-segment model compounds in polyurethane block copolymers. *Macromolecules*, **15**(1), 71–7.

Cherubini, J. (1995), *Polymeric orthotic devices*, US Patent 5,415,623.

Christenson, C. P., Harthcock, M. A., Meadows, M. D, Spell, H. L., Howard, W. L., Creswick, M. W., Guerra, R. E., Turner, R. B. (1986), Model MDI/butanediol polyurethanes: Molecular structure, morphology, physical and mechanical properties. *J Polym Sci: Part B: Polym Phys*, **24**(7) 1401–39.

Crescenzi, V., Manzini, G., Calzolari, G., Borri, C. (1972), Thermodynamics of fusion of poly-β-propiolactone and poly-ε-caprolactone. Comparative analysis of the melting of aliphatic polylactone and polyester chains, *Eur Polym J*, **8**, 449–63.

Dillon, J. G., Hughes, M. K. (1987), *Chemical Characterization of Polyurethane I. GPC, FTIR, and DSC Data*. Rockville, MD: Food and Drug Administration.

Ferguson, J., Petrovic, Z. (1976), Thermal stability of segment polyurethane, *Eur Polym J*, **12**(3), 177–81.

Fourne, R. (1999), Compact melt spinning process for the flexible production of elastic yarns, *Chemical Fibers International*, **49**, 124–5.

Gupta, V. B. (1997), *Manufactured Fiber Technology*, first edn, London: Chapman & Hall.

Hu, J. L., Ji, F. L., Wong, Y. W. (2005a), Dependency of the shape memory properties of a polyurethane upon thermomechanical cyclic conditions, *Polym Int*, **54**, 600–5.

Hu, J. L., Yang, Z. H., Ji, F. L., Liu, Y. Q. (2005b), Cross-linked polyurethanes with shape memory properties, *Polym Int*, **54**, 854–9.

Huang, W. H., Yang, B., An, L., Li, C., Chan, Y. S. (2005), Water-driven programmable polyurethane shape memory polymer: Demonstration and mechanism, *Appl Phys Lett*, **86**, 114–15.

Hyashi, S. (1993), Properties and applications of polyurethane, *International Progress in Urethanes*, **6**, 90–115.

Imada, K., Todokoro, H., Umehara, A., Murahashi, S. (1965), Normal vibrations of the polymer molecules of helical conformation. VI. Polytetrahydrofuran and deuterated polytetrahydrofurans. *J. Chem. Phys*, **42**(4), 2807–16.

Imada, K., Tadokoro, H., Umehara, A., Murahashi, S. (1974), Infrared studies on segmented polyurethane-urea elastomers. *J Macromole Sci B Phys*, **10**(4), 591–618.

Kim, B. K., Lee, S. Y. (1998), Polyurethane ionomers having shape memory effects, *Polymer*, **39**, 2803–8.

Kim, B. K., Lee, S. Y., Xu, M. (1996), Polyurethane having shape memory effect, *Polymer*, **37**, 5781–93.

Koberstein, J. T., Leung, L. M. (1992), Compression-molded polyurethane block copolymers. 2. Evaluation of microphase compositions. *Macromolecules*, **25**, 6205–13.

Krol, P. (2008), *Linear polyurethanes: Synthesis methods, chemical structures, properties and applications*, Leiden: Boston VSP.

Lendlein, A., Kelch, S. (2002), Shape-memory polymers, *Angewandte Chemie International Edition*, **41**, 2034–57.

Li, F. K., Zhang, X., Hou, J. A., Xu, M., Luo, X. L., *et al.* (1996), Studies on thermally stimulated shape memory effect of segmented polyurethanes, *J Appl Polym Sci*, **64**, 1511–16.

Lin, J. R., Chen, L. W. (1998), Study on shape-memory behavior of polyether-based polyurethanes, Part II: Influence of soft-segment molecular weight, *J Appl Polym Sci*, **69**, 1575–86.

Liu, J., Ma, D. Z., Li, Z. (2002), FT-IR studies on the compatibility of hard-soft segments for polyurethane-imide copolymers with different soft segments, *Eur Polym J*, **38**, 661–5.

Liu, Y., Chung, A., Hu, J. L., Lu, J. (2007), Shape memory behavior of SMPU knitted fabric, *Journal of Zhejiang University SCIENCE A*, **8**, 830–4.

Luo, N., Wang, D. N., Ying, S. K. (1996), Crystallinity of hard and hydrogen bonding segments in segmented poly(urethane urea) copolymers, *Polymer*, **37**, 3577–83.

Marchant, R. E., Zhao, Q., Anderson, J. M., Hiltner, A. (1987), Degradation of poly(ether-urethane-urea) elastomer: Infrared and XPS studies. *Polymer*, **28**(12), 2032–9.

Mondal, S., Hu, J. L. (2007), Influence of hard segment on thermal degradation of thermoplastic segmented polyurethane for textile coating application, *Polymer-Plastic Technology and Engineering*, **46**(1), 37–41.

Pan, G. H., Huang, W. M., Ng, Z. C., Liu, N., Phee, S. J. (2008), The glass transition temperature of polyurethane shape memory polymer reinforced with treated/non-treated attapulgite (playgorskite) clay in dry and wet conditions, *Smart Materials and Structures*, **17**(4), 045007.

Petrovic, Z.S., Zavargo, Z., Flynn, J.H., Macknight, W.J. (1994), Thermal-degradation of segmented polyurethanes, *J Appl Polym Sci*, **51**(6), 1087–95.

Ping, P., Wang, W., Chen, X., Jing, X. (2005), Poly(ε-caprolactone) polyurethane and its shape-memory property, *Biomacromolecules*, **6**, 587–92.

Schoonover, J. R., Thompson, D. G, *et al.* (2001), Infrared linear dichroism study of a hydrolytically degraded poly(ester Urethane). *Polymer Degradation and Stability*, **74**, 87–96.

Srichatrapimuk, V. W., Cooper, S. L. (1978), Infrared thermal analysis of polyurethane block polymers. *J Macromole Sci B Phys*, **15**(2), 297–311.

Tang, S.L.P., Stylios, G.K. (2006), An overview of smart technologies for clothing design and engineering, *International Journal of Clothing Science and Technology*, **18**, 10828.

Tobushi, H., Hashimoto, T. N., Hayashi, S., and Yamada, E. (1998), Shape fixity and shape recovery in a film of shape memory polymer of polyurethane series, *J Intell Mater Syst Struct*, **9**, 127–36.

Wischke, C., Neffe, A. T., Steuer, S., Lendlein, A. (2009), Evaluation of a degradable shape-memory polymer network as matrix for controlled drug release. *Journal of Controlled Release*, **138**, 243–50.

Yang, B., Huang, W. M., Li, C. (2006), Effects of moisture on the thermomechanical properties of a polyurethane shape memory polymer, *Polymer*, **47**(4), 1348–56.

Yang, J. H., Chun, B. C., Chung, Y.-C., Cho, J. H. (2003), Comparison of thermal/mechanical properties and shape memory effect of polyurethane block-copolymers with planar or bent shape of hard segment, *Polymer*, **44**, 3251–8.
Yao, M., Zhou, J., Huang, S., Shao, L., An, R., Fan, D. (1990), *Textile materials*, 2nd edn., Peking: Textile Industry Publishing House.
Yoshihaya, N., Ishihara, H., Yamada, T. (2003), Relationship between segment structutures and elastic properties of segmented poly(urethane-urea) elastic fibers, *Polym Eng Sci*, **43**, 1740–6.
Zhang, Z. R. (2004), *Shape memory bone fixing splint*, China Patent CN20020150720.
Ziabicki, A., 1976. *Fundamentals of Fiber Formation*, New York: John Wiley & Sons.

12
Future developments in shape memory polymers

DOI: 10.1533/9780857098542.320

Abstract: In spite of the many achievements already accomplished in the field of shape memory polymers (SMPs), there are many challenging problems still to overcome. There is much scope to improve the shape memory properties and provide SMPs with advanced functions. Furthermore, extensive and intensive application research of SMPs has only just begun. This chapter puts forward the future work which authors currently believe to be promising or even crucial in the area of SMPs.

Key words: model study, reversible chemical cross-links, gas-sensitive shape memory effect, medical applications.

12.1 Introduction

The influences of structure and thermo-mechanical conditions on shape memory properties are very complex, especially for T_g-shape memory polyurethanes (SMPUs). The establishment of a model to correlate microstructure and thermo-mechanical conditions with the shape memory properties of T_g-SMPUs is significant. Some attempts have been made (Buckley *et al.*, 2007; Chen and Lagoudas, 2008; Li and Larock, 2000; Liu *et al.*, 2006; Morshedian *et al.*, 2005; Tobushi *et al.*, 2001), but less significant progress has been achieved in modeling the shape memory behavior of polymers, especially in cases of large deformation.

Tobushi *et al.* (2001) presented an empirical model based on the observation of the thermo-mechanical behavior of shape memory polymers (SMPs). In the model, the modulus of the SMPs was considered to linearly decrease with the temperature increase in a narrow region around T_g ($T_g - T_w < T < T_g + T_w$). The stress vs strain profiles in shape memorization were predicted based on a non-linear visco-elastic equation. This model mainly took account of the modulus variation above and below T_g, but noted no further microscopic changes in the shape memorization (Tobushi *et al.*, 2001).

Liu *et al.* (2006) developed a model based on SMPs comprising frozen and active phases, whose variations with temperature account for the shape memory effect (SME) of polymers (Buckley *et al.*, 2007; Chen and Lagoudas, 2008; Liu *et al.*, 2006). They excluded the time effects in the model. The model was, moreover, largely phenomenological and attributed less to the microscopic changes, although the predictions largely coincided with the experimental results.

Buckley *et al.* (2007) demonstrated the shape memory behaviors of polyurethanes based on a linear, visco-elastic model, in which the SME was ascribed to the variation of molecular mobility around the T_g. They obtained the retardation spectrum via micro creep experiments under varying temperatures, and predicted the shape memory behavior. However, the linear model is only appropriate in the case of small deformations. One advantage of SMPs in comparison with shape memory alloys is the SME with large deformation. In some experiments, it was found that the deformation amplitude had significant influences on shape memory behaviors.

Therefore, a model suitable for predicting shape memory behaviors in cases of large deformation should be developed in the future. The model should take account of the effects of temperature, time and deformation on molecular mobility in the course of shape memorization. However, it is reportedly challenging to model the non-linear visco-elastic and visco-plastic behaviors by taking into account all three of these factors simultaneously. The effect of deformation on molecular mobility in particular represents a frontier field of polymer physics.

12.2 T_m-shape memory polyurethane (SMPU) with varying T_m

Many investigations have been made into T_m-SMPUs and much effort has been made to increase the strength of physical cross-links or to incorporate chemical cross-links. However, the shape memory switch temperature could not be adjusted as flexibly as with T_g-SMPUs, because the types of commercially available polyester or polyether diols still able to crystallize in polyurethane copolymers are somewhat limited. The development of a series of polyols with varying T_m would represent a significant future achievement. Via copolymerization, the polyols with varying T_m can be obtained by selecting different monomers and varying their compositions.

12.3 T_g-SMPUs with thermally reversible chemical cross-links

The phase separation degree of T_g-SMPUs is lower than that of thermoplastic polyurethanes (TPUs) and T_m-SMPUs. The higher compatibility of the soft and hard segments of T_g-SMPUs results in the phase separation and thus T_g can be readily influenced by thermal processes. It was discovered that the T_g of T_g-SMPUs could be increased significantly after extrusion from a slit die, leading also to a significant change in the mechanical properties. The lower stiffness difference between soft phase and hard phase in T_g-SMPUs, moreover, is unfavorable for the polymers showing a narrow shape recovery region. Replacing the hard domains with chemical cross-links is a potential solution for the problem, but this will almost certainly lower processability. It has been well-established that some chemical cross-links are thermally reversible, so incorporation of the thermally-reversible chemical cross-links would improve the shape memory properties of T_g-SMPUs, without sacrificing their good thermal processability.

12.4 Two-way shape memory fibers

The SME of the shape memory fibers (SMFs) developed in the project is one-way. This means that the shape change of the SMF is not reversible, and cannot be repeated without using some external force. Several technologies have been reported to develop SMPs with two-way SME. Chung *et al.* (2008) reported a two-way SME in cross-linked shape memory poly(cyclooctene). Cooling-induced crystallization under a tensile load results in elongation, and subsequent heating melts the contracting network yields (shape recovery). However, this does not avoid the problem that a tensile load is necessary for the two-way SME. Furthermore, the two-way SME is difficult to control.

Ahir *et al.* (2006) and Qin and Mather (2009) showed two-way SME through an anisotropic chain conformation change of liquid-crystalline polymers. Unfortunately, this requires a temperature of above 100°C, which makes the material unsuitable for clothing applications. Furthermore, it is difficult to tailor the shape recovery properties of the SMPs for specific applications.

Shaojen *et al.* (2008) research group has achieved two-way SME on SMP composites by laminating an SMP with an elastic polymer. The two-way SME was ascribed to the release of elastic strain of the SMP layer upon heating, and the recovery of elastic strain induced by the bending force of the elastic polymer layer upon cooling. We are also developing bi-component fibers with one component as a SMP and another component as an elastic fiber, and studying the two-way SME of the conjugate fibers.

12.5 Gas-sensitive shape memory BINA-HDI copolymers (PUPys)

The hitherto most widely studied SMEs are thermally-induced, although there have also been some reports on light-induced, magnetic-induced and water-influenced SMEs. In the present study, the thermally-induced SME and moisture-sensitive SME are both observed in the BIN-SMPUs. In this system, the non-covalent molecular switch controls the shape recovery and shape fixation in both the SMEs. The dynamic feature of the non-covalent bond is responsive to the external stimulus.

According to the principle of dynamic combinatorial chemistry or constitutional dynamic chemistry, a non-covalent bond such as the hydrogen bond will be replaced by a more adaptive non-covalent bond after the application of an external stimulus. It is known that the moisture is the water gas or water vapor. Similarly, by applying a chemical gas to the BIN-SMPUs, the newly formed hydrogen bonds between the pyridine rings and chemistry gas, for example acetic acid vapor, will interrupt the original hydrogen bonding. Hence, a chemical gas-sensitive SME can be expected in the BIN-SMPUs.

A preliminary study shows that after the deformed film of BIN-SMPU is immersed in the acetic acid vapor condition, the temporary shape will recover its original form within a few minutes. The recovery process is similar to that of the moisture-sensitive shape recovery process. Since the acetic acid vapor sensitive SME can be achieved in the BIN-SMPU, the future study of the acetic acid vapor-sensitive SME will be very interesting.

12.6 Chemically cross-linked PUPys

According to the nature of net points, SMPs are divided into physically cross-linked SMPs and chemically cross-linked SMPs. Generally, the chemically cross-linked SMPs show a high shape recovery and a higher recovery force. To synthesize the physically cross-linked SMPs, many kinds of chemically cross-linked SMPs are also currently being studied.

In our research, the physically cross-linked pyridine-containing polyurethanes (PUPys) were carefully examined. Although many unique properties have been developed in physically cross-linked PUPys, some properties need to be improved. For example, the final shape recovery under the higher moisture condition or water condition is not sufficiently efficient; shape recovery force is still low in the PUPys with higher N,N-bis(2-hydroxyl ethyl) isonicotinamide (BINA) content. According to Chen *et al.* (2009), it has been noted that the rubber modulus can be improved significantly by means of chemical cross-linking. As a result, the hard phase, acting as net points, should be more stable in the PUPys with chemical net points. Consequently, it is expected that the water-influenced shape recovery can be improved by the chemical cross-linking. Since the study on the cross-linked PUPys is interesting, many new properties are expected to be developed in cross-linked PUPys.

12.7 Multi-stimuli responsive shape memory fibers

At present, the SMFs are responsive only to one specific stimulus signal, namely the thermal stimulus. By incorporating hydrophilic segments into the SMPU, a water/moisture active SME was obtained (Chen *et al.*, 2009; Huang and Yang, 2005). Likewise, by incorporating conductive particles into SMPs, electro-responsive SMPs were prepared (Leng *et al.*, 2009; Meng and Hu, 2010), while light-responsive SMPs have also been developed by copolymerizing light-responsive azobenzene moiety into the SMPs (Lendlein *et al.*, 2005). In future, SMPs with multi-stimuli responsive functions should be developed by integrating the different stimuli-response materials into one polymer, thereby allowing the SMFs to adapt themselves to different environmental conditions with varying temperature, light, humidity, electricity signals and pH value, etc. .

Figure 12.1 shows a smart shirt known as the 'Sphere React Shirt' (Nike, 2007), whose rear vents can open up to allow perspiration and heat to escape when the wearer perspires. The vents automatically close at the dry state. It can be particularly effective to use thermal/moisture-dual responsive shape memory textiles which can change macro-shape or microstructures to achieve functions for moisture and heat management between skin and fabric. For example, if the wearer begins to

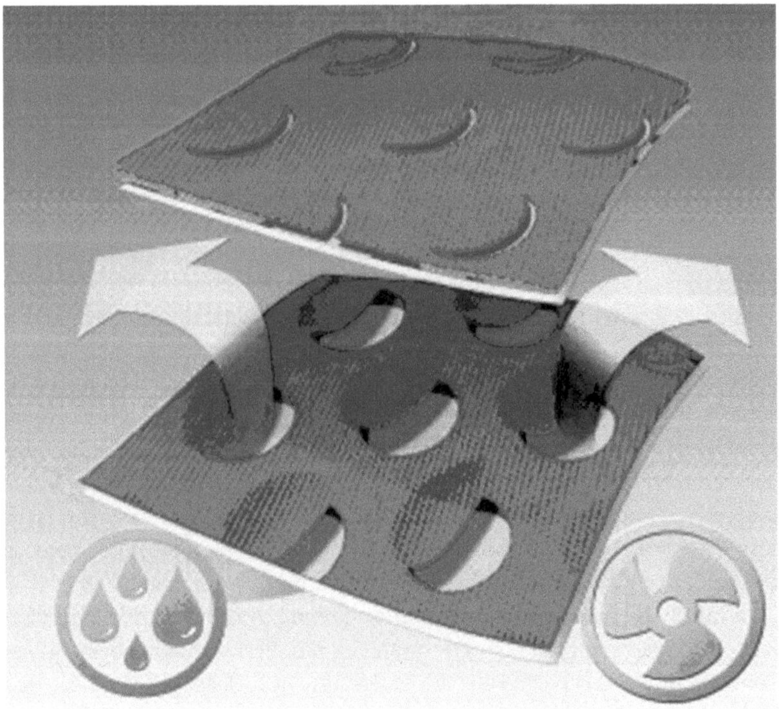

12.1 The back of the Nike 'sphere react shirt' with a smart vent structure (reprinted with kind permission of Nike (2007) company, www.nike.com).

feel hot and starts sweating after excessive muscular activity, the windows in the fabric will open to allow heat and moisture to escape into the environment. In a cold situation, without sweating, the windows close and keep the wearer warm.

12.8 PUPys polymer blends with other polymers

Polymer blending provides an easy method to fabricate new kinds of SMP networks. For example, Zhang *et al.* (2009) prepared a novel tyrene-butadiene-tri-block copolymer (SBS) and polycaprolactone (PCL) blend for the utilization of shape memory materials (SMMs). The phase separation and macromolecular miscibility are important in these blends. In addition, the specific interactions, particularly hydrogen bondings, are of great importance to a large number of polymer-blend systems. Pyridine containing polymers provide a large fraction of hydrogen-acceptor, which can in turn form strong hydrogen bonds with hydrogen-donors such as the carboxyl and phenol groups. For example, poly(2-vinylpyridine) (P2VP), which is inherently weakly self-associated, can form a strong association with polymer containing methacrylic acid groups (Lee *et al.*,

1988). Another example is the miscible polymer blends of poly(styrene-co-methacrylic acid) with copolymers containing vinylpyrrolidone and vinylpyridine groups, which are prepared using hydrogen bonding.

As for the BIN-SMPUs in the present study, the excellent thermally–induced SMEs can be achieved based on the association–dissociation of hydrogen bonding. BIN-SMPUs such as PUPyBDO53 contain a large fraction of pyridine rings, which can provide a large number of hydrogen-acceptors. This allows PUPys to form miscible polymer blends with the polymer-containing methacrylic acid groups. Based on the supramolecular structure of the polymer network after the addition of PUPys, high-quality SMEs could be achieved in polymers possessing no previous SME properties. Other characteristics such as mechanical and damping properties can also be improved in the new polymer blends, meaning the preparation and characterization of SMP blends with PUPy, and the investigation of the interaction of PUPys polymer blends, are particularly important developments.

12.9 Supramolecular liquid crystalline shape memory polymers

Supramolecular liquid crystalline polymers (LCPs) are one type of polymer currently under intensive study. The hydrogen-bonded polymer complex has been particularly easy to achieve in the polymer containing a pyridine ring. For example, the hydrogen-bonded liquid crystalline polyurethane complexes with 4-dodecylcoylbenzoic acid have been carefully studied by Ambrozic and colleagues (Ambrozic *et al.*, 2002; Ambrozic and Zigon, 2005a). The self-organized comb, copolymer-like system is obtained by hydrogen bonds existing between poly (4-vinylpyridine) (P4VP) and 4-nonadecylphenol.

LCPs are capable of forming regions of highly ordered structure, and many unique properties can be developed in liquid crystal elastomers (LCEs). A mesomorphic structure, for example, is observed in the flexible polymer-surfactant systems, due to the presence of hydrogen bonds between the P4VP and the pentadecylphenol. Furthermore, a two-way SME was also reported in the tri-block LCEs, leading to an increasing amount of interest in LCPs in recent years. LCPs or LCEs can provide a route to nano-scale research, as they are an effective polymer for many functional materials.

In the present study, the BIN-SMPUs used contain a large fraction pyridine ring, with strong hydrogen bonds among the urethane groups. A polymer complex can thus be formed in the pyridine ring with the addition of 4-dodecylcoylbenzoic acid. At the same time, the hydrogen bonds present in the urethane group can provide the polymer networks for the formation of a liquid crystal phase. Since it is similar to supramolecular LCEs, this system would be expected to carry some unique properties. However, SMEs are still kept in these supramolecular LCEs. Through the polymer complex, the liquid crystalline properties combine with the BIN-SMPUs, making the study of supramolecular LCEs significant for future development.

12.10 Main-chain pyridine-containing SMPUs

The main-chain type pyridine containing SMPU has so far not been fully investigated. In future studies, the pyridine ring will be attached to the polymer backbone as a pendant. Based on the hydrogen-bonded supramolecular structure, the BIN-SMPUs will have many unique advantages due to their shape memory, mechanical and damping properties.

It was reported that the main-chain type of PUPys containing the pyridine ring on the backbone could be synthesized with 2,6-bis(hydroxylmethyl) pyridine (Ambrozic *et al.*, 2002; Ambrozic and Zigon, 2005a,b). The polymer chain of the main-chain type of PUPys is more regular than that of the side-chain type. Accordingly, the supramolecular structure and morphology will differ significantly from the side-chain type of PUPys. Thus, we would again expect it to show some other unique properties or better SMEs. The investigation of properties will reinforce the understanding of the supramolecular SMPs. In future studies some unique properties and new applications are expected to be found in this new kind of main-chain type of PUPys.

In conclusion, the present study related to the supramolecular SMPs is important for future development. The current investigation, focusing on the BIN-SMPUs, is just the beginning of the research. It is believed that the present work in this field could initiate ever more research directions with regard to SMPs, which will in turn promote the application of SMPs to a greater extent.

12.11 Applications

Being one of the most attractive smart materials, the applications of SMPs cover various areas of everyday life, such as smart fabrics and heat-shrinkable tubes and films, as well as more specialized applications such as self-deployable devices in spacecraft, self-disassembling mobile phones, intelligent medical devices or implants for minimally invasive surgery, thermally-reversible recording, and temperature sensors and actuators (Behl and Lendlein, 2007; Hu *et al.*, 2008; Huang and Yang, 2005; Jeong *et al.*, 2001). In previous studies, the potential applications of the smart breathable textiles in areas such as controlled release, food packaging and gas separation were comprehensively reported by Hu and colleagues (Alteheld *et al.*, 2005; Hu *et al.*, 2008). There is also a wealth of literature on the applications of SMPs in medicine (Fenix *et al.*, 2007; Lendlein and Behl, 2009; Lendlein and Langer, 2002; Sokolowski *et al.*, 2007).

12.11.1 Re-shape applications

It is known that SMPs have the ability to remember their original shapes, i.e., the temporary shape can recover to a permanent shape by applying an external stimulus. In the case of the one-way SME, the permanent shape is the remembered shape, while the temporary shape cannot be remembered. Generally, the permanent shape can be prepared by the conventional process, whereas the temporary shape is achieved by deformation, usually at high temperature. To give an example, permanent shapes are molded from a melt polymer at a high temperature, or from a polymer solution followed by the evaporation of solvent. In order to achieve the SME, the permanent shape should be deformed by increasing the temperature to above the transition temperature, or deformed at room temperature with a higher external force. However, such treatment may harm the potential application of the SMPs.

In our research, it was observed that the BIN-SMPUs show excellent moisture absorption and moisture-sensitive SMEs. Higher moisture absorption will result in the decrease in modulus. This is because the absorbed moisture molecules can destroy the original hydrogen bonds present in both the soft phase and hard phase, particularly under wet conditions, i.e. under water or in an environment with higher relative humidity (RH) levels. As a result, the original shape, shape A, can be deformed easily to a second shape, shape B, when the polymer becomes soft by conditioning in water at room temperature.

The second shape can also be fixed after it is dried. Since the hydrogen bonds acting as physical net points in the hard phase can also be destroyed using higher RH or water, the polymer networks can be reassembled. When the polymer is dried under the second shape, the newly associated hydrogen bonds present in the hard phase will form new physical net points to provide a new polymer network. However, the newly formed hydrogen bonds present in the soft phase will once again serve as the molecular switch. Finally, the second shape can be remembered as a permanent shape. When the second shape is deformed to any other shape, including shape A at a temperature above the transition temperature of soft phase (T_{trans}) but below the transition temperature of hard phase, the new temporary shape can be obtained and fixed by cooling down to below T_{trans}. Thereafter, the second shape, for example shape B, can be recovered in the thermal recovery process. This technology provides an easy method to prepare shape memory products with any potential shape.

The opening process of a reshaped flower is presented in Fig. 12.2. A shape memory flower with eight petals is made from a flat BIN-SMPU film prepared by solution casting. The initial shape is an opened flower. However, by immersion in the water or keeping it wet at RH = 95% and T = 40°C, the petals of the flower become soft within one hour. The initial opened shape is then deformed to a closed shape by applying an external force. By holding the closed shape, the flower is then dried in an oven at 50°C for 24 hours to remove the absorbed moisture, leaving the closed shape memory flower as the new permanent shape. After it is deformed to an opened shape at 80°C and fixed at ambient temperature, the opened shape will become a new temporary shape. The shape memory flower will, upon heating, demonstrate the self-closing process of real petals, as shown in

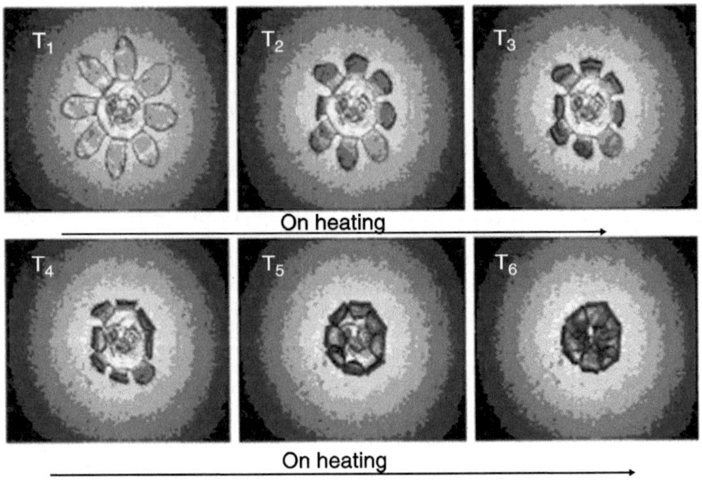

12.2 Shape memory flower closing process upon heating: $T_1 = 25°C$; $T_2 = 35$ various; $T_3 = 40°C$; $T_5 = 50°C$; and $T_6 = 80°C$.

Fig. 12.2. As the temperature increases from $T_1 = 25°C$ to $T_6 = 80°C$, the petals are observed to close step by step. The opened flower eventually completely recovers the closed flower shape.

12.11.2 Two-way shape memory polymer laminates

The two most common SMEs observed in SMMs are the one-way SME and the two-way SME. In general, the one-way SME, when cooled from high temperatures, does not cause a macroscopic shape change. However, the SMMs with the two-way SME can remember two different shapes: one at lower temperatures and another at higher temperatures.

Of all the various kinds of SMMs, it is SMPs which have found wide industrial applications, due to their better versatility of chemical structure, lower cost, easier pretreatment procedure, larger deformation and lower recovery temperature. Since most of the SMPs only show one-way SMEs, researchers are actively developing two-way shape memory properties in the LCEs and the photo-deformation polymers.

In the LCEs, higher deformation of up to 300% can be obtained, but it is still difficult to apply LCEs to industrial applications due to their high manufacturing cost and unstable SME. The photo-deformation polymers, meanwhile, show very low shrinkage if the azobenzene moieties cannot be incorporated into LCPs, rendering their application potential equally limited. Most recently, Chen *et al.* (2008) introduced another kind of two-way SMP in the form of polymer laminated composites. These are prepared by the layer technique with one-way SMPs. In the resulting polymer laminates, not only can the two-way SME be achieved, but the reversible deformation can also be controlled, and a wide range of response temperatures can be achieved.

With two-way SMP laminated composites, two independent layers are required. One is the active layer made with SMPs, while the other is the substrate layer made of a polymer with a high glassy modulus and slightly higher T_g than the T_{trans} of SMPs. Hence, the produced bending force of the substrate layer can be the complement of the recovery force of the active layer in the SMP laminate during the recovery process, particularly in the cooling process. Figure 12.3 shows a typical two-way SME of SMP laminated composites.

From the investigation of the morphology of BIN-SMPUs, it is known that the BIN-SMPUs have a very high glassy modulus, i.e. above 1.0 GPa. The glassy modulus does not decrease before 60°C, and some of the PUPys-MB series of PUPys still show a high modulus above 70°C. Therefore, the BIN-SMPUs are not only used as an active layer after deformation, but also serve as a good candidate for a substrate layer, as they can provide a high contrary recovery force during the cooling process (Chen *et al.*, 2008).

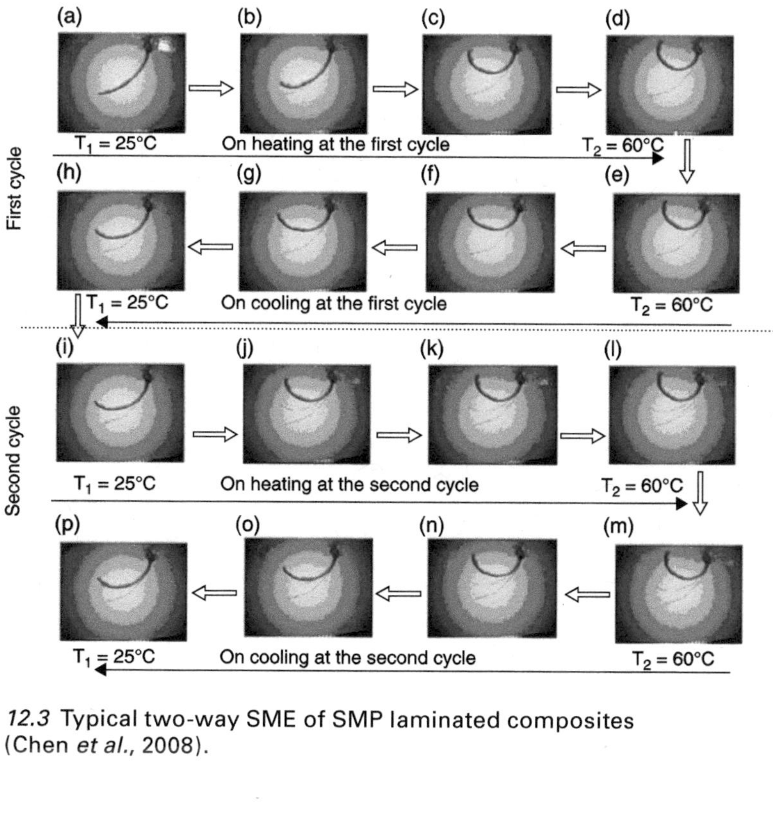

12.3 Typical two-way SME of SMP laminated composites (Chen *et al.*, 2008).

In this case, it is proposed that the laminated polymer composite be prepared to have the thermally-induced SMPU with the T_{trans} below 50°C acting as the active layer, and PUPyMB30 acting as the substrate layer. Upon heating, the laminated polymer composite will bend toward the active layer due to the strain recovery of SMPU above 50°C, and reach its balanced state. However, the bending force of the PUPyMB30 layer upon cooling will draw the polymer composite to an unbending state.

The polymer composites, therefore, show an interesting two-way shape recovery behavior; namely, bending upon heating and unbending upon cooling. Due to the low cost, light weight and easy preparation methods, the two-way SMP laminated composites are expected to promote the application of SMPs in many fields.

12.11.3 Intelligent windows for smart textiles applications

The common SMPs only show the thermally-induced one-way SME. After recovery, an external force is usually required to transform the permanent shape to the temporary shape. Although, as outlined above, two-way SMEs may be achieved in polymer laminated composites, it is still not easy in textiles to utilize a solid film, as the sole SMP fiber or layer cannot fabricate a satisfactorily intelligent smart textile.

However, the BIN-SMPU not only shows the thermally-induced SME, but also demonstrates good moisture-sensitive SME. By combining the thermally-induced SME and moisture-sensitive SME into a textile with two soft layers, an intelligent textile system can be achieved. Figure 12.4 illustrates the intelligent textile system for self-opening and self-closing. It is thus proposed that the textile is composed of two layers. One is the thermally-induced SMP layer, while another one is the deformed moisture-sensitive SMP layer. As the moisture content and temperature of inner clothing increases, the moisture-sensitive SMP layer recovers its original length. The window is then coiled and the channel used for moisture permeation is opened, leaving the body feeling dry and cool. However, when the intelligent textile is dried at the temperature above T_{trans} for the thermally-induced SMPs, the pre-stored strain presented by the moisture-sensitive recovery will be recovered when the temperature is raised to beyond T_{trans}. The thermally-induced shape recovery will result in the self-closing of the window. At the same time, the moisture-sensitive layer is dried, and the T_g is also increased to above ambient temperature. The moisture-sensitive layer can then show a strain recovery with respect to high RH conditions again. This process can be repeated many times.

12.11.4 Medical anti-bacterial applications

In recent years, medical devices that help prevent bacterial colonization have attracted a lot of attention, as control over bacterial infections is an important part

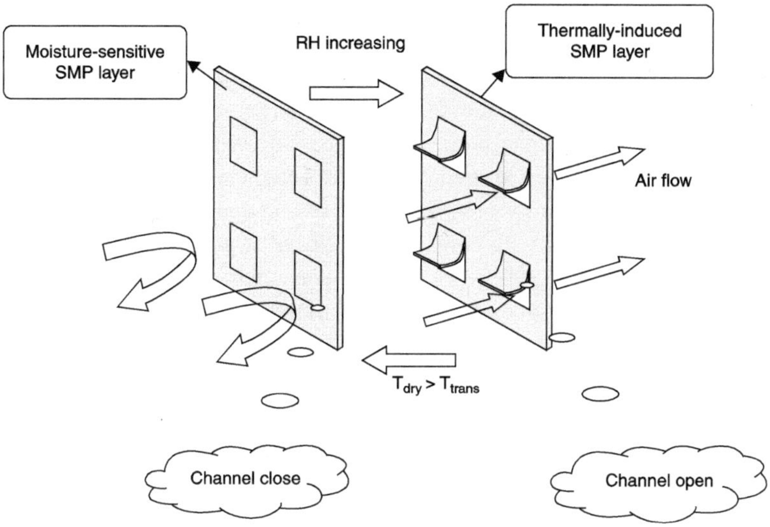

12.4 Illustration of intelligent textiles for self-opening and self-closing.

of any medical procedure. Nowadays, the majority of anti-bacterial materials used to build medical devices are polymeric. Over the past few decades, many kinds of polymeric materials have been developed that show anti-bacterial properties. Among them are, notably for us, some materials related to the pyridinium moieties (Ayfer *et al*., 2005; Yao *et al*., 2008; Zhu *et al*., 2009). Moreover, it was reported that the surface anti-microbial properties are improved by increasing the concentration of the pyridinium groups at the surface. In addition, the bactericidal efficiency would be enhanced with respect to the increase of hydrophilicity and pyridinium concentration at this surface. Furthermore, by incorporating BINA as a chain extender, it is found that biocidal polyurethane possesses high anti-bacterial activity against the Gram-positive *S. aureus* after it is quaternized with 1-iodooctadecane and 1,1,1,2,2,3,3,4,4,5,5,6,6,7,7,8,8- heptadecane-iododecane[264]. More recently, Zhu *et al*. (2009) showed that the ionic polyurethane containing BINA as a chain extender also demonstrates excellent anti-bacterial properties.

PUPys are synthesized to contain a large fraction of BINA moieties ranging from 53% down to 10%; in other words, the pyridine content is much higher than in the previous systems. After quaternizing with alkyl halides such as 1-iodooctadecane (C18) and 1-iodooctane (C8), the BIN-SMPU is expected to show excellent anti-bacterial properties due to its high pyridinium moieties and high hydrophilicity. The anti-bacterial properties will particularly improve when they are electro-spun to nano-fibers. Even the pure PUPys such as PUPyMB30 have anti-bacterial properties due to their high surface area to volume. Hence,

BIN-SMPUs can be further developed to show excellent anti-bacterial properties and, by extension, important potential for medical applications.

SMFs and fabrics in their special forms may find potential applications as biomedical materials due to their soft structures, permeability, and good compatibility with human bodies in comparison with film or bulk counterparts. SMFs may also be used in medical applications where high mechanical strength is required. Shape memory textiles with volume and shape changes may be used for skin-care products with controllable release of perfume, nutrition or a certain drug (Wischke *et al.* 2009). They may also open new opportunities for smart textiles in the field of medicine, thus widening their applications.

12.11.5 Other applications

In addition to the SMEs, many unique properties are also found in the BIN-SMPUs, which are different from traditional SMPs. With these unique properties in mind, many special potential applications are expected. They are, for example, particularly good candidates for the utilization of damping materials because of their high energy loss during the phase transition process between the soft phase and the hard phase. The phase transition will even occur at a lower temperature after conditioning by moisture or a watery environment. The BIN-SMPUs have a very high glassy modulus, and are thus applicable in high performance polymers or fibers. They can be further developed to show a high shape recovery force by improving the rubber modulus and glassy modulus. Finally, as the BIN-SMPUs show excellent moisture absorption, they may present a new option for the production of sweat absorption fibers or fabrics for textile applications.

12.12 References

Ahir, S. V., Tajbakhsh, A. R., Terentjev, E. M. (2006), Self-assembled shape-memory fibers of tri-block liquid-crystal polymers, *Advanced Functional Materials*, **16**, 556–60.

Alteheld, A., Feng, Y. K., Kelch, S., Lendlein, A. (2005), Biodegradable, amorphous copolyester-urethane networks having shape-memory properties, *Angewandte Chemie – International Edition*, **44**, 1188–92.

Ambrozic, G., Mavri, J., Zigon, M. (2002), Liquid-crystalline complexes of polyurethane containing an isonicotinamide moiety with 4-dodecyloxybenzoic acid, *Macromolecular Chemistry and Physics*, **203**, 439–47.

Ambrozic, G., Zigon, M. (2005a), Hydrogen-bonded liquid-crystalline polyurethane complexes with 4-dodecyloxybenzoic acid, *Acta Chimica Slovenica*, **52**, 207–14.

Ambrozic, G., Zigon, M. (2005b), Hydrogen-bonded polyurethane complexes based on 4-alkoxybenzoic acids as the low molar mass components, *Polymer International*, **54**, 606–13.

Ayfer, B., Dizman, B., Elasri, M. O., Mathias, L. J., Avci, D. (2005), Synthesis and anti-bacterial activities of new quaternary ammonium monomers, *Designed Monomers and Polymers*, **8**, 437–51.

Behl, M., Lendlein, A. (2007), Actively moving polymers, *Soft Matter*, **3**, 58–67.
Buckley, C. P., Prisacariu, C., Caraculacu, A. (2007), Novel triol-cross-linked polyurethanes and their thermo-rheological characterization as shape-memory materials, *Polymer*, **48**, 1388–96.
Chen, S., Hu, J., Zhuo, H., Zhu, Y. (2008), Two-way shape memory effect in polymer laminates, *Materials Letters*, **62**, 4088–90.
Chen, S. J., Hu, J. L., *et al.* (2009), Supramolecular polyurethane networks containing pyridine moieties for shape memory materials, *Materials Letters*, **63**, 1462–64.
Chen, S., Hu, J., Yuen, C., Chan, L. (2009), Novel moisture-sensitive shape memory polyurethanes containing pyridine moieties, *Polymer*, **50**, 4424–8.
Chen, Y. C., Lagoudas, D. C. (2008), A constitutive theory for shape memory polymers, Part I: Large deformations, *J Mech Phys Solids*, **56**, 1752–65.
Chung, T., Romo-Uribe, A., Mather, P. T. (2008), Two-way reversible shape memory in a semicrystalline network, *Macromolecules*, **41**, 184–92.
Fenix, Y. K., Zhang, L., Wu, Z. Z. (2007), Biodegradable polyurethanes in medical applications, *Research Journal of Chemistry and Environment*, **11**, 78–83.
Hu, J. L., Lu, J., Zhu, Y. (2008), New developments in elastic fibers, *Polymer Reviews*, **48**, 275–301.
Huang, W. M., Yang, B. (2005), Water-driven programmable polyurethane shape memory polymer: Demonstration and mechanism, *Applied Physics Letters*, **86**, 114105.
Jeong, H. M., Song, J. H., Lee, S. Y., Kim, B. K. (2001), Miscibility and shape memory property of poly(vinyl chloride)/thermoplastic polyurethane blends, *Journal of Materials Science*, **36**, 5457–63.
Lee, J. Y., Painter, P. C., Coleman, M. M. (1988), Hydrogen bonding in blends, Part 4: Blends involving polymers containing methacrylic acid vinylpyridine groups, *Macromolecules*, **21**, 954–60.
Lendlein, A., Langer, R. (2002), Biodegradable, elastic shape-memory polymers for potential biomedical applications, *Science*, **96**, 1673–6.
Lendlein, A., Behl, M. (2009), Shape-memory polymers for biomedical applications, *Smart Materials & Micro/Nanosystems*, **54**, 96–102.
Lendlein, A., Jiang, H., Junger, O., Langer, R. (2005), Light-induced shape-memory polymers, *Nature*, **434**, 879–82.
Leng, J., Lan, X., Liu, Y., Du, S. (2009), Electroactive thermoset shape memory polymer nanocomposite filled with nanocarbon powders, *Smart Materials and Structures*, **18**, 074003.
Li, F., Larock, R. C. (2000), New soybean oil-styrene-divinylbenzene thermosetting copolymers: Shape memory effect, *J Appl Polym Sci*, **84**(8), 1533–43.
Liu, Y., Gall, K., Dunn, M. L., Greenberg, A. R., Diani, J. (2006), Thermo-mechanics of shape memory polymers: Uniaxial experiments and constitutive modeling, *Int J Plasticity*, **22**, 279.
Meng, Q., Hu, J. L. (2010), A brief review of stimulus-active polymers responsive to thermal, light, magnetic, electric, and water/solvent stimuli, *Journal of Intelligent Material Systems and Structures*, **21**, 859–85.
Morshedian, J., Khonakdar, H. A., Rasouli, S. (2005), Modeling of shape memory induction and recovery in heat-shrinkable polymers, *Macromol Theory Simul*, **14**, 428.
Nike (2007), Featured technology, Nike sphere react cool, http://store.Nike.Com/?Country=us&lang_locale=en_us&l=shop,pdp,ctr-inline/cid-100701/pid-240998 – l=shop,pdp,ctr-inline/cid-100701/pid-240998.

Qin, H., Mather, P. T. (2009), Combined one-way and two-way shape memory in a glass-forming nematic network, *Macromolecules*, **42**, 273–80.

Shaojun Chen, Jinlian Hu, Haitao Zhuo, Yong Zhu. Two-way shape memory effect in polymer laminates. *Materials Letters*. Volume 62, Issue 25, 30 September 2008, Pages 4088–4090.

Sokolowski, W., Metcalfe, A., Hayashi, S., Yahia, L., Raymond, J. (2007), Medical applications of shape memory polymers, *Biomedical Materials*, **2**, S23–7.

Tobushi, H., Hayashi, S., Ito, N. (2001), Thermo-mechanical constitutive model of shape memory polymer, *Mech Mater*, **33**, 545.

Yao, C., Li, X. S., Neoh, K. G., Shi, Z. L., Kang, E. T. (2008), Surface modification and anti-bacterial activity of electrospun polyurethane fibrous membranes with quaternary ammonium moieties, *Journal of Membrane Science*, **320**, 259–67.

Zhang, H., Wang, H., Zhong, W., Du, Q. (2009), A novel type of shape memory polymer blend and the shape memory mechanism, *Polymer*, **50**, 1596–601.

Zhu, Y., Hu, J. L., Yeung, Y. (2009), Effect of soft-segment crystallization, hard-segment physical cross-link on shape memory function in anti-bacterial segmented PU ionomers, *Acta Biomaterials*, **19**, 3346–57.

Index